The Cerebral Veins

An Experimental and Clinical Update

Edited by

L. M. Auer and F. Loew

With a Foreword by F. Heppner

Springer-Verlag Wien GmbH

Prof. Dr. Ludwig M. Auer
Department of Neurosurgery
Graz University, Graz, Austria

Prof. Dr. F. Loew
Department of Neurosurgery
Saarland University Hospital, Homburg/Saar
Federal Republic of Germany

With 166 Figures

© 1983 by Springer-Verlag Wien

Originally published by Springer-Verlag Wien New York in 1983

Softcover reprint of the hardcover 1st edition 1983

Library of Congress Cataloging in Publication Data. International Symposium on Cerebral Veins (1st:
1982: Graz, Austria). The cerebral veins. 1. Cerebral veins—Diseases—Congresses. 2. Cerebral veins—
Congresses. I. Auer, Ludwig M., 1948– . II. Loew, F. (Friedrich), 1920– . III. Title. RC388.5.I5157.
1982. 616.8'1. 83-14667

ISBN 978-3-7091-4126-7   ISBN 978-3-7091-4124-3 (eBook)

DOI 10.1007/978-3-7091-4124-3

Foreword

Research in the morphology—angioarchitecture and ultrastructure—of cerebral veins has been widely neglected in past decades; investigation was mainly focussed on the arterial side of brain circulation. This circumstance has certainly had a negative impact on the development of knowledge in clinical medicine about cerebral venous disease. Cerebral venous pathology and its consequence is, however, a frequent problem in clinical neurosurgery, both with regard to operative techniques and conservative management. Therefore, it is not surprising that the initiative to collect, for the first time, data on our present knowledge in basic research of cerebral veins, their structure and function under normal and pathological circumstances, came from clinicians.

Regarding the cerebral veins the clinician has primarily in view the dysfunctions originating from embryogenetic malformations, phlebitic obstruction, tumourous shunts, or traumatic lesions. But in addition to that, particular attention should be paid to the microstructure of the venous vessel walls, their barrier function, and the venous vasomotor system. Studying these interrelationships has for a long time been both fascinating and of immediate interest to me.

Not too long ago, the physiology of the cerebral blood flow was dominated by the opinion that these vessels had no ability to alter their lumen actively, that they were nothing else but passive tubes with a pressure-dependent flow (Max Schneider). Neurosurgeons were among the first to become aware of how absurd this conception was—not only for teleological reasons but quite simply in view of spastically contracted arteries following aneurysmatic bleeding, or a cortical area becoming paler during epileptic discharge. Aiming to disprove this bizarre thesis we began in Graz, in 1955, in coordination with Marx, Lechner, and Diemath, to investigate the cerebral surface in experimental animals and in patients whose cortex was exposed for therapeutic reasons. Of interest was the behaviour of veins in the presence of oedematous swelling of the brain—a situation where the venous blood became pale and just as brick-red as it is in the arterioles; the capillaries disappeared and the veins contracted spastically. This was also the case in humans. Blood in the subarachnoid space showed, incidentally, the same mechanism. These studies were started without knowledge of the work of Hakim in Boston nor of Gurdjian's in Detroit; they used similar equipment and published their findings in 1957 and 1958, respectively.

Vivid interest in the brain's terminal blood flow has kept us active for nearly thirty years and initiated the international meeting of September

1983, which brought together in Graz distinguished researchers from all over the world. The contents of this volume, based on this symposium, is dealt with under the headings of anatomy, barrier mechanisms with regard to ICP, to neurogenic regulation, and finally to the clinical appearance of phlebogenic brain disturbances. It is an important aim of this book to provide basic material for consideration by clinicians, and to stimulate clinical research in this field. One of its main impacts may become more frequent investigation of cerebral blood volume under various pathological conditions in patients, since it has become evident that the regulation of cerebral blood volume depends not necessarily only on the factors regulating cerebral blood flow, but on separate mechanisms.

I should like to express my sincere thanks to all who cooperated in the composition of this book.

Graz, August 1983   F. Heppner
                                                    Head and Chairman,
                                            Department of Neurosurgery,
                                                    Graz University

Contents

Cerebral Blood Flow, Blood Volume and Intracranial Pressure

Pharmacological Effects

Cerebral Veins Under Various Pathological Circumstances

Diagnosis and Treatment of Cerebral Venous Diseases

Index of Contributors

Morphology of Cerebral Veins

Laboratoire d'Anatomie, Faculté de Médecine, Besançon, France

Cortical Veins of the Human Brain

H. M. DUVERNOY

Summary

The cortical veins were injected by two different techniques: indian ink and low viscosity resin.

The pial or superficial cortical veins form a dense network covered by arterial ramifications. The junction between intracortical and pial veins was particularly studied. We equally studied venous anastomoses which were of variable importance depending upon the gyri; no arteriovenous anastomoses existed on the cortical surface.

The deep or intracortical veins were divided into five groups according to their degrees of cortical penetration. The principal veins (group 5) are remarkable by their important diameter and their large territory. In the cortex, though venous anastomoses appear to be absent, the presence of precapillary arterio-venous anastomoses is suspected. The intracortical veins (groups 4 and 5) have a regular disposition. They seem to form cortical vascular units in association with the arteries which surround them.

Keywords: Cortical veins; human; angioarchitecture.

Introduction

Recent progress in the exploration of cerebral vascularization, as well as advances in neurosurgical techniques, make the knowledge of fine and precise anatomy of the cerebral vessels an absolute necessity. This certitude pushed us to embark upon the study of the cortical vascularization, the results of which were recently published [9]. In this report, however, only the observations concerning cortical veins will be discussed.

Material and Methods

Twenty-five brains were examined. The different techniques of intravascular injections already described in detail in the above-cited paper will only be treated briefly in this report.

Intra-arterial injection of india ink proved to be the method most frequently utilized. This method consists of injecting 400 cc of a solution of india ink and 5% gelatin. The injection is made directly into the internal carotid and vertebral arteries of the brain, either in situ or after necropsy. The principal advantage of this method is the filling of the entire vascular network—arteries, capillaries and veins. Throughout the entire regions subsequently rendered transparent by Spalteholz's technique, one may follow both the pial and the intra-cortical vessels throughout their entire trajectory. This particular technique, which leaves the surrounding tissues intact, also provides the possibility of practising silver impregnation (Bodian technique) which permits the correlation between nervous tissue and vascular network.

In studying the veins of the brain, most authors utilize the counter-current injection of the venous network, beginning at the collecting venous trunks. This retrograde injection, however, is by no ways complete, and we have always preferred total injection by the arteries.

The injection of a low viscosity resin (MERCOX) was frequently utilized. After tissue destruction and examination by a scanning electron microscope (SEM) we obtain a precise cast of the arteries and veins; the capillary network is equally visible. In comparison with india ink, this technique provides more details of the morphology of the blood vessels which are not altered by fixation. The SEM examination also provides a spectacular three-dimensional vision. We think, in fact, that these two techniques of injection are complementary, and together permit a more precise study of the vascularization of any particular region.

Results

Observations concerning cortical pial veins (surface veins) and intra-cortical veins will be successively discussed.

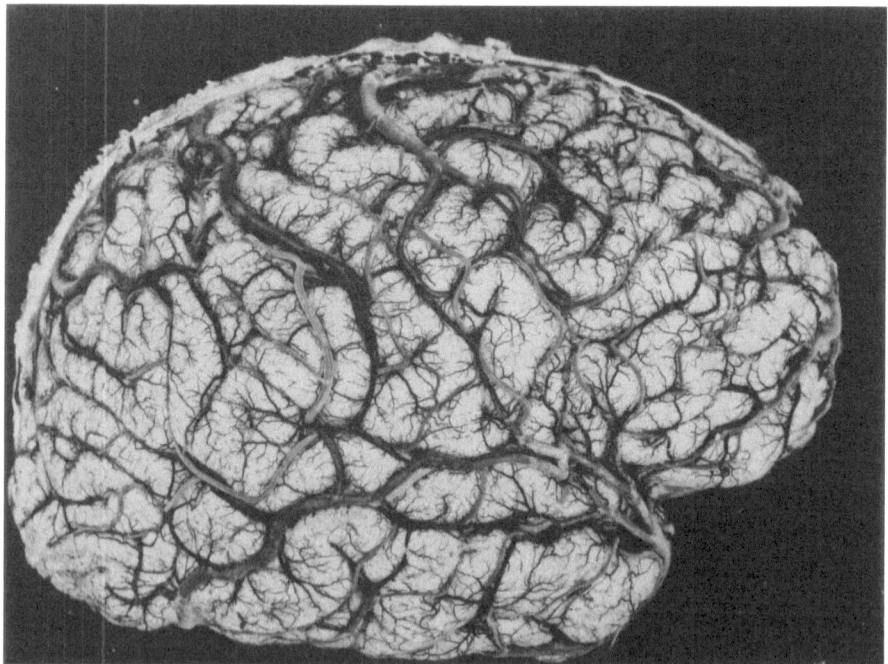

Fig. 1. Over-all view of superficial vessels. Right hemisphere. Indian ink injection. Arteries are grey and veins are black

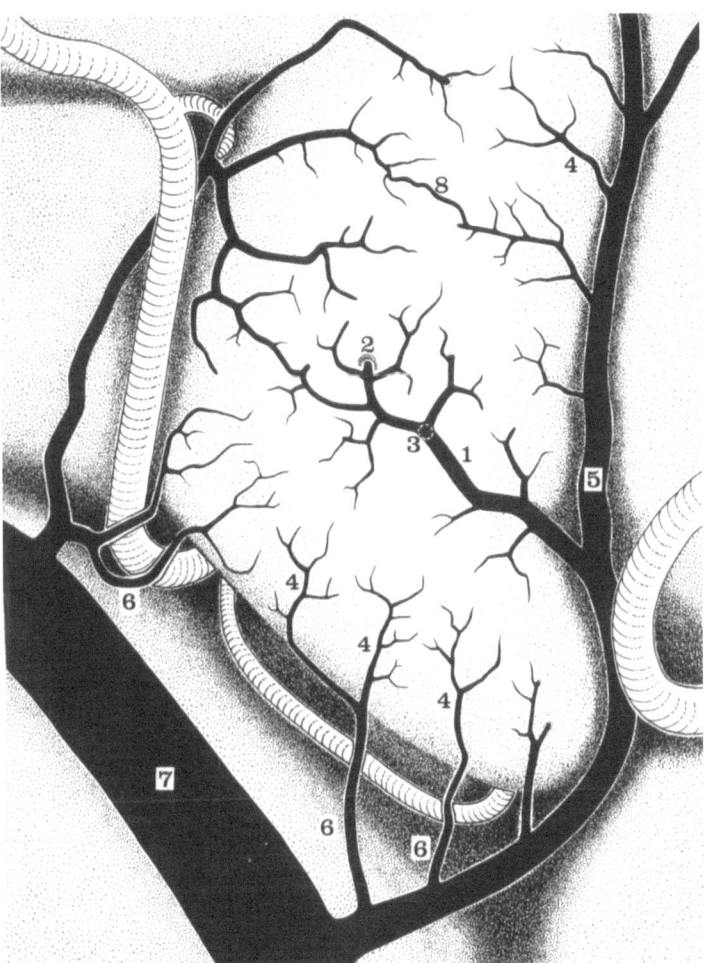

Fig. 2. Drawing of the general arrangement of pial veins. The pial arteries have been removed. *1* Central vein of the gyrus; note its angular path. *2* Junction point of an intracortical vein at the pial vein extremity. *3* In this case, the junction point of an intracortical vein is not visible. Note however, the sudden diameter increase of the pial vein downstream. *4* Peripheral veins. *5* Marginal vein. *6* The collecting veins of the pial network cross over the arteries and the sulcus to flow into the superficial cerebral vein. *7* Superficial cerebral vein. *8* Pial venous anastomoses

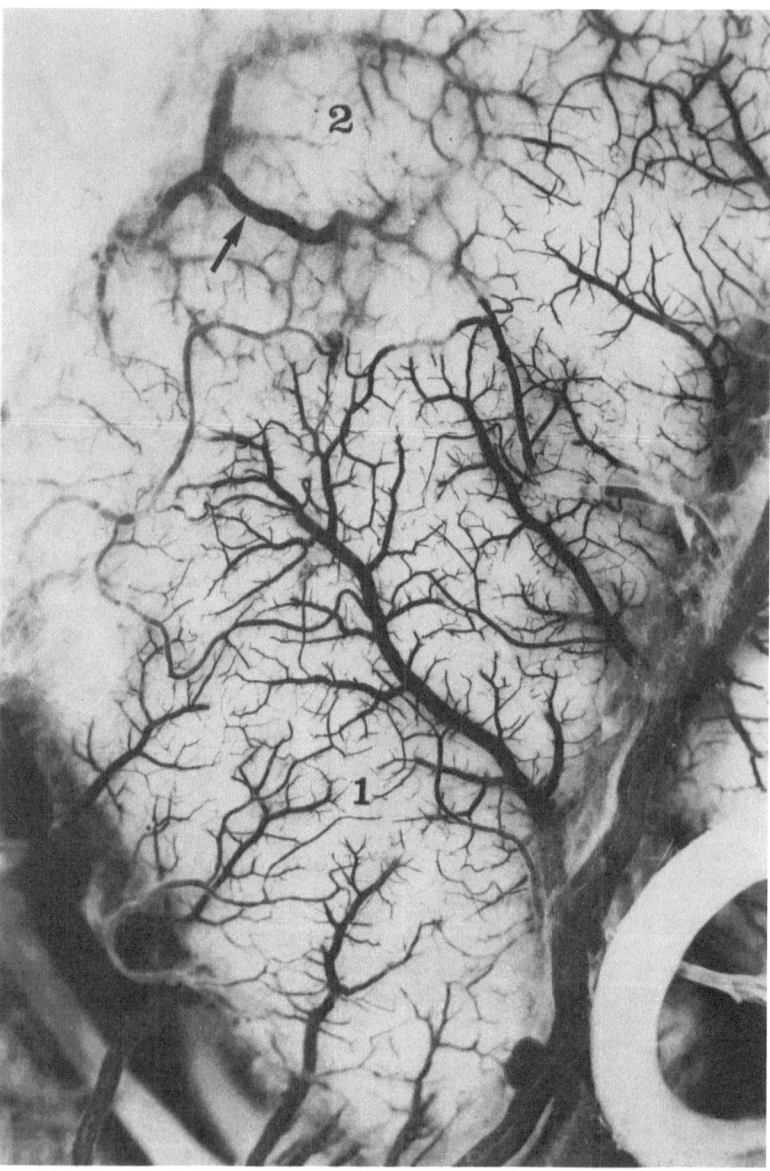

Fig. 3. Supramarginal gyrus (× 7,5). *1* After removal of arachnoid, the arterial and venous pial network is visible. *2* In this area arachnoid has been left intact and the vascular network is seen in transparency. A collecting vein adhering to the arachnoid is clearly visible (arrow)

I. Cortical Pial Veins

The thin, delicate venous network covering the surface area of the gyri will be studied; the large superficial cerebral veins are therefore excluded (Fig. 1). On microscopic examination, the pial vessels can be seen beneath the arachnoid matter. Once this layer is excised, the density of the pial vessels appears in its entirety (Fig. 3).

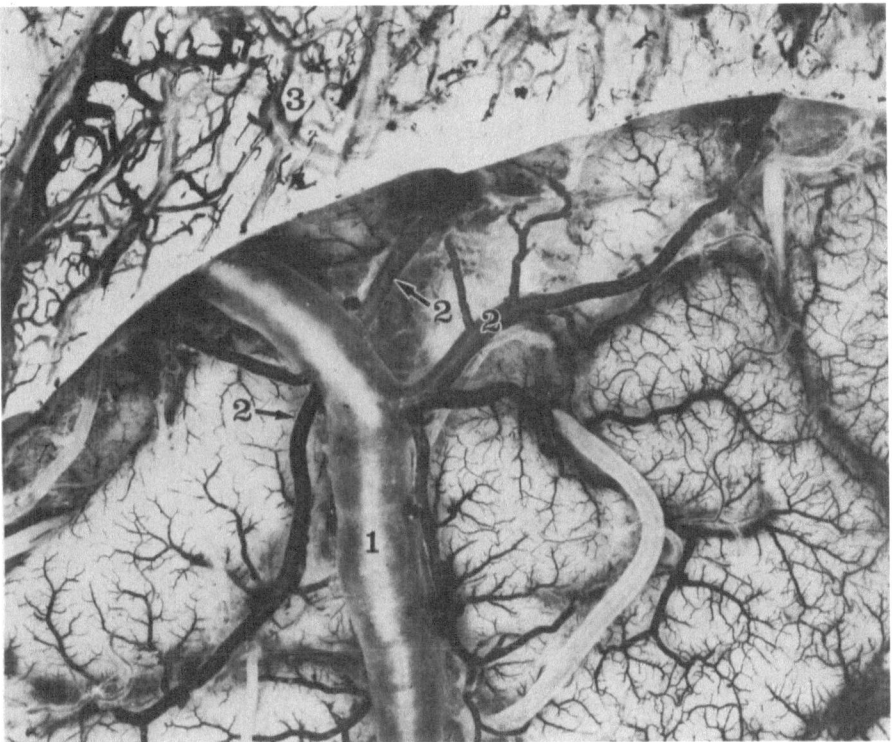

Fig. 4. Parietal lobe (× 2,8). *1* Superficial cerebral vein. *2* Collecting veins cross over the sulci to flow into the superficial cerebral vein. *3* Dura mater has been left intact

The venous network adheres to the nervous tissue; it is generally covered by fine arterial ramifications (Fig. 9). At the periphery of the lobule, however, the collecting veins which drain the pial network detach themselves from the cortex to flow into the superficial cerebral veins; in general they cross over the arteries (Figs. 2, 4, and 5). The relationship to the arachnoid is also variable: at the center of the lobule the arterioles, more superficial than the venules, adhere to the arachnoid by a few fibrous bridges; at the periphery, however, the veins which drain the pial network, cross over arteries and sulci entering directly into contact with the arachnoid (Fig. 3). This adhesion to the arachnoid is so intimate that their dissection is difficult to perform. A strong traction applied to the arachnoid may damage their walls. Furthermore, in the case of cerebral edema these veins appear highly distended (Fig. 6).

After excision of the arachnoid and the pial arteries, it is possible to detail the various aspects of the pial veins.

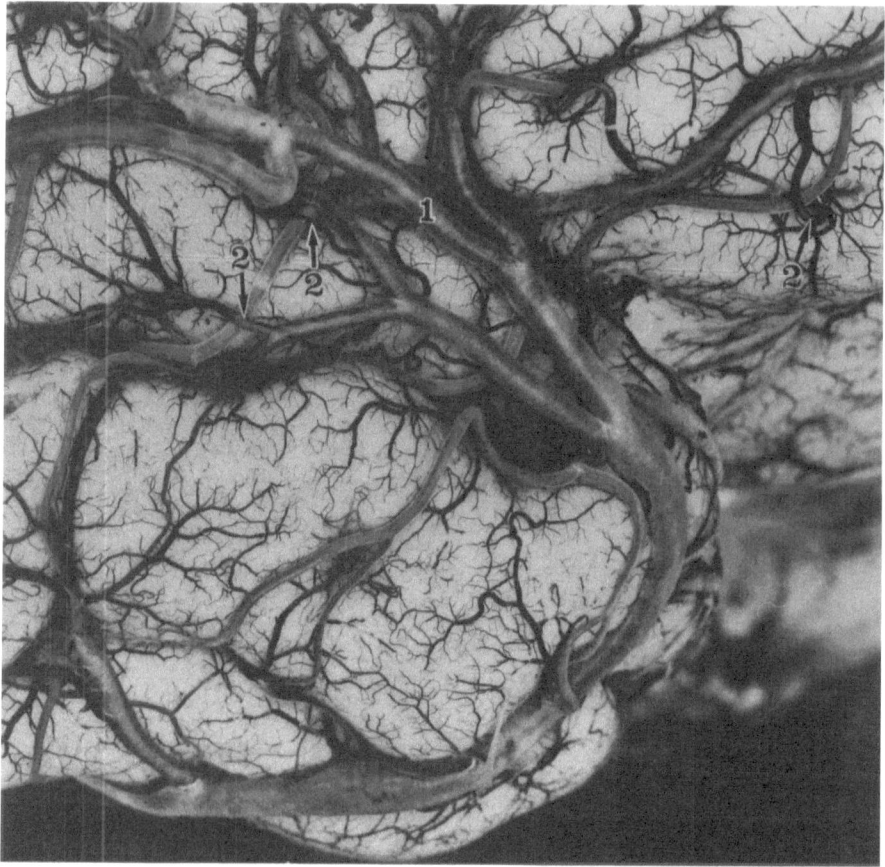

Fig. 5. Anterior pole of the temporal lobe (× 3). *1* The network of the superficial middle cerebral vein overlays the lateral fissure. *2* Collecting veins cross over the arteries to flow into cerebral veins

General disposition of pial veins (Fig. 2)

Two types of pial veins are generally encountered at the surface of the gyrus: the central veins and the peripheral veins (Figs. 2, 9, and 10). Frequently, each gyrus contains one or several veins of large diameter (280–300 μ), whose abundant arborisations occupy the center, whereas, at the

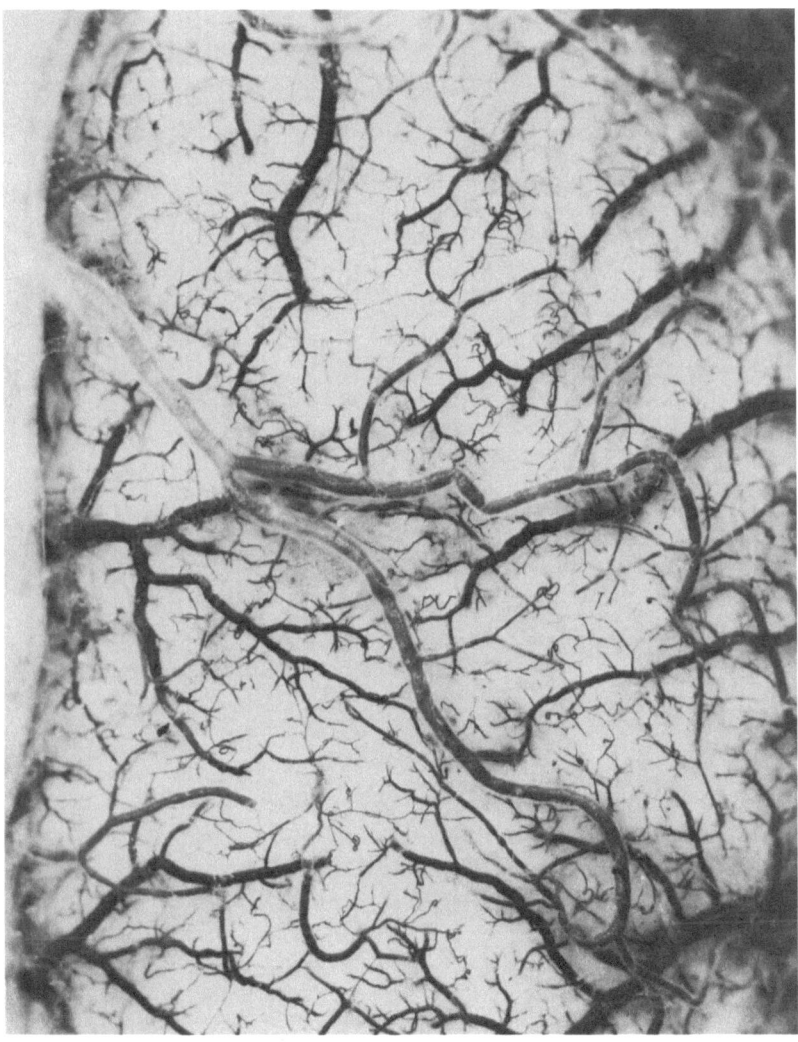

Fig. 6. Gyrus occipitalis secundus (× 22). Cerebral edema; irregular appearance of pial cortical vessels

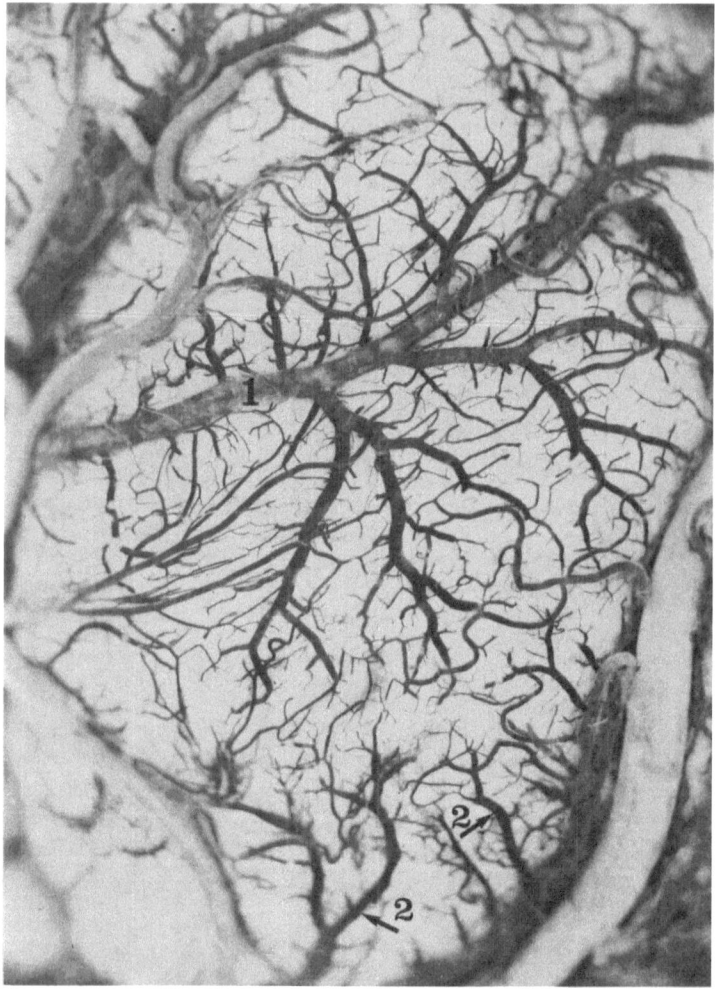

Fig. 7. Lingual gyrus (× 7,5). *1* Central vein whose numerous arborizations cover the greatest part of the gyrus surface. *2* Small peripheral veins

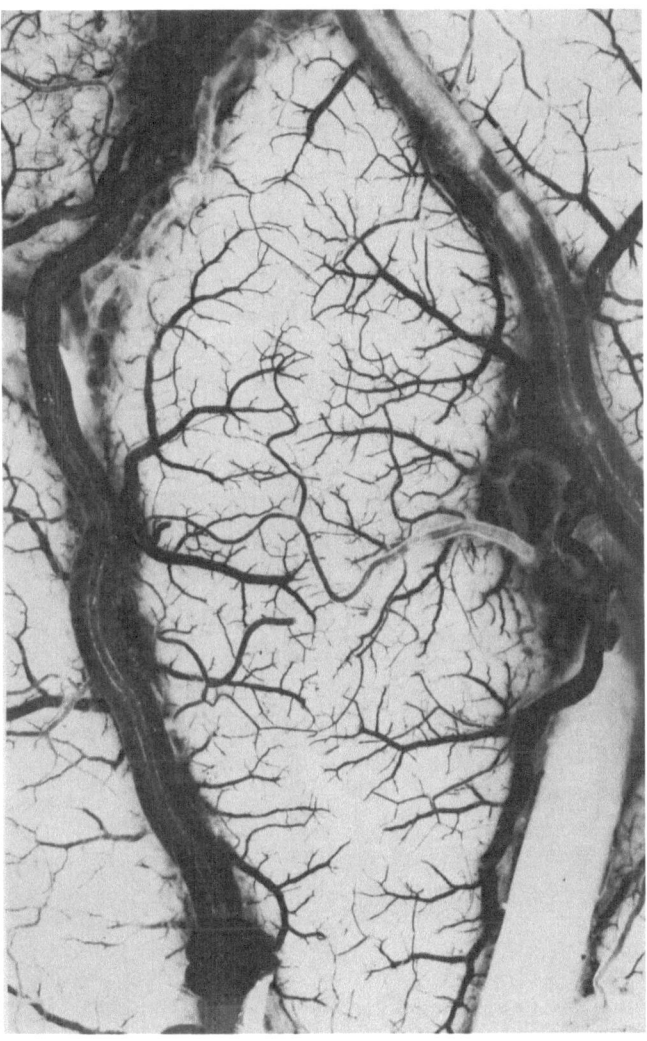

Fig. 8. Parietal lobe (× 5). On this elongated lobule only peripheral pial veins are visible without notable anastomosis

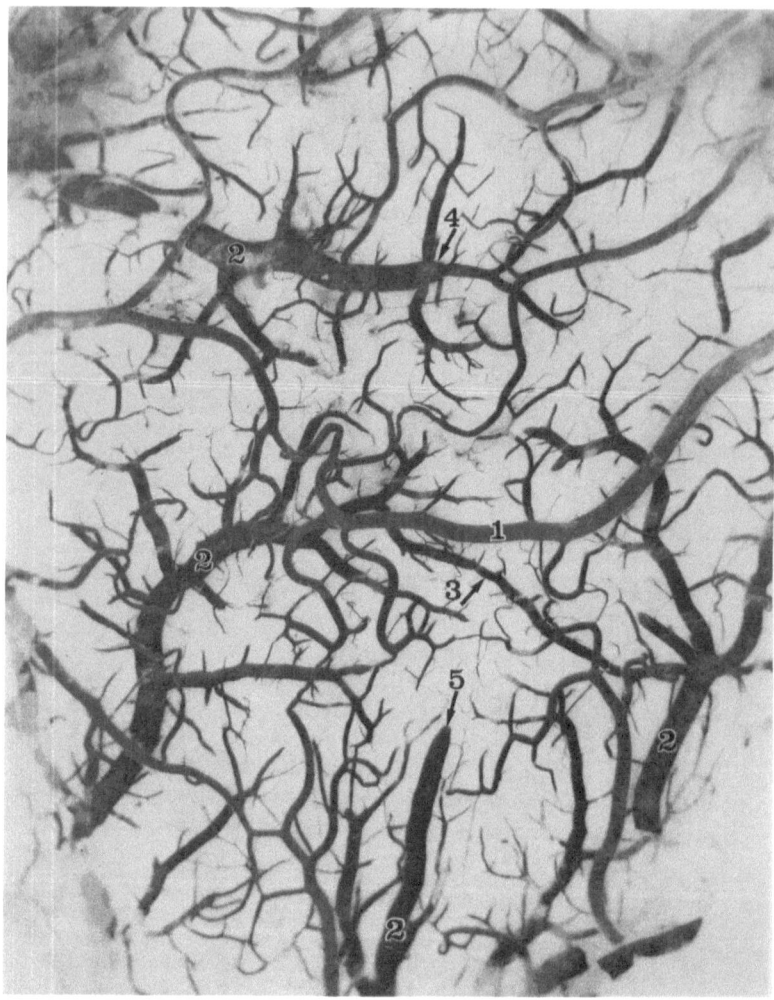

Fig. 9. Paracentral lobe (× 9). Relationship between arteries and veins. *1* Central artery: its ramifications cover the pial venous network. *2* Pial veins. *3* Anastomosis between two pial veins. *4* Diameter increase of a pial vein marking the hidden junction between a large intracortical vein and the superficial vein. *5* Emergence of an intracortical vein reaching the extremity of the pial vein

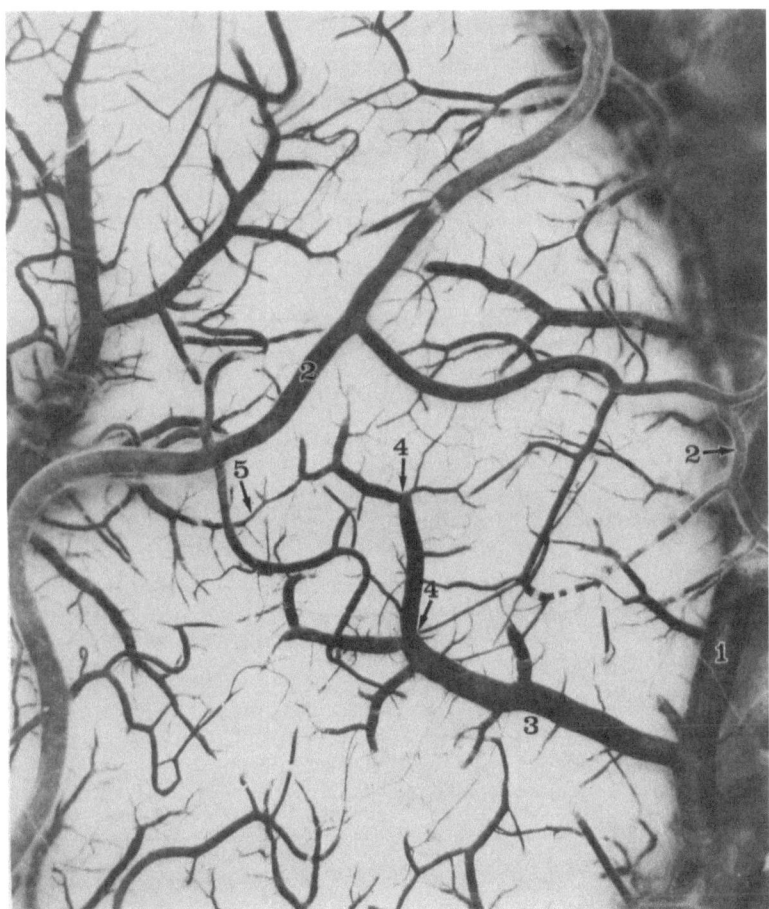

Fig. 10. Gyrus occipitalis secundus (× 15). *1* Marginal vein. *2* The arterial network covers the pial veins. *3* Central vein. *4* Enlargements of venous diameter indicate the junction of intracortical veins. *5* Venous anastomosis

periphery, veins of a smaller diameter (130 μ) are disposed in a parallel formation and run into a marginal vein which frequently borders the gyrus. The two types of veins are usually found in the same gyrus. The central type is sometimes predominant (Fig. 7). At other times, on the contrary, the peripheral type is the only one seen, especially in elongated gyri (Fig. 8).

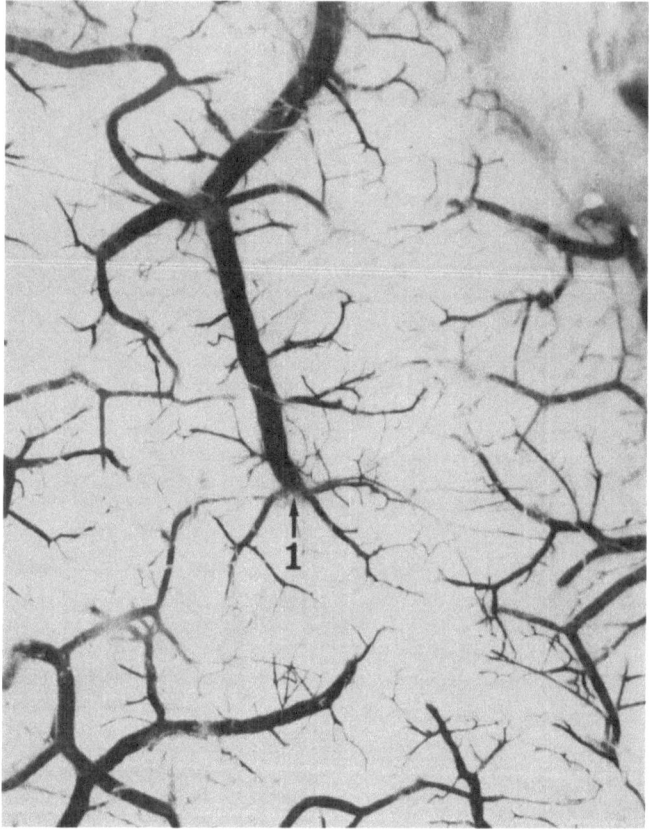

Fig. 11. *1* Emergence of a large intracortical vein; numerous small pial veins flow into the principal vein at its point of emergence (× 15)

The central and peripheral veins have an angular course receiving their secondary branches at the angles (Figs. 2 and 10).

It may be important to locate the precise point where the intracortical veins emerge from the nervous tissue and reach the superficial network. The intracortical veins of large diameter may flow into the extremity of the pial vein (Figs. 2, 9, and 11); in this case, the junction between the two vessels will be highly visible; but more often, the deep vein runs into the superficial vein at the side which adheres to the nervous tissue. In this case the junction is not visible to the observer; upon attentive examination, however, the observer may discover a sudden increase in diameter of the pial vein at this particular point (Figs. 2, 9, and 10).

The intracortical veins of smaller dimension, very numerous in fact,

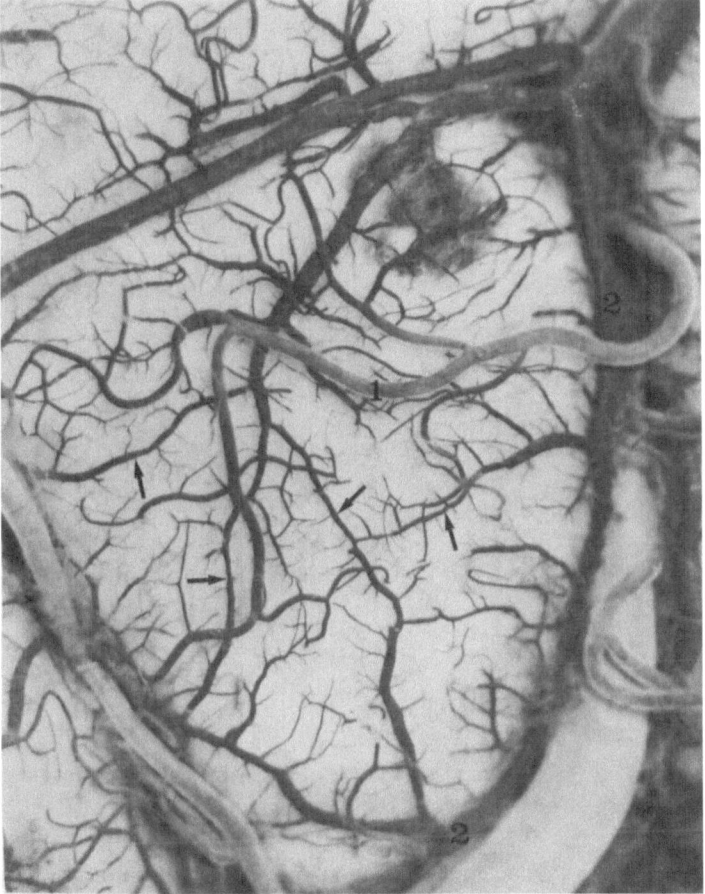

Fig. 12. Supramarginal gyrus (× 10). *1* The arterial network overlays the pial veins. *2* Marginal vein. Arrows indicate numerous venous anastomoses

constitute, after their emergence, a fine pial network which is frequently poorly injected. Its importance is therefore often underestimated. These thin venules sometimes converge upon the point of emergence of the large diameter vein (Fig. 11).

The pial venous network is generally anastomotic. The importance of these anastomoses is variable. Sometimes voluminous veins cross the gyrus and unite two marginal veins to each other; in this case, the distinction between central and peripheral veins is impossible (Fig. 12); in other cases, the pial veins are independant and no anastomose is visible at the surface of the gyrus (Fig. 8). This inconstancy contrasts with the importance of the

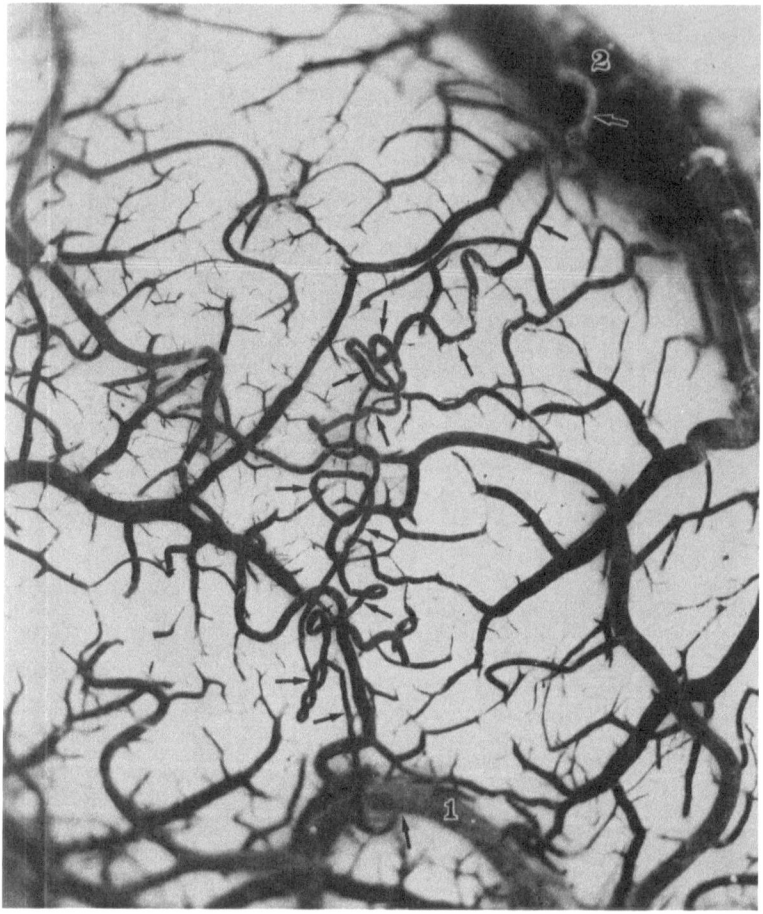

Fig. 13. Gyrus occipitalis primus (× 12,5). Arterio-venous anastomosis (arrows) joining an artery (*1*) to a vein (*2*). It has a tortuous path and crosses over the pial network

arterial anastomoses. On the brains examined, we never found arterio-venous anastomoses. In one case, however, we observed one vessel with an irregular course uniting an artery with a vein. But this anastomosis was adherent to the arachnoid in its entire course and without contact with the pial network (Fig. 13).

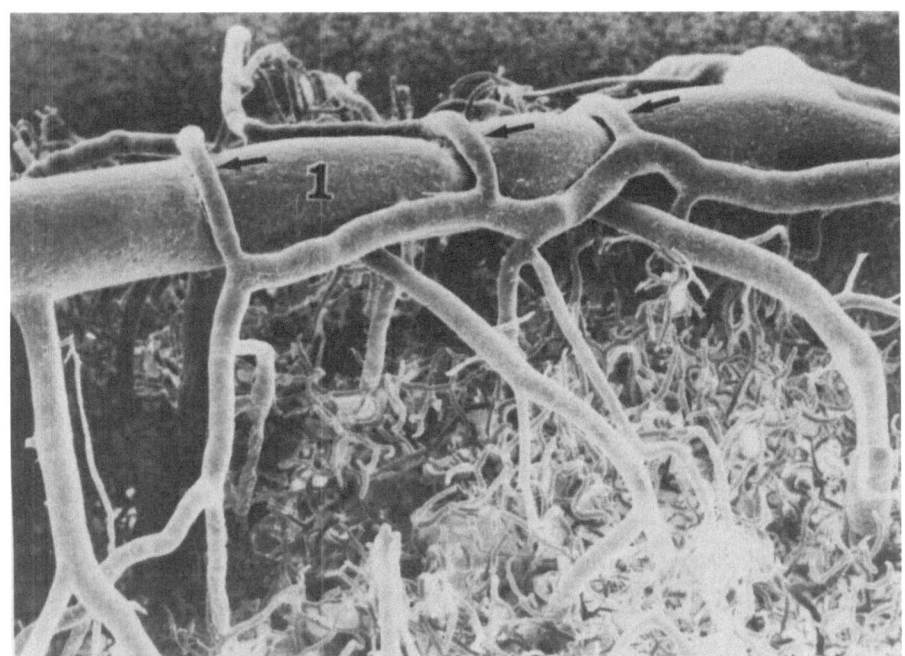

Fig. 14. Pial vascular network (SEM; ×75). *1* Marginal vein. Arrows indicate arteries crossing over the vein and leaving grooves in the venous wall

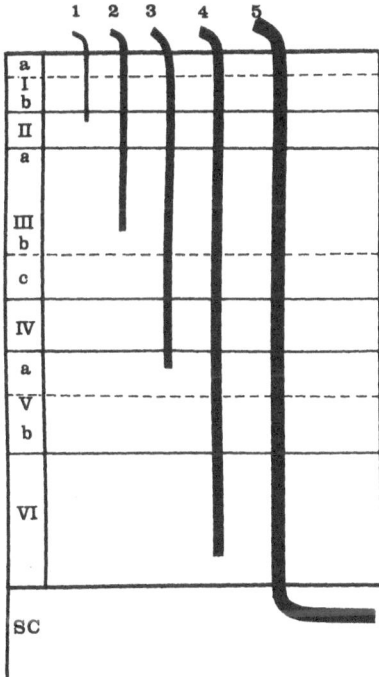

Fig. 15. Degrees of intracortical veins penetration into the cortex. *1, 2, 3, 4,* and *5* represent five groups of intracortical veins (or intracortical arteries). The disposition of cortical cellular layers is shown on the left. *I* Molecular layer. *II* External granular layer. *III* Pyramidal layer. *IV* Internal granular layer. *V* Ganglionic layer. *VI* Multiform layer. *SC* Subcortical white matter

II. Intracortical Veins

In keeping with the description of pial vessels, two types of veins have been described emerging from the cortex: large diameter intracortical veins few in number, and smaller but more numerous veins of much lesser diameter. Close examination of the cortex enables us to divide the intracortical veins into five groups according to their degree of penetration into

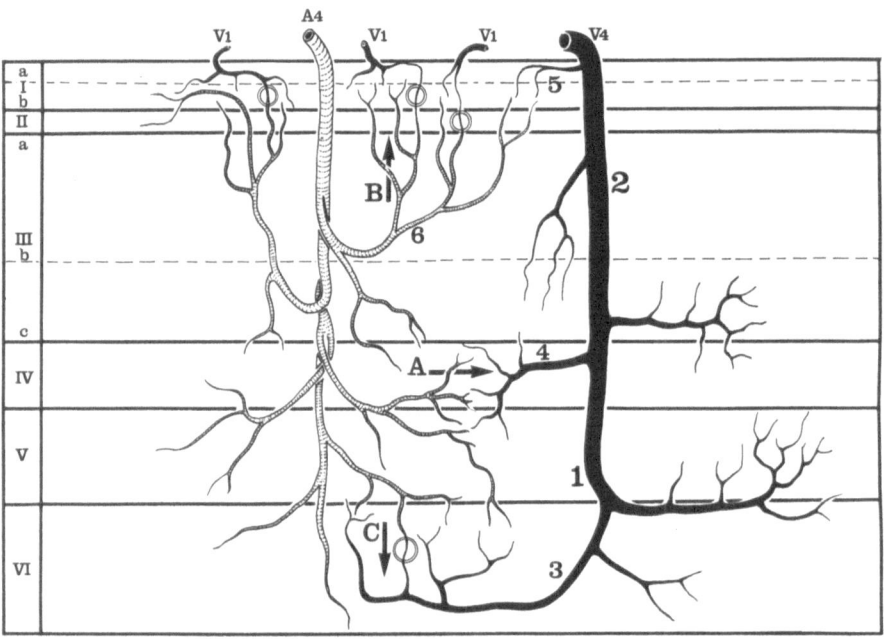

Fig. 16. This drawing indicates the morphological features of typical intracortical artery and vein of group 4 (A 4 and V 4). Vein of group *4*: *1*: its deep path is irregular and *2*: its superficial path is rectilinear. *3, 4, 5*: deep, intermediate and superficial branches. Artery of group *4:6*: recurrent branches. *A* Middle intracortical blood pathway. *B* Superficial intracortical blood pathway collected by short intracortical veins (V 1) and by superficial branches of large intracortical veins (*5*). *C* Deep intracortical blood pathway collected by deep branches of large intracortical veins (*3*). Circles represent possible precapillary arterio-venous anastomoses. The disposition of cortical cellular layers is shown on the left

the cortex. The five venous groups (V 1–V 5) correspond with groups A 1– A 5 of the intracortical arteries (Fig. 15).

Veins of small and medium diameter (20–45 μ, V 1, V 2, V 3): veins of group V 1 do not extend beyond the molecular and external granular layers of the cortex; the group 2 veins enter into contact with the pyramidal layer, and those of group 3 with the internal granular layer.

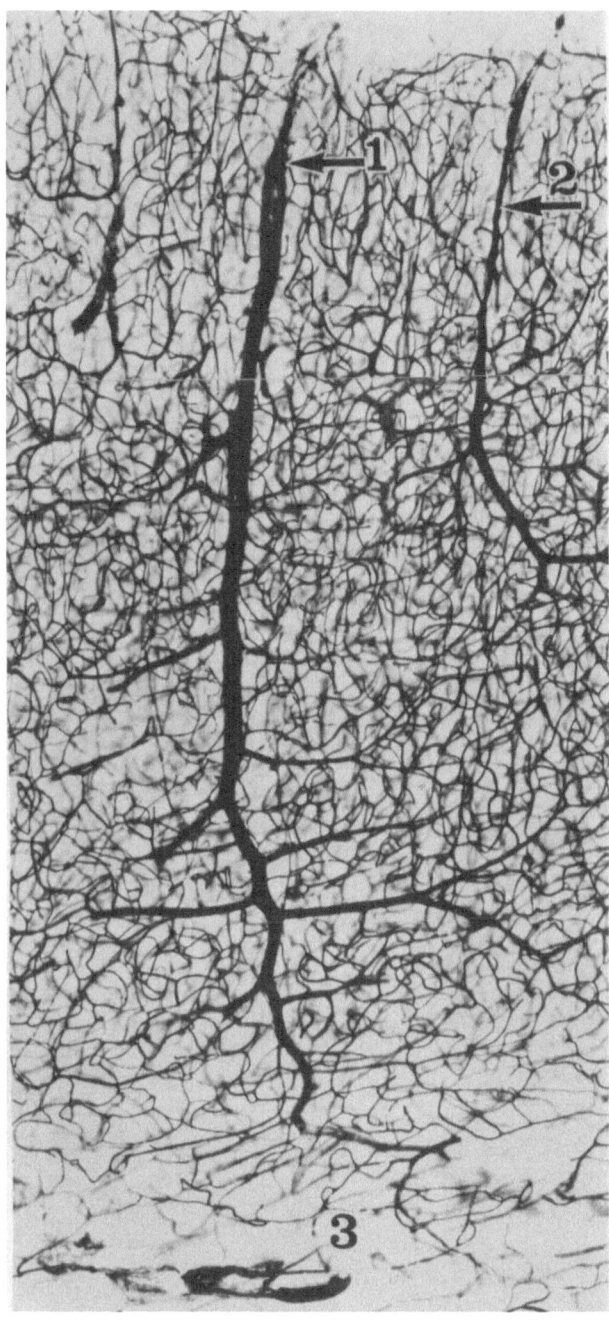

Fig. 17. Middle temporal gyrus (× 64). *1* Typical intracortical vein (V 4); it has a straight path through the superficial layers of the cortex and an angular path through the deep cortical layers. *2* Short cortical vein (V 3) originating in the middle cortical layers. *3* Subcortical white matter

H. M. Duvernoy:

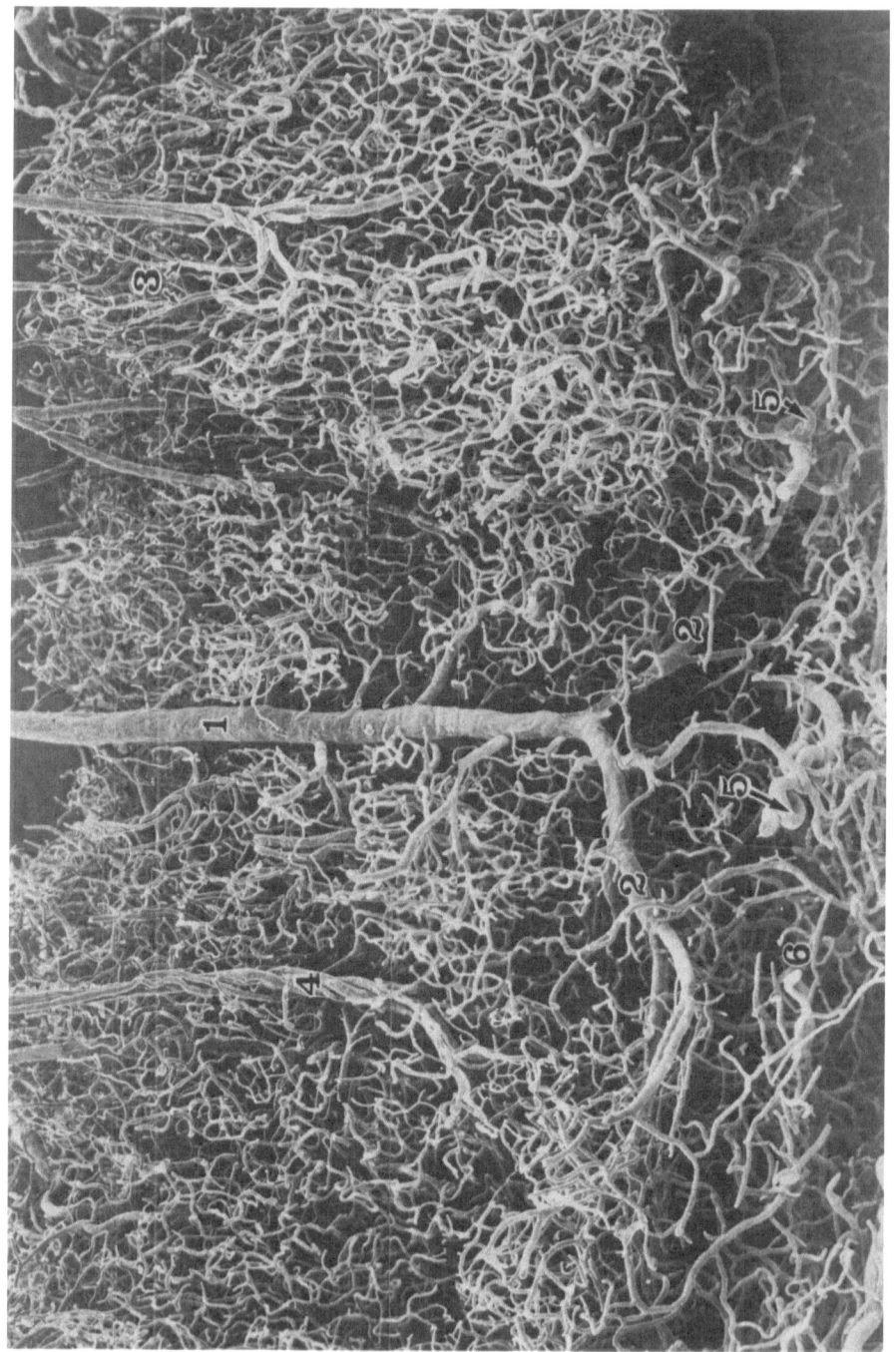

Fig. 18. Section of temporal cortex. General view of cortical vessels (SEM; × 45). *1* Principal intracortical vein (*V 5*). *2* Its large deep branches. *3* Recurrent branches of intracortical artery. *4* Coiling of branches of an intracortical artery. *5* Arterial glomerular loop formation in the subcortical white matter (*6*)

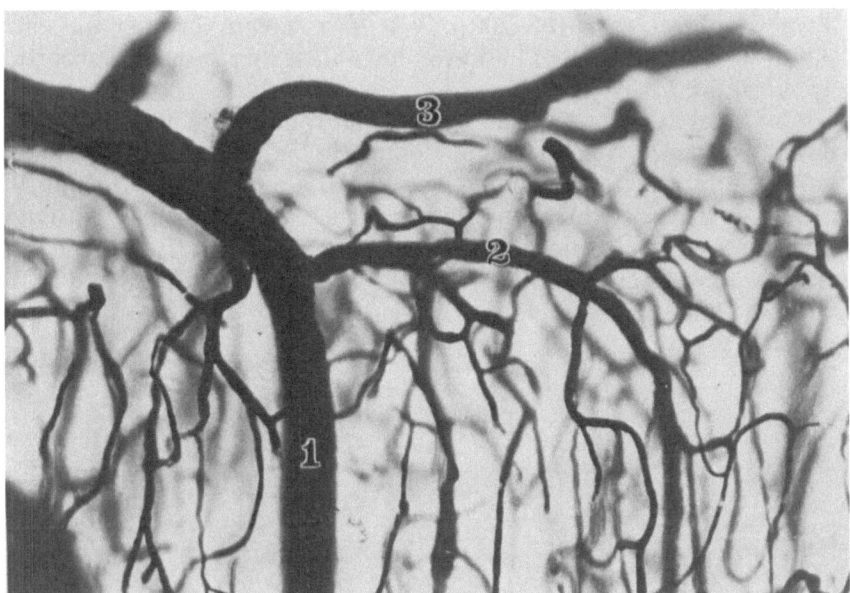

Fig. 19. Middle temporal gyrus (× 140). *1* Principal intracortical vein. Superficial branches in the molecular layer (*2*) and at the cortex surface (*3*)

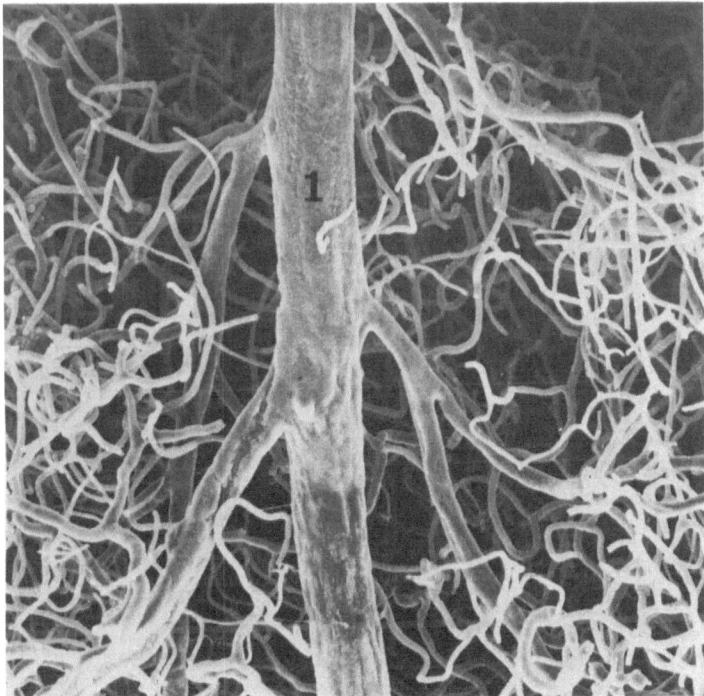

Fig. 20. Pole of the temporal lobe (SEM; × 85). *1* Principal intracortical vein. Intermediate branches reach the main trunk at acute angles

Large diameter veins (65–200 μ, V 4, V 5) spread through the entire thickness of the cortex (V 4) and even have their origin in the subcortical white substance (V 5).

Large diameter veins have an uniform morphology. The group 4 vein, taken as an example, runs an irregular course in the deep cortical layers, and a more regular trajectory in the superficial layers (Figs. 16 and 17). Generally, it presents three types of branches: *deep branches,* extremely ramified, drain a vast cortical area; *intermediate branches,* fixed to the main trunk at an acute angle, and *superficial branches* which are small in size. It is

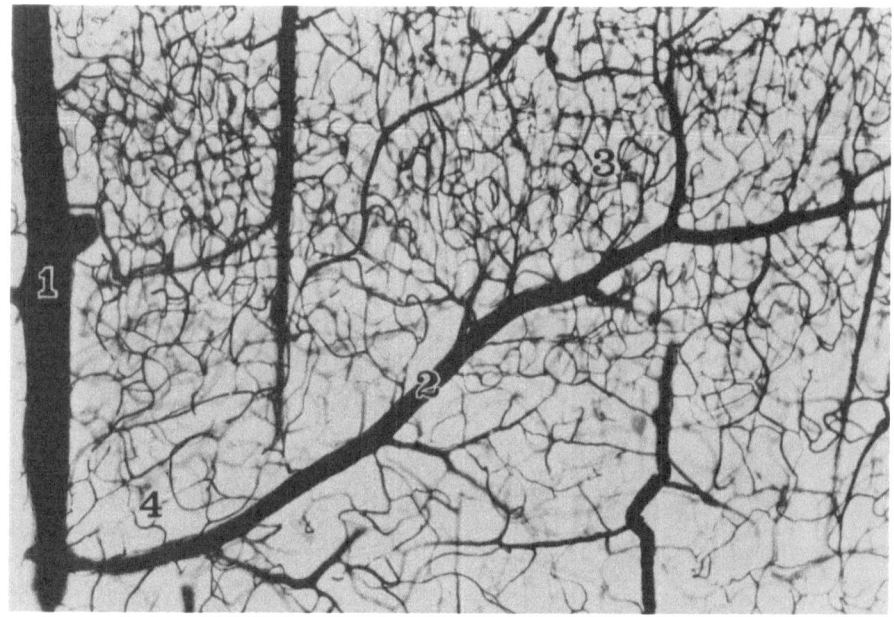

Fig. 21. Paracentral lobule (× 60). *1* Principal intracortical vein. *2* Large deep branch draining the innermost cortical layers (*3*) through the subcortical white matter (*4*)

thus possible to see that the size of the venous arborizations decrease from deep layers towards the superficial areas of the cortex. Generally, the arteries present an opposite arrangement: the very numerous superficial branches curve towards the surface (recurrent arteries) and irrigate a large territory, whereas the deep branches are much smaller (Figs. 2 and 18).

The veins V 5 which have a similar morphology to the preceding group, differ, however, by their origin in the white matter, but even more so by their considerable diameter (Fig. 18). We will refer to them as the *principal veins of the cortex.* If the superficial and intermediate branches bear a

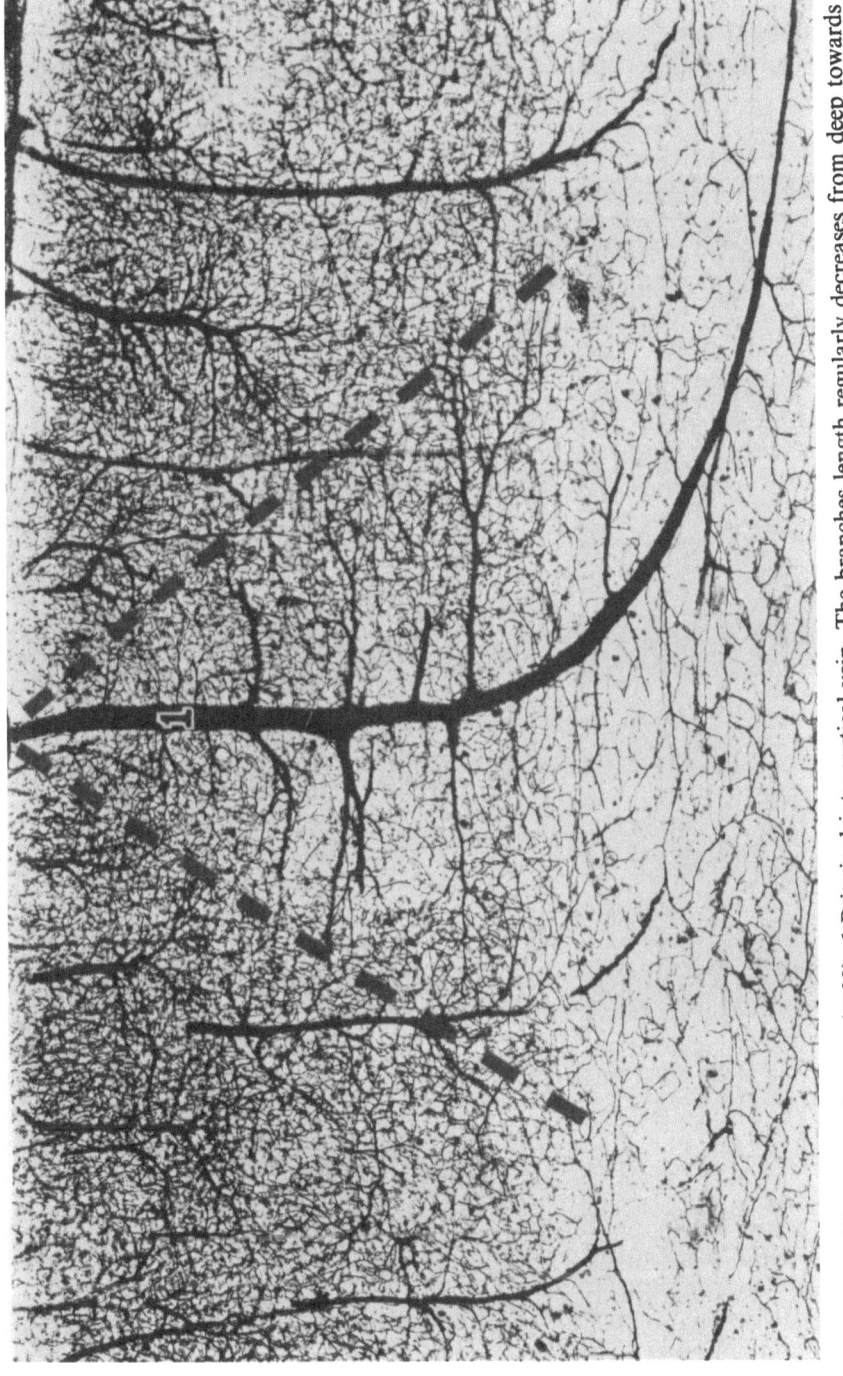

Fig. 22. Middle temporal gyrus (×30). *1* Principal intracortical vein. The branches length regularly decreases from deep towards superficial cortical regions; thus, the vascular territory of a principal vein has a conical appearance (dotted line)

ressemblance to those of V 4 (Figs. 19 and 20), the deep branches are notable by their surface area (Fig. 21); they flow into the principal trunk in stellar formation and drain an important cortical territory (Figs. 23 and 24). Due to the fact that the length of the branches regularly decreases from deep

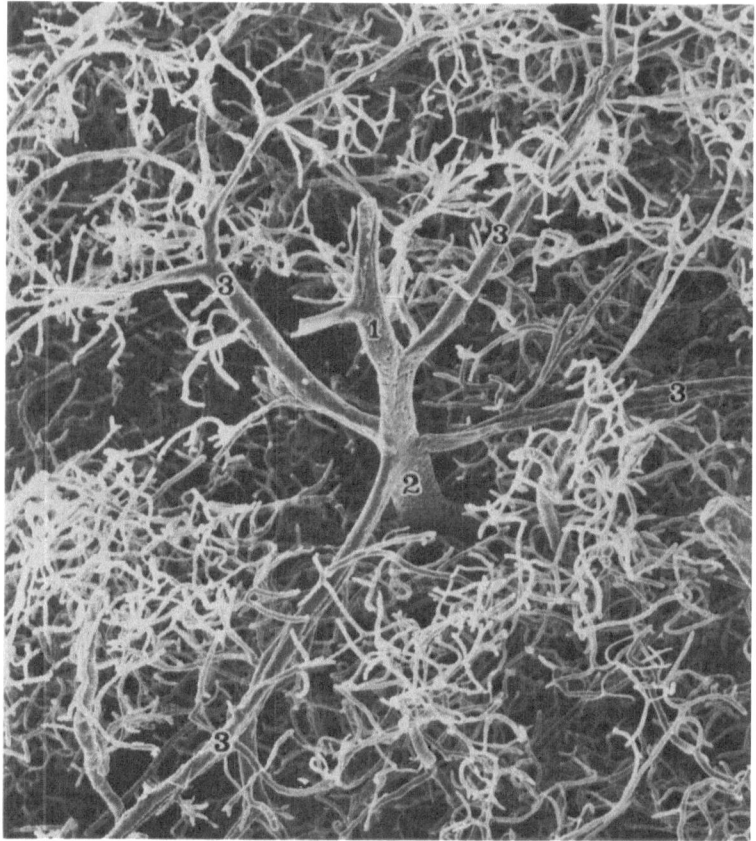

Fig. 23. Middle temporal gyrus (SEM; × 55). Tangential aspect of the deep cortical surface after removal of the white matter. *1* Subcortical segment of a principal cortical vein. *2* Venous trunk penetrating into the deep cortical layers. *3* Long deep branches with stellar aspect, drain a large cortical area

towards superficial cortical regions, the vascular territory of the principal vein has a conical appearance (Fig. 22).

Small diameter veins (V 1, V 2, V 3), unlike the others, do not have an uniform morphology. Their secondary branches frequently converge upon the origin of the vein; it is important to note large increase in vein diameter at the junction of these branches with the principal trunk (Figs. 25 and 26).

The smallest veins (V 1) are very numerous and often poorly injected (Figs. 28 and 29).

Generally, the arborization of the intracortical veins is extremely dense, giving the cortical vascular network a classical bush-like appearance[10, 20]

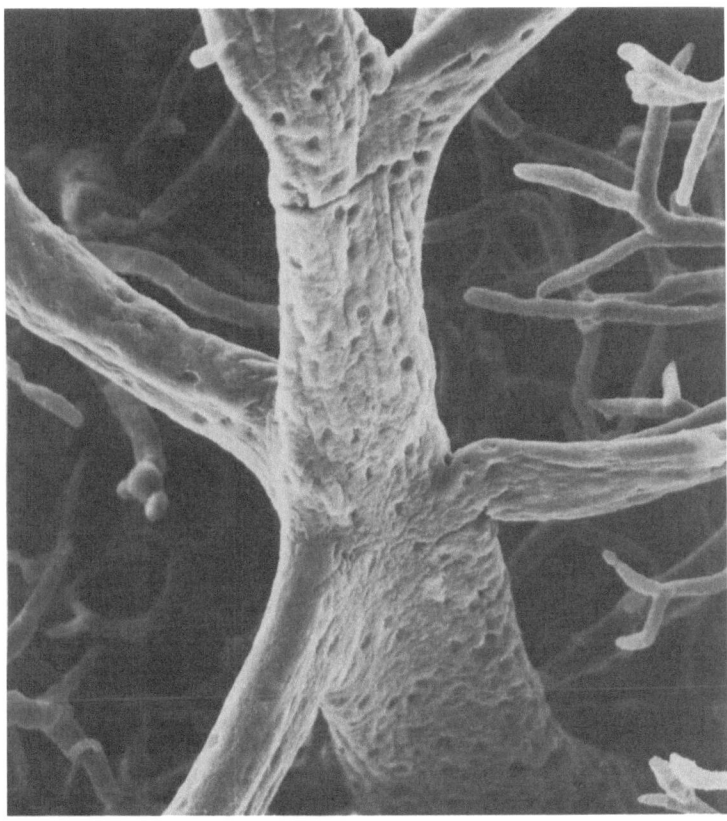

Fig. 24. Enlargement of Fig. 23 showing the junction of the deep branches into the principal vein (SEM; × 260)

(Fig. 27). The orientation of the arterial and venous ramifications suggests the possibility of three cortical vascular pathways (Fig. 16): an *intermediate vascular pathway* where arterial inflow and venous drainage are at the same level, a *superficial vascular pathway* fed by the recurrent arteries and drained towards the surface by the short veins (groups V 1–V 2) or by superficial arborizations of large diameter veins, and a *deep vascular pathway* which is drained by the voluminous deep branches of the large diameter veins V 4 and V 5.

Precapillary arterio-venous anastomoses appear to exist in the deep and superficial cortical vascular pathways (Figs. 16, 31, and 32). These anastomoses frequently unite the arborizations of the recurrent arteries with the small veins of groups 1 and 2. On the contrary, however, we have observed only a few arterial anastomoses and no venous anastomosis at all.

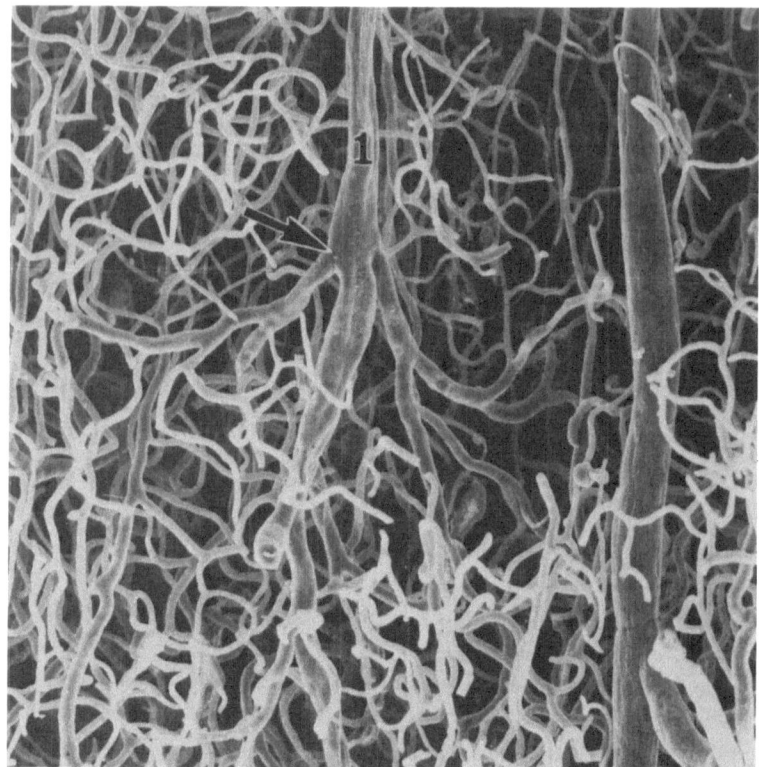

Fig. 25. Morphological aspect of a small intracortical vein (V 2). Arrow indicates diameter enlargement at the junction of secondary branches (SEM; × 100)

There appears to exist a special organization of arteries and veins visible in the deep cortical layers. On tangential sections, one will note the regular arrangement of large diameter veins (V 4 and V 5) (Fig. 33); each vein constitutes a vascular unit arround which several arteries run in a ring-like formation (Figs. 34, 35, and 36). One large vein may be surrounded by several arterial rings; sometimes, however, two large veins may be centered by the same arterial ring. This particular organization explains why there exists in general four times more arteries than veins in the deep layers of the cortex.

Discussion

1. Pial Cortical Veins

The venous network is generally covered by arterial ramifications (Figs. 9 and 10) as described by several authors[5,11,21]. We have arrived at the

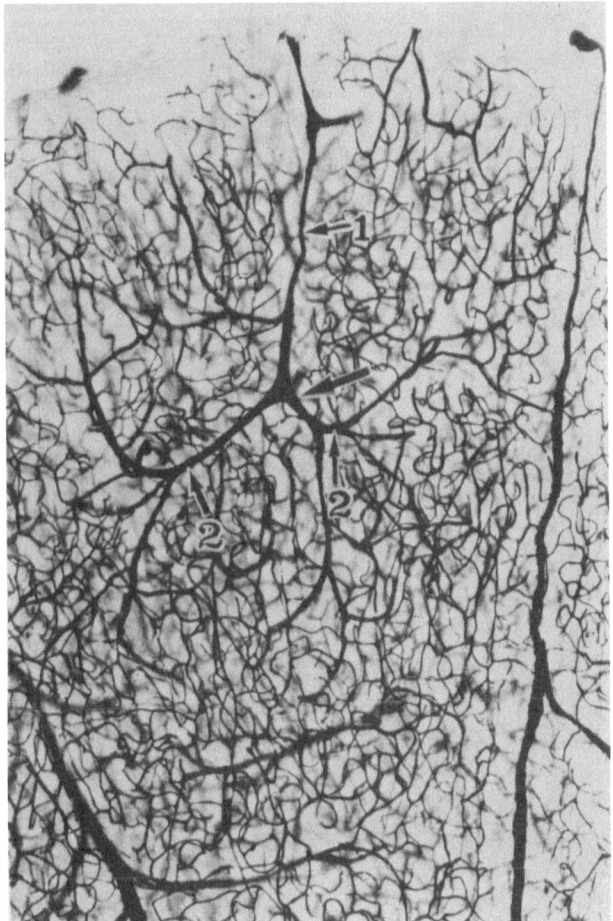

Fig. 26. *1* Morphological aspect of a small intracortical vein (V 3) *2* radiating branches converging towards the principal trunk (arrow); note the diameter enlargement at the junction point

same conclusions in studying other regions of the brain[8]. We note, however, that at the periphery of the gyri, the collecting veins cross over the arteries, flow into the superficial cerebral veins, and finally adhere intimately to the arachnoid (Figs. 2–5). In the central region of the gyri however the pial veins are separated from the arachnoid by the arterial network.

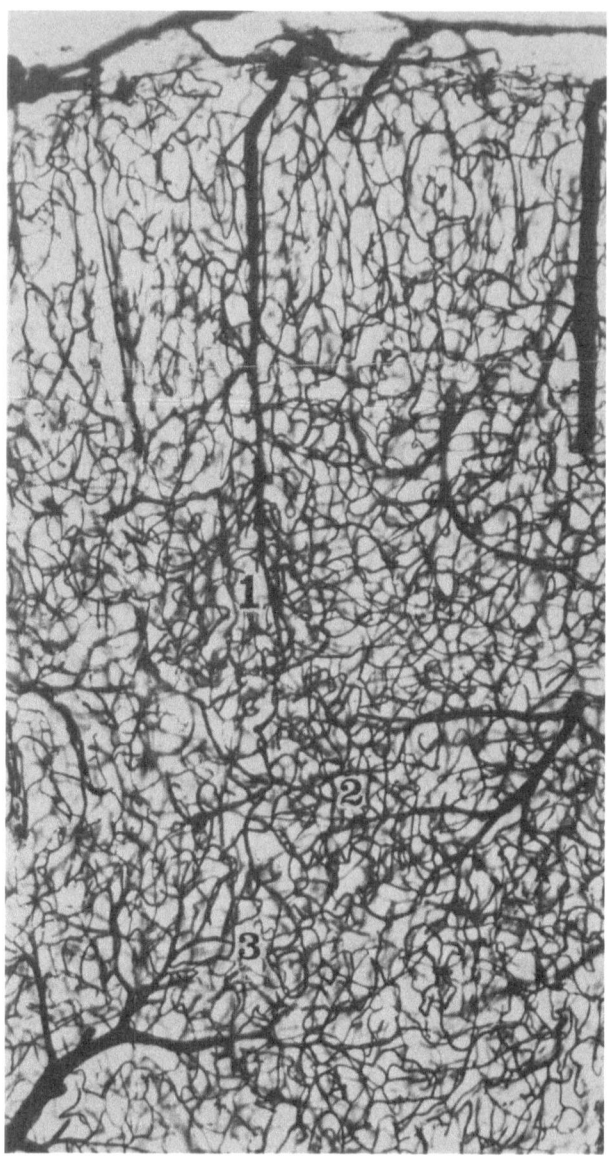

Fig. 27. Middle temporal gyrus (× 60). Section of the cortex showing the bush-like venous network (*1, 2, 3*) formed by ramifications of V 3, V 4, and V 5 veins

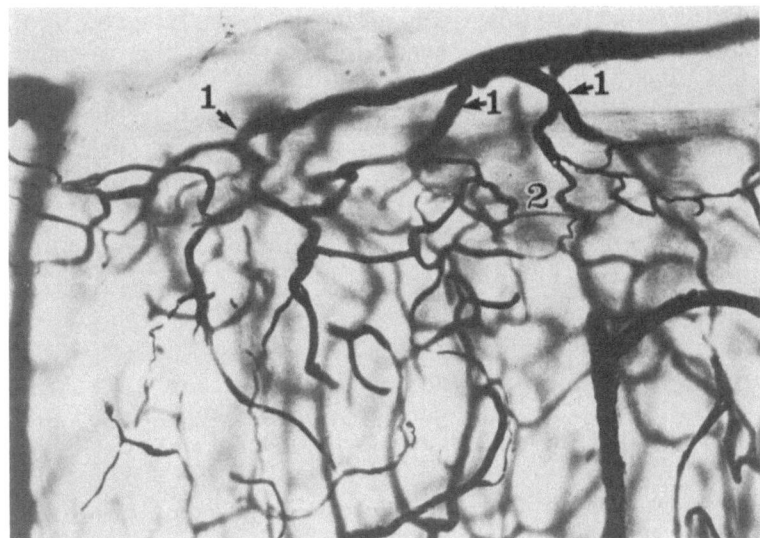

Fig. 28. Middle temporal gyrus (× 140). *1* Small intracortical veins (V 1). *2* Ramifications in the molecular layer

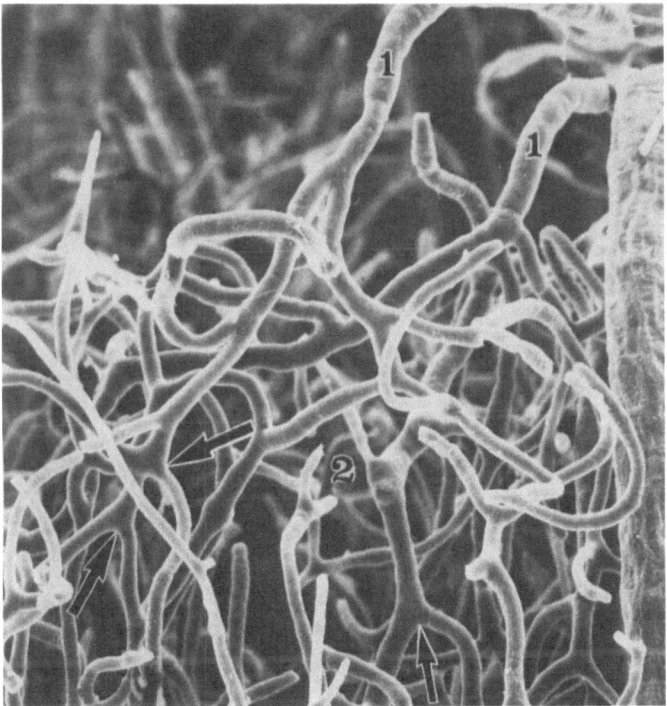

Fig. 29. Middle temporal gyrus (SEM; × 245). *1* Small intracortical veins (VI) whose ramifications are situated within the superficial cortical layers (*2*); arrows indicate diameter dilatations at the junction of secondary branches (SEM)

The blood appears to flow freely in the pial veins. However these veins are sometimes compressed by overcrossing arteries; this therefore may diminish the venous blood flow (Fig. 14).

The question of pial vessel anastomoses has often been debated. While arterial anastomoses are numerous and recognized as such by several authors[3, 5, 12, 16, 18], venous anastomoses are variable in number depending upon the gyri[21] (Figs. 2, 8, 9, and 12). Our observations concord with those of other researchers[4, 7, 12, 21] concerning both the absence of arterio-venous anastomoses and that of capillary network at the cortical surface.

2. Intracortical Veins

The difference between intracortical arteries and veins has long been a hot subject of discussion among researchers[6, 14, 19, 20, 21, 24]. While the

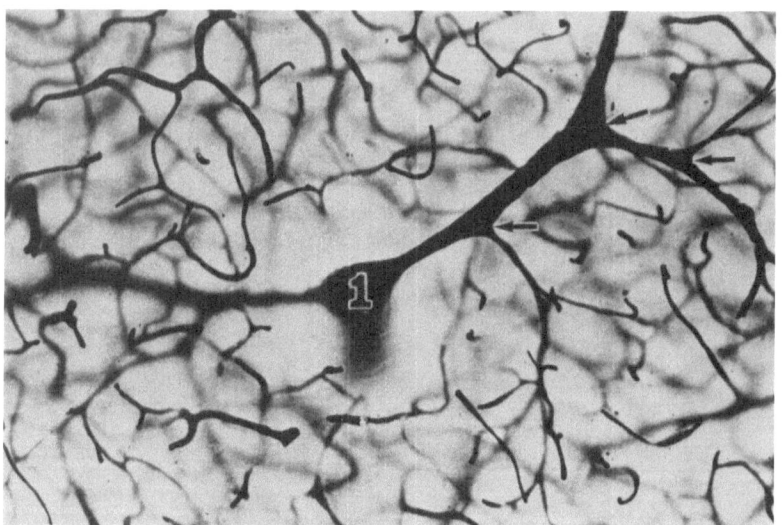

Fig. 30. Superior frontal gyrus. Tangential section ($\times$ 130). *1* Intracortical vein; arrows indicate typical triangular enlargement at the union of secondary branches

deeper veins (groups 4 and 5) are readily recognizable, the problem is more arduous with the more superficial veins (groups 1, 2, and 3). Among the methods of identifications, an important criterion is the triangular shaped increase in diameter at the junction of venous embranchements (Figs. 29 and 30). Important as this may be, absolute certitude can only be obtained in our opinion in following the easily recognizable pial vessels from their onset to their junction with the intracortical vessels.

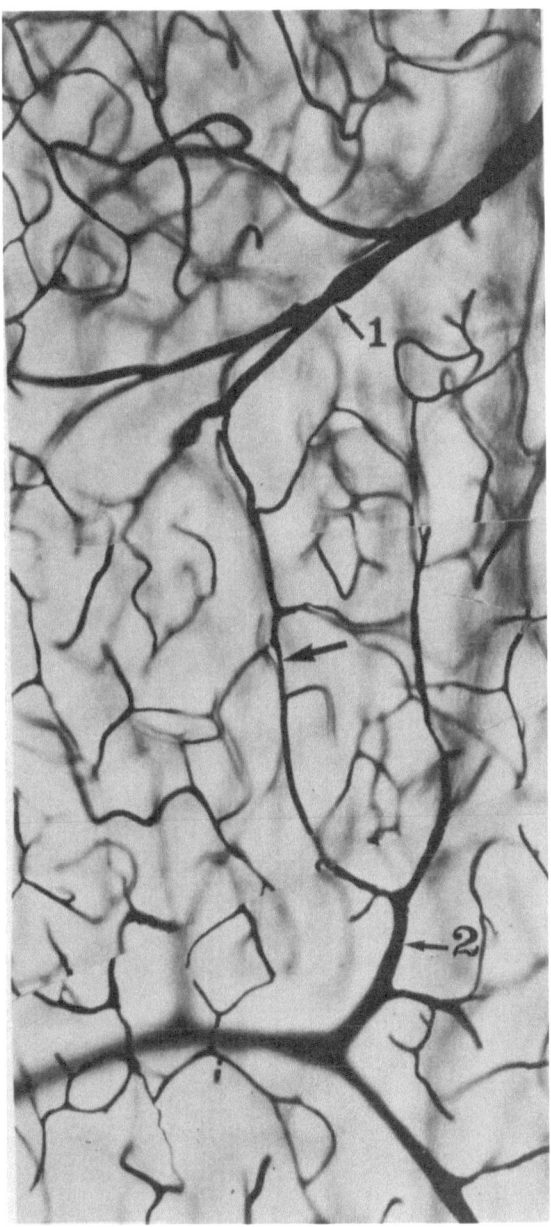

Fig. 31. Middle temporal gyrus. Deep cortical layers (× 130). Arrow indicates probable precapillary arterio-venous anastomosis between an arterial branch (*1*) and a deep venous branch (*2*)

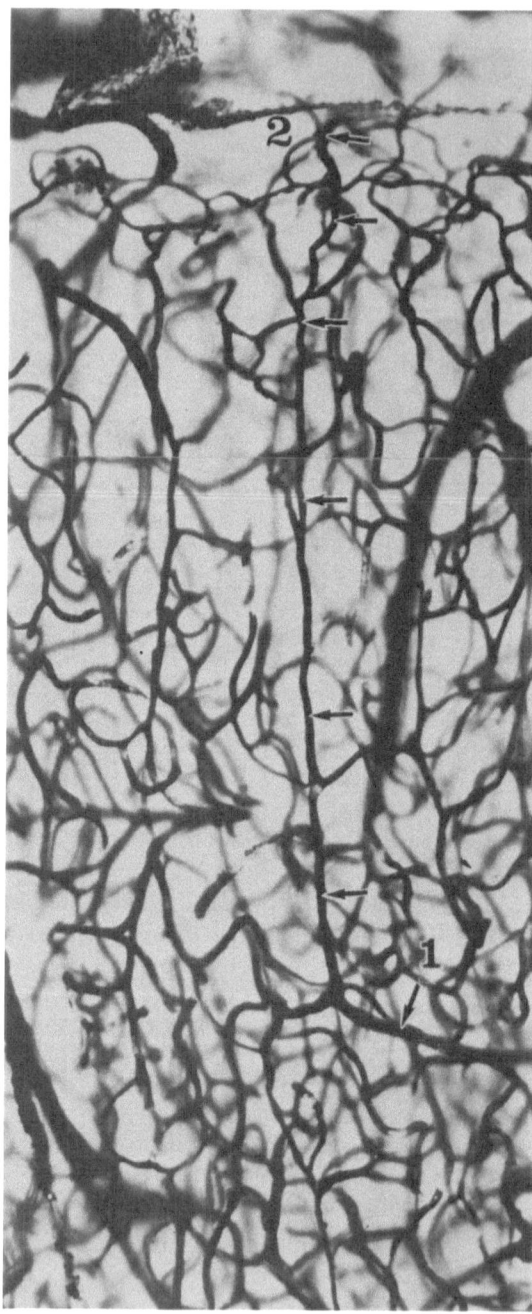

Fig. 32. Middle temporal gyrus. Superficial cortical layers (×140). Arrows indicate a probable precapillary arteriovenous anastomosis joining a recurrent artery (*1*) to a type 1 small cortical vein (*2*)

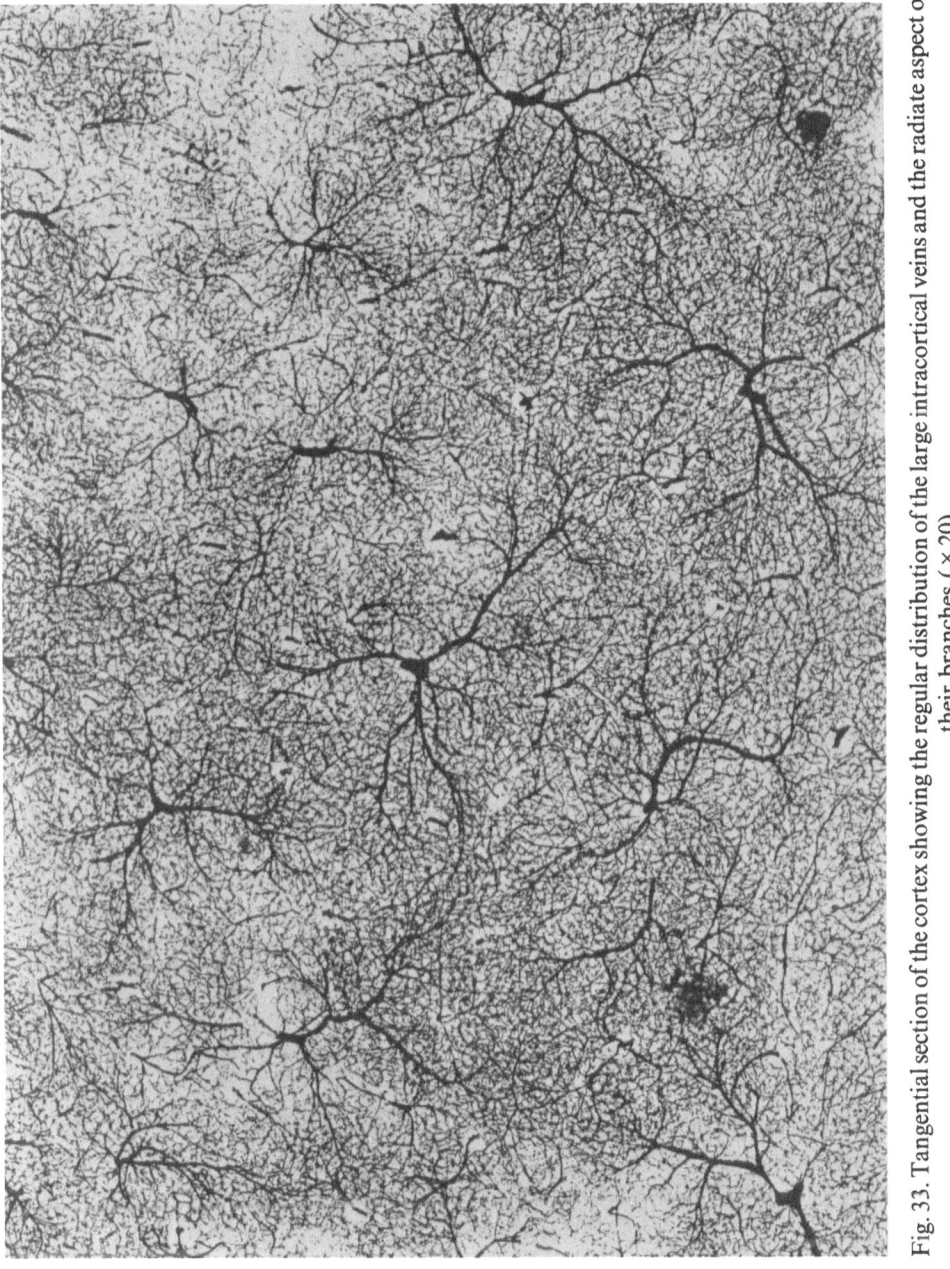

Fig. 33. Tangential section of the cortex showing the regular distribution of the large intracortical veins and the radiate aspect of their branches (× 20)

The simultaneous study of intracortical arteries, capillary network, and veins provides a more precise notion of the intracortical circulation which seems to be divided into three pathways as already mentioned (Fig. 16). The presence of intracortical vascular anastomoses may modify this circulation. In reality no important anastomosis, either arterial or venous, has been observed.

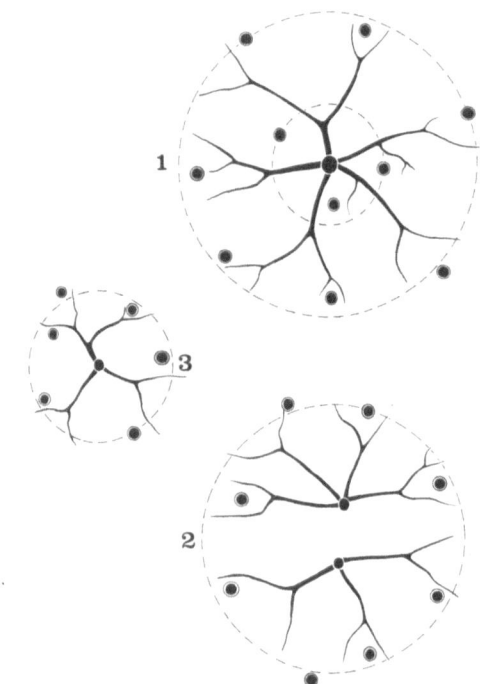

Fig. 34. Drawing of tangential section of venous units. Both arteries and veins are shown in black, but arteries have double circles. *1* Large venous unit surrounded by two arterial rings. *2* Two adjacent veins center this unit. *3* Small venous unit surrounded by only one arterial ring

Precapillary anastomoses however are still the object of controversy. While we have encountered along with other authors [1, 4, 13, 15, 22] a few intra arterial precapillary anastomoses, we have never observed any venous anastomosis of the same type.

Although the precapillary arterio-venous anastomoses are difficult to affirm anatomically, they are probably present in both the superficial and deep cortical layers (Figs. 16, 31, and 32). These shunts already described by several researchers [14, 17] may momentarily deprive certain capillary territories of sufficient blood flow.

Geometric disposition of intracortical vessels was first observed in rat brain by Wolf[23] and Bär[2]. We have observed in the human brain the presence of vascular units each of which is formed by a central vein surrounded by an arterial ring (Figs. 33–36). If these hypothesis are verified,

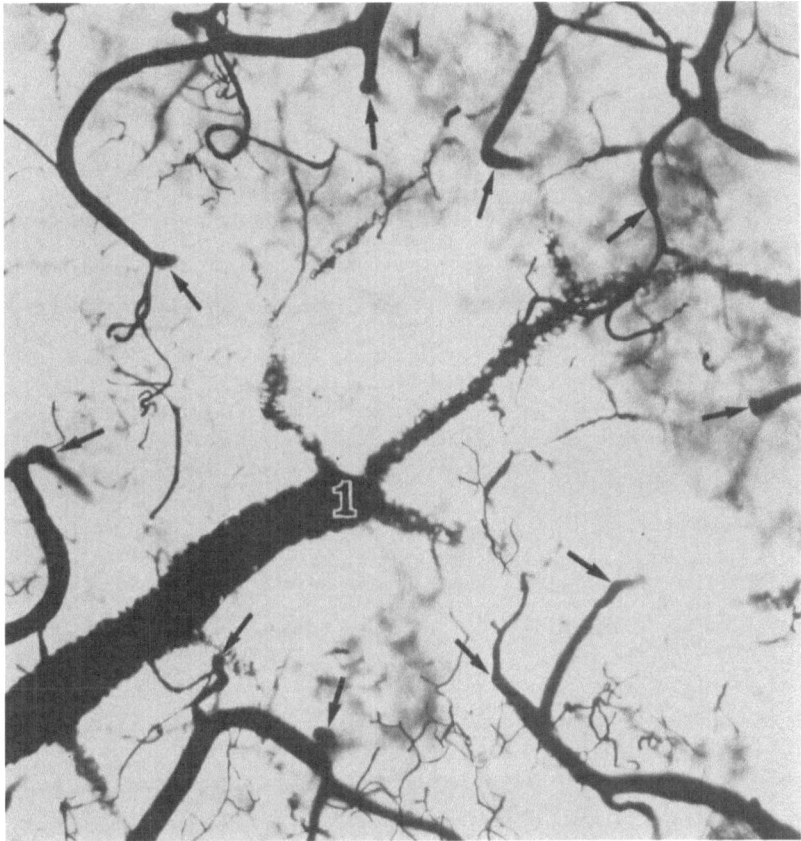

Fig. 35. Venous unit at the cortical surface (after clearing). *1* Emergence of a principal intracortical vein. Arrows indicate cortical arteries which penetrate at right angle in a ring arround the emerging vein (× 50)

it would be interesting to compare these vascular units with neuronal modules classically described in the cortex.

In conclusion we recognize the fact that our knowledge of the anatomy of cerebral vessels remains incomplete. More precise notions are indispensable in order to fully understand the cortical circulation and its pathology.

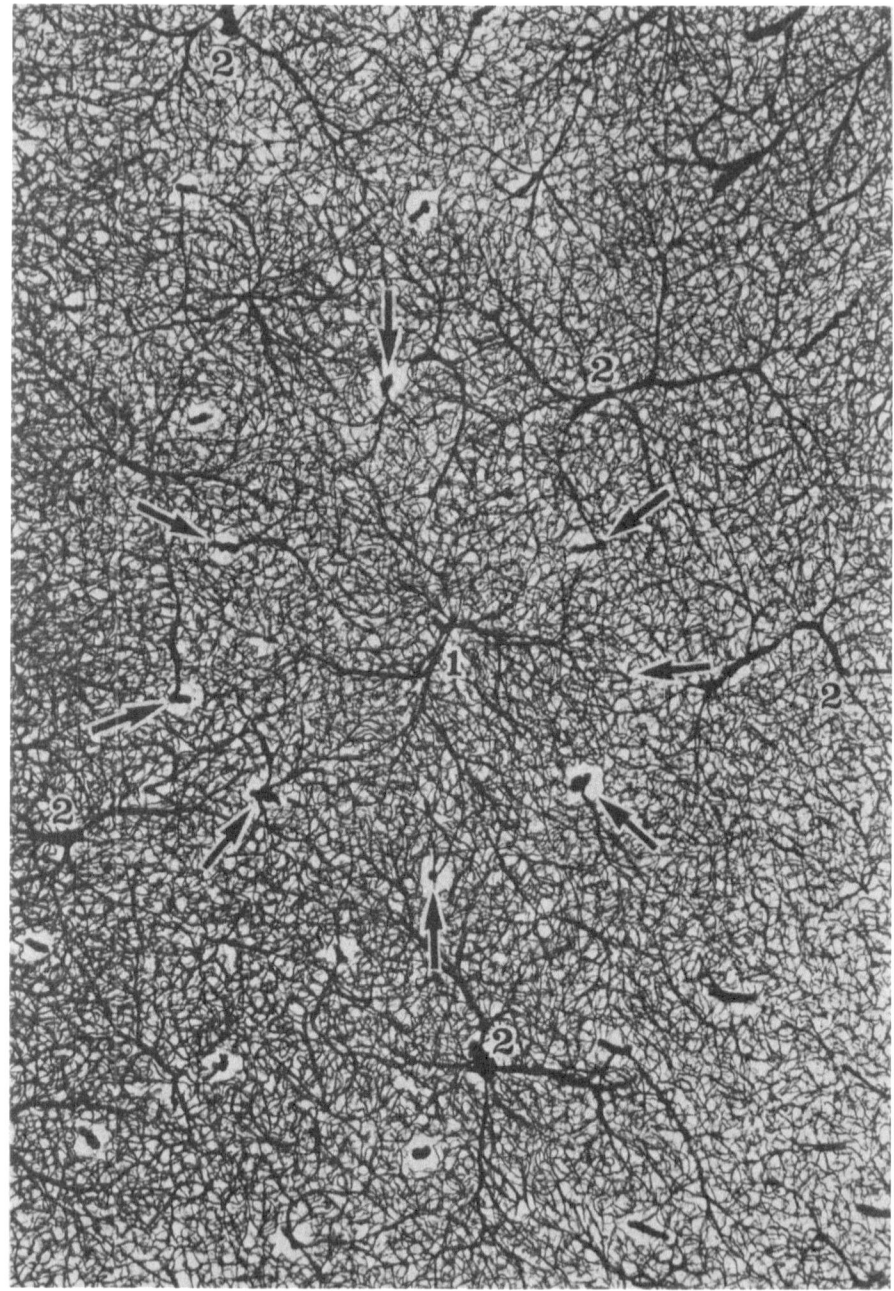

Fig. 36. Tangential section of the cortex. *1* Venous unit with its arterial ring (arrows). *2* Adjacent venous units (× 38)

References

1. Anderson, B. G., Anderson, W. D.: Shunting in intracranial microvasculature demonstrated by SEM of corrosion-casts. Amer. J. Anat. *153*, 4 (1978), 617—624.

2. Bär, Th.: The vascular system of the cerebral cortex. In: Advances in Anatomy Embryology and Cell Biology, Vol. 59. Berlin-Heidelberg-New York: Springer. 1980.

3. Batson, O. V.: Anatomical problems concerned in the study of cerebral blood flow. Fed. Proc. *3* (1944), 139—144.

4. Campbell, A. C. P.: The vascular architecture of the cat's brain. A study by vital injection. In: The Circulation of the Brain and Spinal Cord. Proc. Ass. Res. Ment. Dis. New York, 1937, Vol. XVIII, pp. 69—93. Baltimore: Williams and Wilkins. 1938.

5. Charpy, A.: In: Traité d'Anatomie Humaine: Système nerveux. Tome troisième, deuxième fascicule (Poirier, P., Charpy, A., eds.), pp. 373—610. Paris: Masson. 1902.

6. Courville, C. B.: Vascular patterns of the encephalic gray matter in man. Bull. Los Angeles Neurol. Soc. *23*, 1 (1958), 30—43.

7. Duret, M.: Recherches anatomiques sur la circulation de l'encéphale. Arch. Physiol. Norm. Pathol., Deuxième série *1* (1874), 316—353.

8. Duvernoy, H.: Human Brainstem Vessels. Berlin-Heidelberg-New York: Springer. 1978.

9. Duvernoy, H., Delon, S., Vannson, J. L.: Cortical blood vessels of the human brain. Brain Res. Bull. *7* (1981), 519—579.

10. Fang, C. H.: The studies of cerebral arterioles and capillaries in normal and pathologic states in man. In: Pathology of Cerebral Microcirculation (Cervós-Navarro, J., ed.), pp. 431—441. Berlin-New York: De Gruyter. 1974.

11. Florey, H.: Microscopical observations on the circulation of the blood in the cerebral cortex. Brain *48* (1925), 43—64.

12. Forbes, H. S.: The cerebral circulation. 1: Observation and measurement of pial vessels. Arch. Neurol. Psychiat. *19* (1928), 751—761.

13. Hale, A. R., Reed, A. F.: Studies in cerebral circulation. Methods for the qualitative and quantitative study of human cerebral blood vessels. Amer. Heart J. *66* (1963), 226—242.

14. Hasegawa, T., Ravens, J. R., Toole, J. F.: Precapillary arteriovenous anastomoses. "Thouroughfare channels" in the brain. Arch. Neurol. *16* (1967), 217—224.

15. Kennady, J. C., Taplin, G. V.: Shunting in cerebral microcirculation. Amer. Surg. *33*, 10 (1967), 763—771.

16. Lierse, W.: Über die Beeinflussung der Hirnangioarchitektur durch die Morphogenese. Acta Anat. *53* (1963), 1—54.

17. Ravens, J. R.: Anastomoses in the vascular bed of the human cerebrum. In: Pathology of Cerebral Microcirculation (Cervós-Navarro, J., ed.), pp. 26—38. Berlin-New York: De Gruyter. 1974.

18. Ruckes, J.: Die arterielle Vascularisation der Pia mater des Neugeborenen. Frankfurt. Z. Pathol. *76* (1967), 227—234.

19. Saunders, R. L. de C. H., Bell, M. A.: X-ray microscopy and histochemistry of the human cerebral blood vessels. J. Neurosurg. *35* (1971), 128—140.

20. Saunders, R. L. de C. H., Feindel, W. H., Carvalho, V. R.: X-ray microscopy of the blood vessels of the human brain. Part II. Med. Biol. Illus. *15* (1965), 234—246.

21. Scharrer, E.: Arteries and veins in the mammalian brain. Anat. Rec. *78* (1940), 173—196.

22. Van den Berg, R., van der Eecken, H.: Anatomy and embryology of cerebral circulation. Prog. Brain Res. *30* (1968), 20—25.

23. Wolff, J. R.: An ontogenetically defined angioarchitecture of the neocortex. Arzneimittel-Forsch. *26* (1976), 1239.
24. Wünscher, W., Werner, L.: Zur Methodik angioarchitektonischer Untersuchungen. Psychiatr. Neurol. Med. Psychol., Leipzig *13* (1961), 1—6.

Author's address: Prof. Dr. H. M. Duvernoy, Faculté de Médecine, Laboratoire d'Anatomie, Place St. Jacques, F-25030 Besançon Cedex, France.

Institute of Neuropathology of the Free University of Berlin, Berlin

Ultrastructure of Venules in Human and Cat Brain

W. Roggendorf* and J. Cervós-Navarro

Summary

We investigated 70 meningeal and intracerebral venules from normotensive patients and 85 feline venules.

The basic construction of the vascular wall of human and feline venules is similar. They do not have a well defined wall composed of endothelium, media, and adventitia as is characteristic for arteries and arterioles. The human venous endothelium additionally contains cytoplasmic lipoid vacuoles and nuclear inclusions. The perithelium in postcapillary venules contains more branched periendothelial cells than in the capillary. In the collecting venules the perithelium is more complete and compact due to an increase of periendothelial cells. Only the human brain has muscular venules intracerebrally.

In the meninges of both humans and cats, collecting venules and veins are more plentiful, but occasional capillaries and short segments of postcapillary venules are also present. In humans, additional muscular vessels, although with an incomplete layer, are present. Meningeal vessels and veins differ from those in the intracerebral compartment with respect to the perivascular space and their innervation. Meningeal vessels do not display a particular space filled with proteo-glycane material, but they do have bundles of nerve fibres.

Keywords: Venules; human and cat brain; ultrastructure.

Introduction

The morphology of the cerebral microcirculation has been described rather comprehensively by terms such as the "capillary bed", the "microvasculature" and the "terminal system", which includes arterioles, capillaries and venules. Although many authors mentioned venules within the scope of the entire cerebral and meningeal vascular system[6, 7, 10, 12, 13]. The ultrastructure of intracranial venules described by Takahashi[17], Roggendorf *et al.*[16] and Lange and Halata[11] is limited to animal material. The aim of these present studies was twofold: to present ultrastructural results from human intracerebral venules and to compare the intracerebral and meningeal venules in humans and animals.

---

* Additional affiliation at the Washington University, St. Louis, Mo., U.S.A.

Methods and Material

We investigated eight autopsy brains and seven surgical brain biopsies of normotensive patients, aged 26–70 years who had no evidence of other vascular disease. After a death to fixation interval of 100–200 min, we sampled specimens from the frontal, temporal and parietal lobes and pontine base. The material was fixed by immersion for 4 hours in buffered Glutaraldehyde, postfixed in Osmiumtetroxyde and prepared for electronmicroscopy as usual. Besides 70 meningeal and intracerebral venules from humans, we used material from 85 feline venules from our previous study[16].

Venules in the terminal system were identified according to Rhodin[14]: 1. Post capillary venules as micro-vessels with a diameter of about 8–30 μ continuous with venous capillaries. These vessels show a gradual increase in the number of pericytes with increasing luminal diameter. 2. Collecting venules as vessels with a diameter between 30–50 μ, with one complete layer of pericytes and a complete layer of veil cells. 3. Muscular venules as vessels between 50–100 μ in diameter. Nearly all periendothelial cells are smooth muscle cells that overlap and sometimes form two layers. Myoendothelial junctions occur but are rare.

Results

In the present feline as well as in the human material, small venules (post-capillary venules) were frequent whilst larger venules (collecting venules) were rare. Only occasionally in our human material could we detect large muscular venules.

Endothelium

The structure of the endothelium does not differ in either type of venule and is similar to that of capillaries. Cellular organelles, filaments, golgi apparatus, rough endoplasmatic reticulum and mitochondria are sparse, Weibel-bodies and multi-vesicular bodies are rare. In human intracerebral venules, however, we detected large lipid droplets often in close association with the nucleus (Fig. 3 a). Intranuclear inclusion bodies were be found in the endothelium and in periendothelial cells (Fig. 1 b). There is no regular distribution of filaments.

Perithelium

Cells surrounding the venous endothelium, which can be classified neither as pericytes nor as smooth muscle cells, are expediently called periendothelial cells. Subendothelial elastic fibre material or collagenous fibres are absent. In *post capillary venules* periendothelial cells are more branched than in capillaries and lie adjacent to endothelial cell membranes in order to form zonulae adherents (Fig. 1 a). Regarding the periendothelial cells, there are no differences between human and feline venules. In segments where processes of periendothelial cells are absent, the basal lamina of the endothelium merges with that of the surrounding nervous tissue and thus there is no real perivascular space (Fig. 1). The subcellular structure of periendothelial cells is described in detail by Rhodin[14] and by Forbes *et al.*[5]. In the *collecting venules* the periendothelial cells show few ramifications. These cells increase in number and appear to be more compact (Fig. 2). In addition, the periendothelial cells of collecting venules contain more filaments and the plasma membrane has semilunar or irregularly shaped dense areas with no obvious connections with the filaments. Some larger intracerebral venules contain an incomplete sheath of typical smooth muscle cells corresponding to muscular venules (Fig. 3 c).

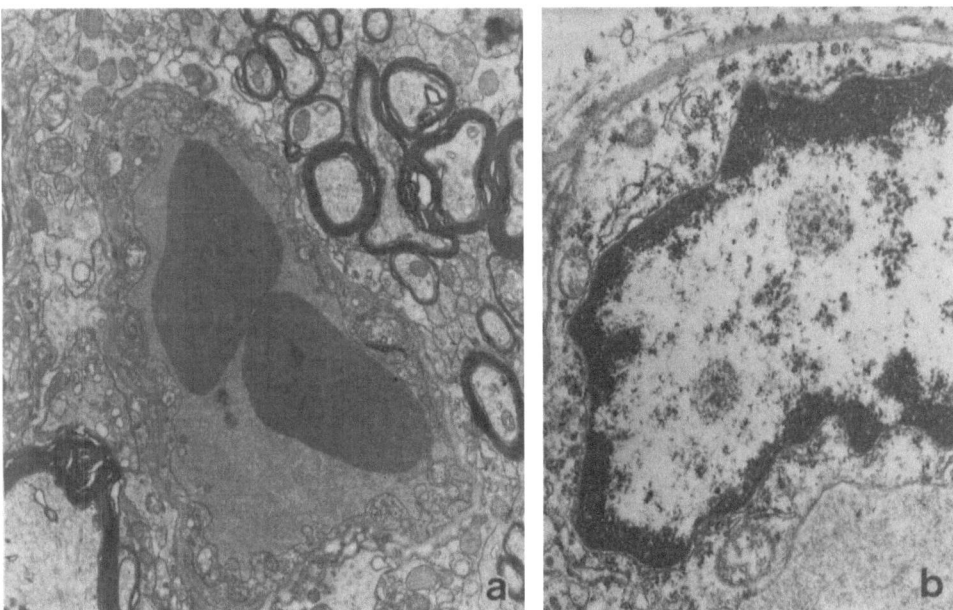

Fig. 1 a. Postcapillary venules from the human pons. Small processes of periendothelial cells showing incomplete sheathing of the vessel circumference. × 11,200

Fig. 1 b. Nuclear inclusion bodies in an endothelial cell of a human brain venule. × 28,000

The volume and nature of the perivascular space around intracerebral venules vary considerably depending upon the location in the brain. In cortical postcapillary venules, the endothelial basal lamina merges with that of the surrounding nerve tissue, so that no perivascular space exists. Outside the cortex the perivascular space occasionally contains collagen fibres, connective tissue and basic substance. In the collecting venules, the narrow perivascular space contains collagen, adventitial cells and cells with numerous lysosomes. These lysosome-containing-cells are much more prominent in human material (Fig. 3 a). The surrounding glial sheath contains both in human and feline material, dense zones shaped either as trapezoides or as symmetrical triangles (Fig. 3 b).

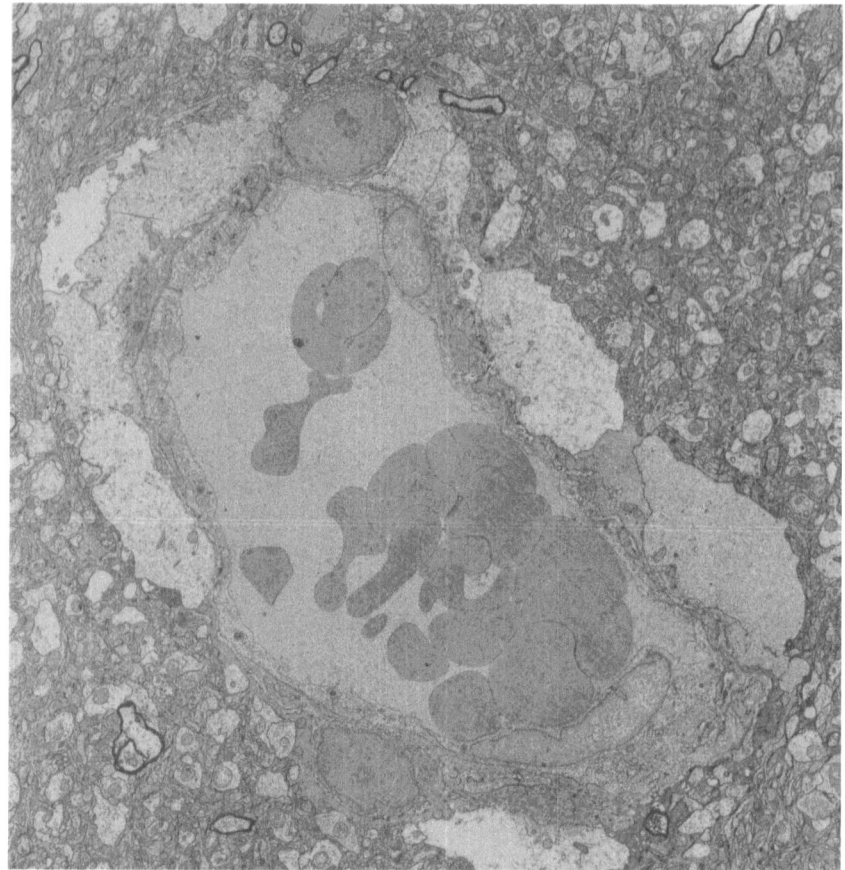

Fig. 2. A collecting venule from the parietal lobe of the cat brain showing a nearly continuous sheath of periendothelial cells containing prominent nuclei. × 3,699
[From L. Roggendorf *et al.* Cell Tiss. Res. *192* (1978) 465]

Meningeal Venules

In the meninges, collecting venules and veins are most plentiful. Occasional capillaries and short sections of postcapillary venules are also present. Collecting venules display the same irregular construction without muscular media. There is no difference between human und feline material.

The narrow appendices of periendothelial cells incompletely surround the vascular circumference so that, over certain distances, the endothelium is separated from the subarachnoid space only by a basal lamina (Fig. 4a). Occasionally unmyelinated nerve fibres occur, which ramify from the sheath via the Schwann cells. These fibres contain vesicles with and without dense cores.

Moreover, in our human material we were able to detect small cell processes which resemble typical smooth muscle cells. In contrast to feline

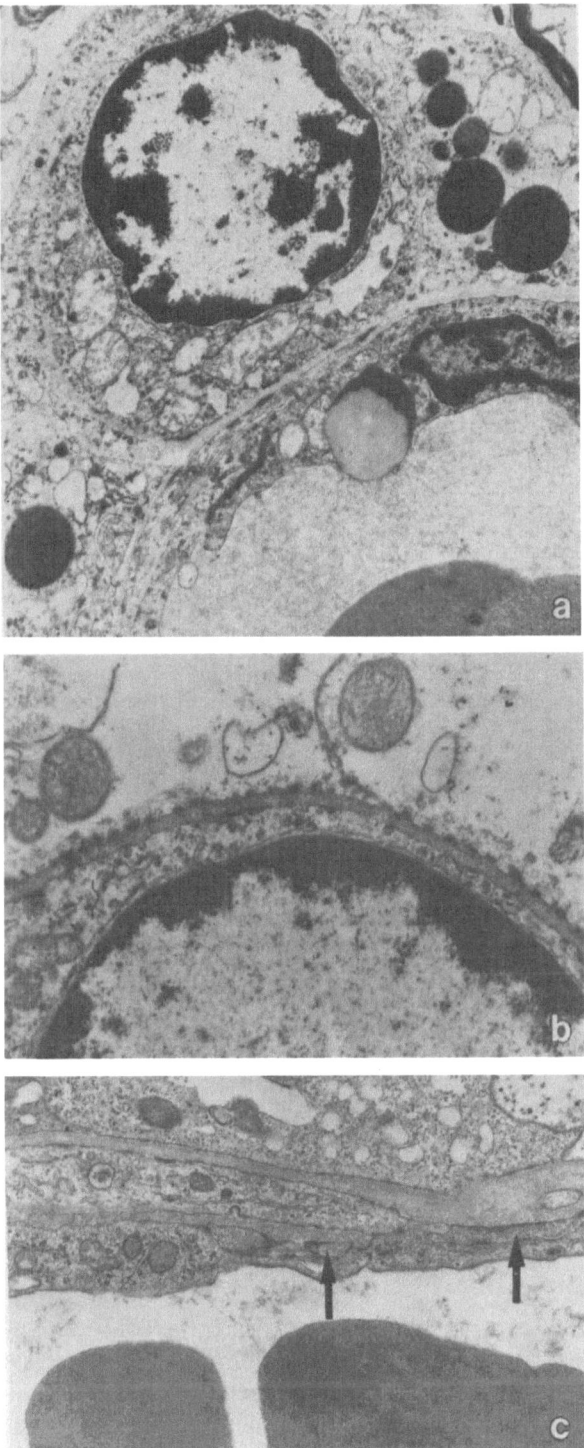

Fig. 3 a. Postcapillary human venule with a large lipid vacuole in the endothelium, lysosomes in the phagocytic cells in the perithelium. × 12,400

Fig. 3 b. Postcapillary venule from the parietal lobe of a cat, dense zones are especially conspicuous around the deformation of the perivascular space. × 3,500

Fig. 3 c. Wall of collecting venule in human brain with a nearly complete sheathing of the endothelium by adventitial cells and processes of smooth muscle cells (arrows). × 25,000

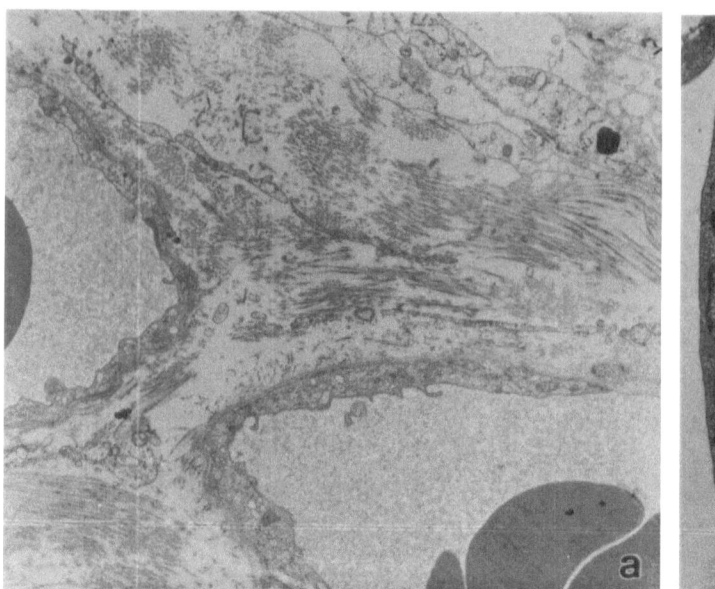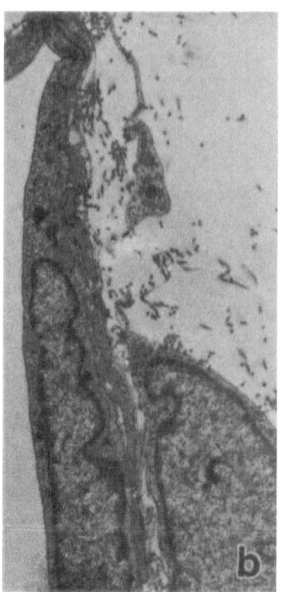

Fig. 4a. Two postcapillary venule segments in the meninges of a human brain. There are only small processes of periendothelial cells The meninges contain large numbers of collagen bundles. × 5,600

Fig. 4b. Muscular venule of the human meninges with a nearly complete smooth muscle layer surrounding the endothelium. Small processes of meningeal cells. × 9,620

endothelium, the endothelium of human material contains large translucent vacuoles (Fig. 4b).

Discussion

Venules do not have a vascular wall consisting of endothelium, media and adventitia as is characteristic for arteries and arterioles. This is an important criterion that distinguishes venules from the rest of the microcirculation. Muscular venules are only detectable in human brains. Such an anatomical difference may reflect differences in permeability during inflammation, capacitance and hydraulic hydrance.

Endothelium

The endothelial organelles of venules differ only slightly in various portions of the brain from those of other organs. Weibl-bodies described by Rhodin[14] and Hammersen[8, 9], as endothelial specific organelles are rare in intracerebral venules. The endothelial cells in human tissue frequently contain large lipid vacuoles and intranuclear inclusion bodies, which were not observed in other species[3, 9, 14]. The functional role of these lipid vacuoles is unclear.

Perithelium

Since the vascular wall of venules do not have a well defined structure with the three compartments of endothelium, media and adventitia as is characteristic for arterioles, we favour the term "perithelium", for describing those structures surrounding the endothelium and the basal lamina. The contradictory views relating to the morphological origin and function of the periendothelial cells were discussed recently[16]. Dense collagenous bundles are essentially structural elements of the perivascular space around venules in the central nervous system. The frequency and distribution of collagen bundles are different with regard to different venous compartments, and have a different localization in the CNS. Dense collagen bundles are found frequently in the postcapillary venules in the brain stem, whereas collagen bundles in postcapillary venules in the cat cortex are rare. These findings correspond to the results of Drommer and Schulz[4]. Collagen is more prominent in venules than in capillaries[1] and in arterioles[2, 3, 11, 15]. In the human venules an age dependent increase in collagen was shown not only in postcapillary venules but also in collecting venules and muscular venules. Dense areas at the glial cell membrane are described in detail by Vaquera-Orte et al.[18]. These authors assume that dense areas may play a functional role during calibre changes in intracerebral vessels.

Meningeal Venules and Veins

Meningeal venules and veins differ from those in the intracerebral compartments with respect to their perivascular space. Meningeal venules do not display a particular space filled with a proteoglycane material. Although the periendothelial cells in meningeal venules contain parallel adjusted filaments which are more numerous than in collecting venules, they still do not correspond to the typical smooth muscle cells of muscular venules. Only in our human material could we find well developed smooth muscle cells. Both the increased number of pinocytotic vesicles in the meningeal venules of the human and cat brain, and the large vacuoles in our human material might play a role in the absorption of spinal fluid together with the Pacchionian granules.

Acknowledgement

This work was supported by a Feodor-Lynen-Fellowship of the Alexander v. Humboldt foundation, Bonn.

References

1. Cervós-Navarro, J.: Elektronenmikroskopische Befunde an den Kapillaren der Hirnrinde. Arch. Psychiat. Z. Ges. Neurol. *204* (1963), 484—504.
2. Dahl, E.: The fine structure of intracerebral vessels. Z. Zellforsch. *145* (1973), 577—586.
3. Drommer, W.: Feinstruktur der normalen Arteriolen und ihre Alterationen nach experimentellem Colitoxinschock im zentralen Nervensystem des Schweines. Acta Neuropath. (Berl.) *22* (1972), 29—41.

4. Drommer, W., Schulz L.-Cl.: Feinstruktur der normalen Kapillaren und Venolen im Rückenmark des Schweines. Anat. Anz. *128* (1971), 232—247.

5. Forbes, M. S., Rennels, M. L., Nelson, E.: Ultrastructure of pericytes in mouse heart. Amer. J. Anat. *149* (1977), 47—70.

6. Frederickson, R. G., Low, F. N.: Blood vessels and tissue space associated with the rat. Amer. J. Anat. *125* (1969), 123—145.

7. Hager, H.: Elektronenmikroskopische Untersuchungen über die Feinstruktur der Blutgefäße und perivaskulären Räume im Säugetiergehirn. Acta Neuropath. (Berl.) *1* (1961), 9—33.

8. Hammersen, F.: Anatomie der terminalen Strombahn. München-Berlin-Wien: Urban und Schwarzenberg 1971.

9. Hammersen, F.: Bau und Funktion der Blutkapillaren. In: Handbuch der allgemeinen Path. III/7, pp. 151—153. Berlin-Heidelberg-New York: Springer. 1977.

10. Hogan, J. M., Feeney, L.: The ultrastructure of the retinal blood vessel. J. Ultrastruct. Res. *9* (1963), 10—28.

11. Lange, W., Halata, Z.: Comparative studies on the pre- and postterminal blood vessels in the cerebellar cortex of Rhesus monkey, cat, and rat. Anat. Embryol. *158* (1979), 51—62.

12. Maynard, E. A., Schultz, R. L., Pease, D. C.: Electron microscopy of the vascular bed of rat cerebral cortex. Amer. J. Anat. *100* (1957), 409—433.

13. Nelson, E., Blinzinger, K., Hager, H.: Electron microscopic observations on the subarachnoidal and perivascular spaces of the syrian hamster brain. Neurology (Minneap.) *11* (1961), 285—295.

14. Rhodin, J. A. G.: Ultrastructure of mammalian venous capillaries, venules and small collecting veins. J. Ultrastruct. Res. *25* (1968), 452—500.

15. Roggendorf, W., Cervós-Navarro, J.: Ultrastructure of arterioles in the cat brain. Cell Tiss. Res. *178* (1977), 495—515.

16. Roggendorf, W., Cervós-Navarro, J., Lazaro-Lacalle, M. D.: Ultrastructure of venules in the cat brain. Cell Tiss. Res. *192* (1978), 461—474.

17. Takahashi, M.: Fine structure of the rat intracranial veins. Acta Anat. Nippon *43* (1968), 238—254.

18. Vaquera-Orte, J., Cervós-Navarro, J., Martin-Girón, F., Becerra-Ratia, J.: Fine structure of the perivascular limiting membrane. In: Cerebral Microcirculation and Metabolism (Cervós-Navarro, J., Fritschka, E., eds.), pp. 129—136. New York: Raven Press. 1981.

Authors' address: W. Roggendorf, M.D., Institute of Neuropathology of the Free University of Berlin, D-1000 Berlin.

Institute of Neuropathology of the Free University Berlin, Berlin

Pathological Changes in the Vessel Wall of Intracerebral Venules: an Ultrastructural Study

J. Cervós-Navarro and W. Roggendorf*

Summary

The first part of our study deals with the effect of hypertension and aging on human intracerebral venules. In the second part we investigated the spastic constriction of intracerebral venules caused by electrical stimulation of the cat brain.

We investigated three brains and four surgical brain biopsies of normotensive patients aged 50 to 70 years who had no evidence of vascular disease. We also studied eight human brains and one surgical biopsy from patients 51 to 73 years old who had a history of at least two years of a raised blood pressure. We sampled specimens from the temporal, parietal, and frontal lobes and pontine base. Venules show focal as well as concentric fibrosis and in some cases typical smooth muscle cells containing numerous myofilaments and dense zones. The changes in normotensive patients during aging show, besides degenerative changes such as fibrosis, an increase in cellular components in the perivascular space containing lipid-like inclusions. Since these changes were apparently more common and more advanced in hypertensives patients, we conclude that hypertension may accelerate the aging of the microvasculature in the brain.

To investigate spastic changes in the intracerbral venules, the brains of two groups, 6 cats in each, were subjected to direct current or electroshock. Tissue samples for electron microscopy were taken from the parietal lobe and the caudate nucleus. Electric stimulation induced a spastic vasoconstriction thereby markedly deforming the entire vessel wall which then appeared to have a starlike profile with only a small lumen. The subendothelial space was undulate. The mechanism as to how spastic changes occur, even in venules in which there is only a minimal amount of contractile elements is still unknown.

Keywords: Intracerebral venules; hypertension; aging; ultrastructure.

Introduction

The ultrastructure of intracerebral venules in different animal species has been described previously[15,21]. Recent data regarding the normal

---

* Additional affiliation at the Washington University, St. Louis, Mo., U.S.A.

human intracerebral and meningeal venules were reported in the chapter proceeding this communication (p. 39). Pathological changes are limited in brain venules despite their importance in a variety of cerebral vascular diseases, especially with regard to permeability and capacitance.

Spontaneously hypertensive rats served as a suitable model to investigate the ultrastructure of the intracerebral venous microcirculation under various conditions including aging[4,9,19]. Electron microscopical observations regarding the effect of hypertension on the human brain were predominantly focussed on arterioles and capillaries[6,16]. Therefore our study primarily deals with the effect of hypertension and aging on human intracerebral venules.

Since there are recent data indicating that contractile filaments may play an important functional role not only in the arterial part of cerebral vessels but also in the capillaries and venules[1,10] the question arises, if there are morphological changes in the venous vessel wall. A suitable experimental condition has been established to investigate intracerebral vessels with spastic constriction produced by electrical stimulation[3,11,17]. Therefore we also investigated the venous compartment of the microcirculation when stimulated by electroshock and direct current.

Material and Methods

We investigated 3 brains and 4 surgical brain biopsies of normotensive patients aged 50 to 70 years who had no evidence of vascular disease. We also studied 8 human brains and 1 surgical biopsy from patients 51 to 73 years old, who had a history of at least 2 years of high blood pressure. The most important criteria to separate both groups were, besides the pathological anatomical findings, the heart weight and the heart to body weight ratio. The body weight in the normotensive patients was 64 to 81 kg and the heart weight ranged from 290 to 410 g. In the hypertensive group the body weight ranged from 46 to 77 kg and the heart weight from 294 to 820 g. After a death to fixation interval of 100 to 200 min, we obtained specimens from the temporal, parietal, and frontal lobes and from the pons. The specimens were fixed by immersion for 4 hours using 4.5% buffered Glutaraldehyde and prepared for electron microscopy as usual. We investigated a total number of 60 human intracerebral venules.

For investigation of spastic constriction we used 21 adult cats. After anaesthesia a catheter was inserted into the aorta and into the inferior vena cava. Arterial blood pressure was monitored and remained at a normal level throughout the experiments. Blood gases were kept normal (pCO$_2$ 35 to 45 mmHg, pO$_2$ 92 to 110 mmHg pH 7.35). Two groups were investigated: 1. Electric convulsive treatment was applied in 6 cats after craniotomy, two metal electrodes being attached to the temporal muscle on each side. An alternating current of 220 V 50 Hz was applied for 500 ms. 2. Electrical stimulation was applied by direct current (DC) in 6 cats. One electrode was attached to the right temporal muscle and the left electrode was an Agar-gel-silver electrode with a plastic tube (DC: 20—100 V for 2 to 30 s corresponding to 20 to 150 ma) brought into direct contact with the venous vessel. The remaining animals were used as controls, being treated as the experimental groups but without electric stimulation. 20 min after treatment specimens from the parietal cortex and subcortical white matter and from the caudate nucleus were excised for electron microscopy. The tissue was fixed in 5% buffered Glutaraldehyde and processed as usual.

Results

Effects of Chronic Hypertension and Aging on Intracerebral Venules

A constant finding in the hypertensive patients was that 2 to 4 venules were found in adjacent position. There is a different basic principle of construction in such convoluted vessels. On the one hand venules have a complete vascular wall consisting of endothelium and perithelium. On the other hand, venules share the periendothelial sheath. In all these cases the perivascular space contains an increased amount of collagen.

The endothelium in venules with chronic hypertension exhibits only minor changes. There is an increase in the total cell volume with prominent filaments and tubuli (Fig. 2 a). The most striking features in the perithelium are large amounts of collagen mixed with fine granular material, resulting in widening of the perivascular space. There are two types of collagen distribution in the perivascular space: 1. There is a focal fibrosis predominantly in postcapillary venules, where only a segment of the entire vessel circumference exhibits an increased amount of collagen (Fig. 1 a, b). 2. A concentric fibrosis is more prominent in collecting venules, where the entire circumference of the vessel contains large amounts of collagen (Fig. 2 b). The periendothelial cells contain an increased number of filaments (6 nm). Therefore, it is occasionally difficult to differentiate between periendothelial cells and primitive smooth muscle cells. However, true smooth muscle cells were found as well as periendothelial cells. The number of contact zones between endothelium and periendothelial cells is normal in number and appearance.

The perivascular glial membrane is characterized by multiple dense areas, sometimes condensed into a lamina (Fig. 2 b).

There are no specific differences in venules between old normotensive individuals and those with hypertension. Fibrosis and widening of the perivascular space, as well as convoluted vessels were more prominent in hypertensive individuals.

Spastic Constriction of Intracerebral Venules

In the following description we refer to spastic contraction when marked deformation of the entire vessel wall leads to a completely obstructed vessel, in contrast to physiological contraction, in which the basic configuration of the vessel is retained. Slightly constricted venules was a frequent finding in both experimental groups, but venules with spastic contraction were rare. In our material we observed spastic venules in less that 5% of all the cross sections of venous vessels. Electrical stimulation induced a spastic a spastic vasoconstruction strongly deforming the entire vessel wall, which appeared to have a starlike profile with a small lumen (Fig. 3 a). Other venules showed a strongly scalloped basal lamina and in some indentations in the wall. Moreover, the periendothelial cells showed a bizarre and marked deformation of the cell processes (Fig. 3 b). Occasionally, we were able to observe a parallel arrangement of filaments in the periendothelial cells, whilst the other compartments of the periendothelial cells were deformed.

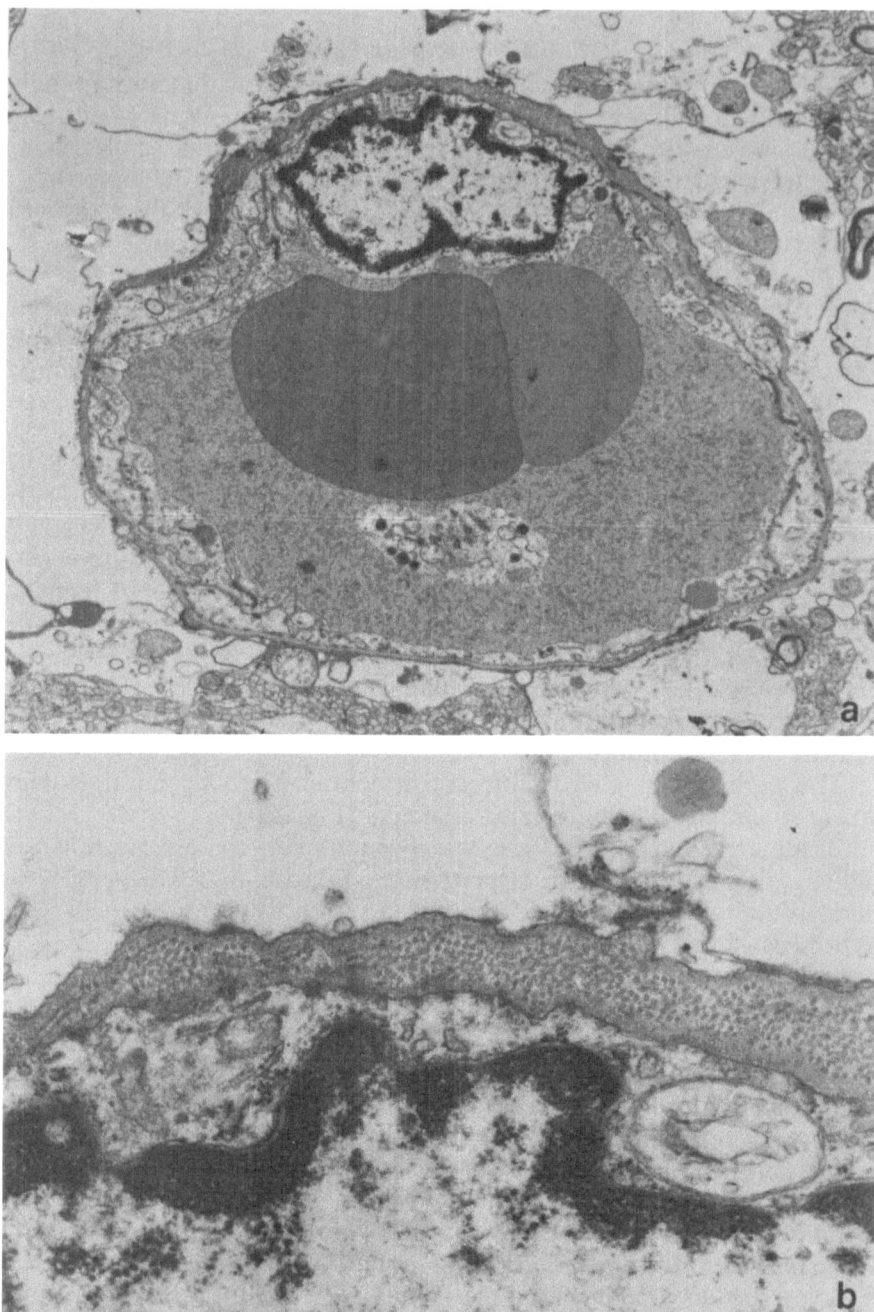

Fig. 1 a. Postcapillary venule of human parietal lobe with segmental fibrosis. × 16,200

Fig. 1 b. Detail from Fig. 1 a. Large numbers of collagen bundles in the perivascular space.
× 33,000

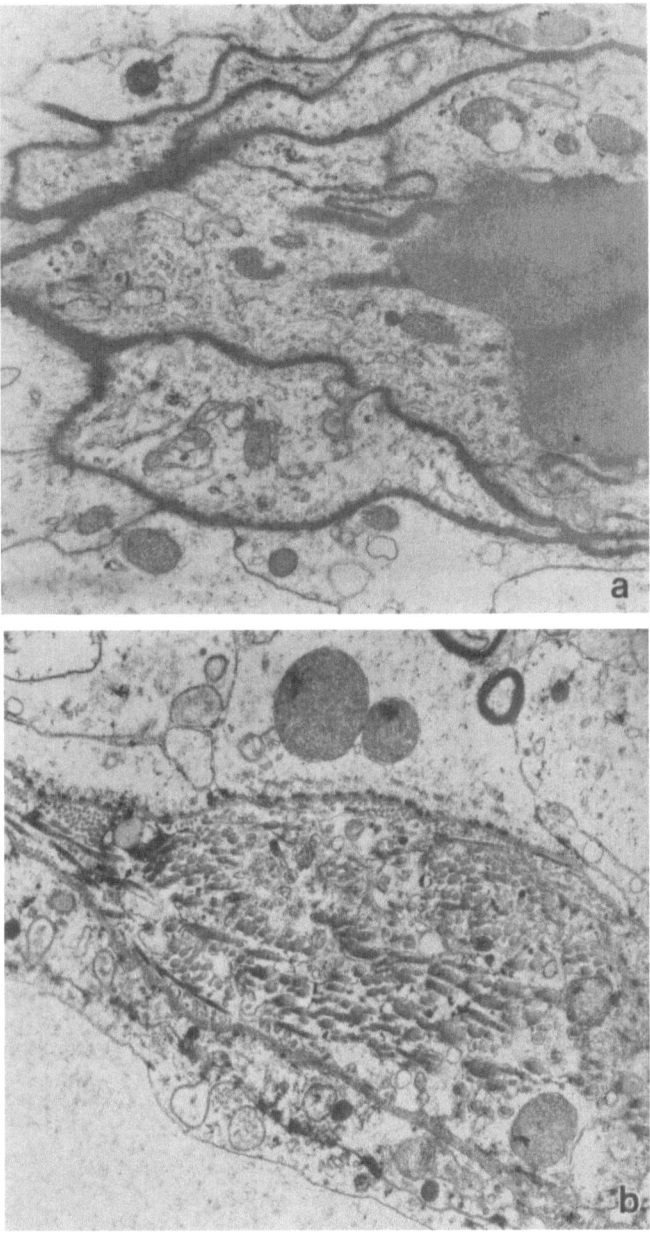

Fig. 2 a. Postcapillary venule from the human pons. The endothelial cells contain large numbers of filaments tubuli. × 10,800

Fig. 2 b. Collecting venule from the human frontal lobe. Marked enlargement of the perivascular space by different sized collagen bundles. Multiple dense areas at the perivascular glial membrane. × 12,400

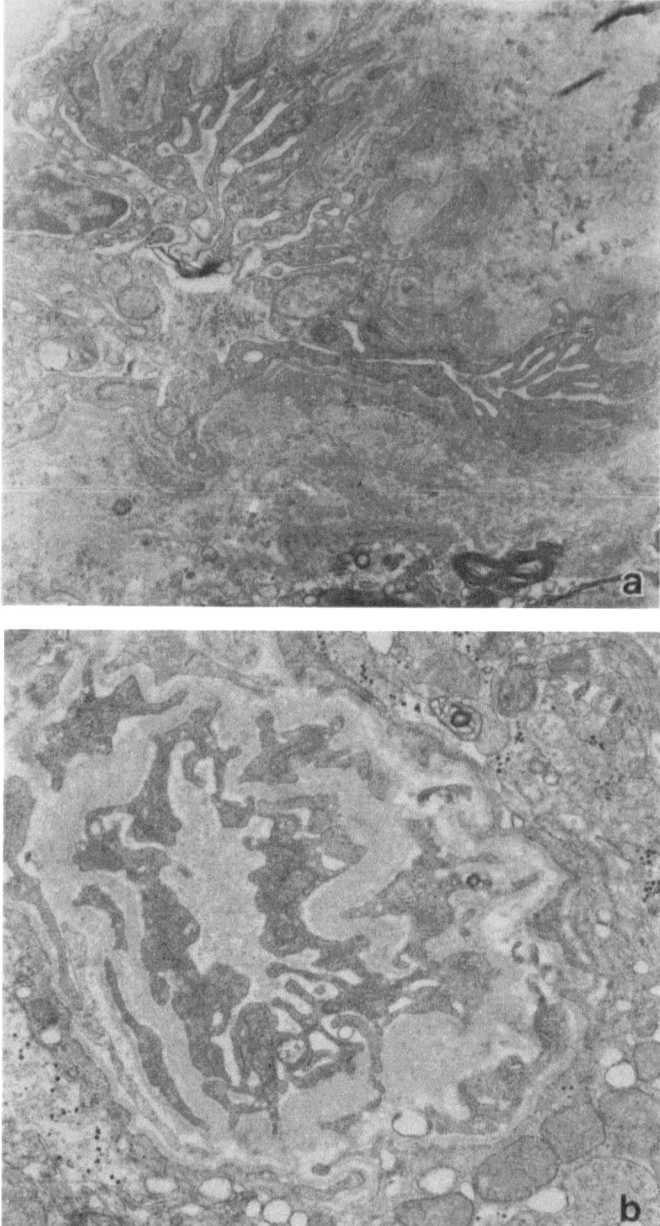

Fig. 3 a. Spastic constriction of a collecting venule of the parietal lobe from a cat. Starlike profile of the vessel with only a small lumen. Dense filaments in the abluminal part of periendothelial cells. × 8,500

Fig. 3 b. Postcapillary venule with spastic constriction. The periendothelial cells show a bizarre and marked deformation. Enlargement of the perivascular space. × 6,400

Discussion

Intracerebral Venules During Chronic Hypertension and Aging

In both groups occasionally two four vessels were seen in one perivascular space. These findings could only be discussed by comparison with previous investigations using light microscopy of the same material after Indian ink perfusion[8]. These authors have shown that, with increasing age and during hypertension, convoluted and dilated vessels are prominent in the venous part of microcirculation as well. Similar results were reported in the brains of normotensive senile patients[14].

The subcellular structure of cerebral vessels differs only slightly between normotensive elderly individuals and hypertensive patients. These findings correspond to the results of Eto[4] and Wiener et al.[23] in spontaneously hypertensive rats and rats with renal hypertension. Their assumption that an increased density of filaments might play a functional role during hypertension was supported, however, by findings in the portal veins of spontaneously hypertensive rats[12].

The most important finding was the increase in the amount of collagen bundles in the perivascular space and between the perithelium and endothelium. These results are in contrast to observations in spontaneously hypertensive rats[9] where collagen bundles in the perivascular space were less obviously. Our findings in the human tissue support the early light microscopic observations by Gellerstedt[7], Scholz and Nieto[18], Baker and Iannone[2]. These authors suggested that fibrosis in venules is an unspecific age-dependent finding which might be influenced or accelerated by hypertension or other basic diseases e.g. diabetes. In some cases the perivascular space was filled with osmeophilic amorpheous material, numerous vesicles and debris. Similar observations are described during varicosis in larger veins and in the perivascular space of the human brain capillaries[6,20]. These are considered to be general signs of regressive changes without significance for aging or chronic hypertension.

Spasm of Intracerebral Venules

Our observations of the intracerebral venules show a characteristic group of morphological changes accompanying spastic constriction. In the endothelial cells clusters of filaments were seen in the abluminal cytoplasm. The basal lamina was undulate. The deformed periendothelial cells occasionally showed a circular arrangement of filaments. The vessel diameter was markedly decreased and in some cases the vessel lumen was obstructed. These findings are in agreement with electron microscopical observations of small arterioles undergoing physiological contraction and spasm both in intracerebral and in mesentric vessels[17,13]. Their were no significant changes in the organelles, in generalized vacuolization or oedema. These observations are in agreement with the studies of Fein et al.[5] and Tanabe et al.[22] which indicated that there are no changes due to spasm in the first hour after stimulation. The most important difference between spastic changes in venules as opposed to arterioles, would seem to be the frequency and their appearance. A reduction in diameter was seen in meningeal veins and venules but lasted only for seconds. However, in intracerebral venules with smaller diameters a powerfull constriction could be shown. Whether this difference in leptomeningeal and intracerebral venous vessels is due to a larger variability in the distribution of contractile elements, or whether this indicates a physiological phenomenon different from that found in larger venules and veind should be clarified by further ultrastructural studies. The systematic investigation of the precise mechanism of intracerebral spasm could be based, at least in part, on experimental conditions derived from this investigation with electrical stimulation of the intracerebral venules and veins.

Acknowledgement

This work was supported by a Feodor-Lynen-Fellowship of the Alexander v. Humboldt foundation, Bonn.

References

1. Auer, L. M., Johansson, B. B., Kuschinsky, W., Edvinsson, L.: Sympathoadrenergic activity of cat pial veins. J. Cerebr. Blood Flow Metabol. *1*, Suppl. 1 (1981), 311—312.
2. Baker, A. B., Iannone, A.: Cerebrovascular disease: II. The smaller intracerebral arteries. Neurology (Minneap.) *9* (1959), 391—396.
3. Cervós-Navarro, J., Matakas, F., Roggendorf, W., Christmann, U.: The morphology of spastic intracerebral arterioles. Neuropath. and Applied Neurobiol. *4* (1978), 369—379.
4. Eto, T., Yamamoto, T., Omae, T.: An electron microscope study on permeability of cerebral venules in the rats with hypertensive encephalopathy. Arch. histol. jap. *38* (1975), 299—306.
5. Fein, J. M., Flor, W. J., Cohan, S. L., *et al.*: Sequential changes of vascular ultrastructure in experimental vasospasm. Myonecrosis of subarachnoid arteries. J. Neurosurg. *41* (1974), 49—58.
6. Garcia, J. H., Ben-David, E., Conger, K. A.: Arterial hypertension injures brain capillaries. Stroke *12* (1981), 410—413.
7. Gellerstedt, N.: Zur Kenntnis der Hirnveränderungen bei der normalen Altersinvolution. Uppsala: Almquist & Wiksell. 1933.
8. Iglesias-Rozas, J. R., Meencke, H. J., Cervós-Navarro, J.: Microangio-architecture of the cerebral cortex during chronic hypertension in man. In: Cerebral Ischemia and Arterial Hypertension (Mossakowski, M., Zelman, J. B., Kroh, H., eds.). Warsaw: Polish Medical Publishers. 1978.
9. Knox, C. A., Yates, R. D., Chen, I., Klara, P. M.: Effects of aging on the structural and permeability characteristics of cerebrovasculature in normotensive and hypertensive strains of rats. Acta Neuropath. (Berl.) *51* (1980), 1—13.
10. LeBeux, Y. L., Willemont, J.: Actin like filaments in the endothelial cell of adult rat brain capillaries. Exp. Neurol. *58* (1978), 446—454.
11. Matakas, F., Cervós-Navarro, J., Roggendorf W., Christmann, U., Sasaki, S.: Spastic constriction of cerebral vessels after electric convulsive treatment. Arch. Psychiat. Nervenkr. *224* (1977), 1—9.
12. Peiper, U., Klemt, P., Popov, R.: The contractility of venous vascular smooth muscle in spontaneously hypertensive or renal hypertensive rats. Basic Res. Cardiol. *74* (1979), 21—34.

13. Phelps, P. C., Luft, J. H.: Electron microscopical study of relaxation and constriction in frog arterioles. Amer. J. Anat. *125* (1969), 399—429.
14. Raven, J. R.: Anastomoses in the vascular bed of the human cerebrum. In: Pathology of Cerebral Microcirculation (Cervós-Navarro, J., Matakas, F., Betz, E., Grčević, N., eds.), p. 26. New York: Raven Press. 1974.
15. Roggendorf, W., Cervós-Navarro, J., Lazaro-Lacalle, M. D.: Ultrastructure of venules in the cat brain. Cell Tiss. Res. *192* (1978), 461—474.
16. Roggendorf, W., Cervós-Navarro, J., Iglesias-Rozas, J. R.: Contractile elements of human brain vessels in chronic hypertension. In: Cerebral Microcirculation and Metabolism (Cervós-Navarro, J., Fritschka, E., eds.), pp. 443—453. New York: Raven Press. 1981.
17. Roggendorf, W., Cervós-Navarro, J.: Ultrastructural characteristic of spasm in intracerebral arterioles. J. Neurol. Neurosurg. Psychiatr. *45* (1982), 120—125.
18. Scholz, W., Nieto, D.: Studien zur Pathologie der Hirngefäße I. Fibrose und Hyalinose. Z. ges. Neur. Psych. *162* (1938), 675—693.
19. Singh, D. N. P., Jeria, M., Melax, H., Cookson, F., Yoshida, O.: Ultrastructural changes of capillaries in the cerebral cortex of spontaneously hypertensive rats. IRCM Med. Sci. 9 (1981), 485—486.
20. Staubesand, J.: Matrix Vesikel und Mediadysplasie. Ein neues Konzept zur formalen Pathogenese der Varikose. Phlebol. Proktol. *7* (1978), 109—140.
21. Takahashi, M.: Fine structure of the rat intracranial veins. Acta Anat. Nippon *43* (1968), 239.
22. Tanabe, Y., Sakata, K., Yamada H. *et al.*: Cerebral vasospasm and ultrastructural changes in cerebral arterial wall. J. Neurosurg. *49* (1978), 229—238.
23. Wiener, J., Giacomelli, F.: The cellular pathology of experimental hypertension. VII. Structure and permeability of the mesentric vasculature in angiotensin-induced hypertension. Amer. J. Pathol. *72* (1973), 221—240.

Authors' address: J. Cervós-Navarro, M.D., Institute of Neuropathology, Free University Berlin, D-1000 Berlin.

[1] Laboratory of Neuropathology of the Institute of Pathological Anatomy, [2] Clinic of Neurosurgery of the University of Graz, Austria, and [3] Institute of Surgical Neurology, Research Institute of Brain and Blood Vessels, Akita, Japan

Morphological Analysis of Contractile Elements in Pial and Intraparenchymal Veins After in vivo Perfusion Fixation

G. F. WALTER[1], L. M. AUER[2], and I. SAYAMA[3]

Summary

After perfusion-fixation pial arteries and veins showed a statistically significant dilatation calling in question the usual definition of veins with regard to their diameters. Immunohistochemical investigations revealed that arteries and veins, but not capillaries, contained smooth muscle myosin. The need for a correlation between immunohistochemical and electron microscopic results by the application of immunocytochemical methods on the ultrastructural level is stressed.

Keywords: Cerebral veins; contractile elements; perfusion fixation.

Introduction

Although many studies have been published concerning the morphology of the cerebral veins as part of the entire cerebral vascular system[5,8,11-15,17-19,21], only a few investigations have dealt with the morphological peculiarities of pial and intraparenchymal veins and venules in detail[17-19]. Rhodin[17] used the following terminology: "postcapillary venules" with diameters of between 8–30 μm, "collecting venules" with diameters of between 30–50 μm and "muscular venules" with diameters of between 50–100 μm. It has, however, never been clearly defined whether the above diameters refer to vessel diameters measured from a microscopic picture, or whether they were adjusted to the in vivo situation by some calculation procedure. It is possible that two factors might influence the vascular dimensions, when tissue fixation with a usual fixative is performed: first the fixative itself, secondly, after cessation of the normal perfusion pressure the vessels collapse, and fixation presents vessel diameters smaller than their natural size in vivo.

Moreover, data available on contractile elements and/or smooth muscle cells in cerebral veins when related to their calibre are not satisfactorily clear. However, precise knowledge would be of interest in the light of recent findings on active venous constriction[3].

In the present series of experiments, therefore, the diameter variations of pial arteries and veins during perfusion-fixation were followed by means of a multichannel videoangiometer. The structure of photographed lepto-meningeal and neighbouring cortical vessel walls was investigated for contractile elements.

Material and Methods

Experiments were performed in 10 cats of either sex with a body weight of 1.5 to 3 kg. The animals were anaesthetized with 30 mg/kg pentobarbital, intubated endotracheally, relaxed with 60 μg/kg pancuroniumbromide and ventilated with a 3:1 mixture $N_2O:O_2$. One femoral vein and artery were cannulated for continuous blood pressure monitoring, control of blood gases (AVL gas check) and drug administration. One common carotid artery and its bifurcation were exposed: the lingual artery was cannulated and a loop put around the common carotid artery. Thereafter, the animal was put into the prone position and the head fixed in a stereotaxic headholder. Ipsilateral to the carotid artery preparation, a cranial window was made as described earlier[1]. After opening of the dura, the bone defect was closed with a cover glass and sealed with acrylic glue.

Pial vessels under the cranial window were photographed at fourty-fold magnification with a Zeiss operating microscope. Then the preparation was transferred under the intravital microscope. Under steady state conditions (normoxia, normocapnia, normotension) pial vessel diameter variations were recorded continuously using the multichannel videoangiometer[2] over a period of five minutes. Then the loop around the common carotid artery was closed and retrograde perfusion through the lingual artery was started: 5% glutaraldehyde solution stained with Evans blue was infused under continuous pressure control, adapting the perfusion pressure to the animal's normal systemic blood pressure. This perfusion procedure was performed for 20 minutes. After the perfusion fixation, the cranial glass shield was removed and the cortex underneath incised in a ring shape the same as the craniotomy using a sharp cylinder or a microscalpel. Then the animal was decapitated.

The in vivo fixed brain tissue was post-fixed by immersion in 3% buffered glutaraldehyde. Some tissue was embedded in paraffin and stained with H.E. and Masson's trichome. Other portions were post-fixed in 1% osmium tetroxide and, after dehydration in graded alcohols, embedded in Epon. Ultra-thin sections were examined with the Philips EM 400 electron microscope and compared with semi-thin sections from the same block which were stained with Toluidine blue.

For immunohistochemical investigations, unfixed brain tissue from two other cats was frozen and cut in a cryostat. The specimens were FITC-stained with anti-chicken gizzard muscle myosin and anti-thymus myosin using the unlabelled antibody enzyme technique. The extraction and purification of the antigens, the raising of the antisera in rabbits and the testing of their specificity have been previously described elsewhere[6,10]. In control stainings, no immunoreactivity was obtained.

Results

Perfusion-fixation with glutaraldehyde always resulted in well unfolded pial arteries and veins, both in light microscopy of conventionally paraffin-embedded tissue and in Epon-embedded semi-thin sections. The intraparenchymal blood vessels did not collapse either. The vessel walls of pial veins and venules were extremely thin.

From four cats, the diameters of 46 pial veins continuously recorded using the multichannel videoangiometer were calculated. The resting diameters ranged between 44 μm and 224 μm (mean 85.7 ± 7.9 μm). During perfusion-fixation with 5% glutaraldehyde, the vessels dilated by $22.9 \pm 1.6\%$ to an average diameter of 126.6 ± 9.3 μm $\pm$ SEM (statistical significance $p < 0.01$ using variance analysis). 43 pial arteries with resting diameters between 50 μm and 202 μm (mean 138.2 ± 8.1 μm) dilated by $27.5 \pm 2.6\%$ to 176.0 ± 11.1 μm (Fig. 1). The MAP before the start of

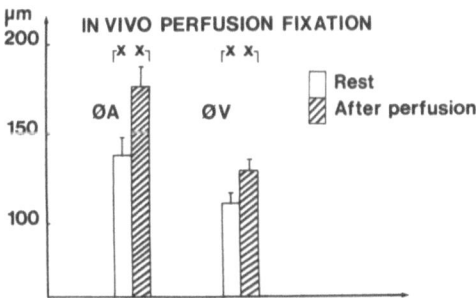

Fig. 1. Absolute dilatation of pial vessels (μm, ordinate) during in vivo perfusion fixation. $\emptyset$A = pial arteries, $\emptyset$V = pial veins

perfusion-fixation was 150 ± 9 mmHg on the average, the PaO_2 102.4 ± 4.8 mmHg and the $PaCO_2$ 28.4 ± 0.9 mmHg.

In morphological investigations, the arteries showed the typical parallel arrangement of little packets of smooth muscle cells which could be seen both in electron microscopy and in light microscopy (Fig. 2). The smooth muscle cells exhibited a distinct immunoreactivity with both anti-chicken gizzard muscle myosin and anti-thymus myosin.

Using the anti-thymus myosin, a considerable background fluorescence labelling of both capillaries and veins occurred (Fig. 3 b). With regard to these results, special attention was paid to the immunoreactivity of the veins with anti-chicken gizzard muscle myosin and a positive reactivity was found in pial veins with an external diameter of about 50 μm and more, and in intraparenchymal veins with an external diameter of about 15 μm and more (freeze sections), but never in capillaries (Fig. 3 a). This immunohistochemical technique did not permit the determination of the nature of the fluorescent cells unequivocally, but the general impression was that the endothelial cells were free of fluorescence. In electron microscopy, pial and intraparenchymal veins with the above mentioned diameters showed pericytic cells within their vessel walls, containing filaments of approximately 10 nm diameter (Fig. 3 c). These filaments seemed to have some focally dense areas. Similar filaments without focal densities were sometimes found in the flattened endothelial cells of the pial veins.

G. F. Walter *et al.*:

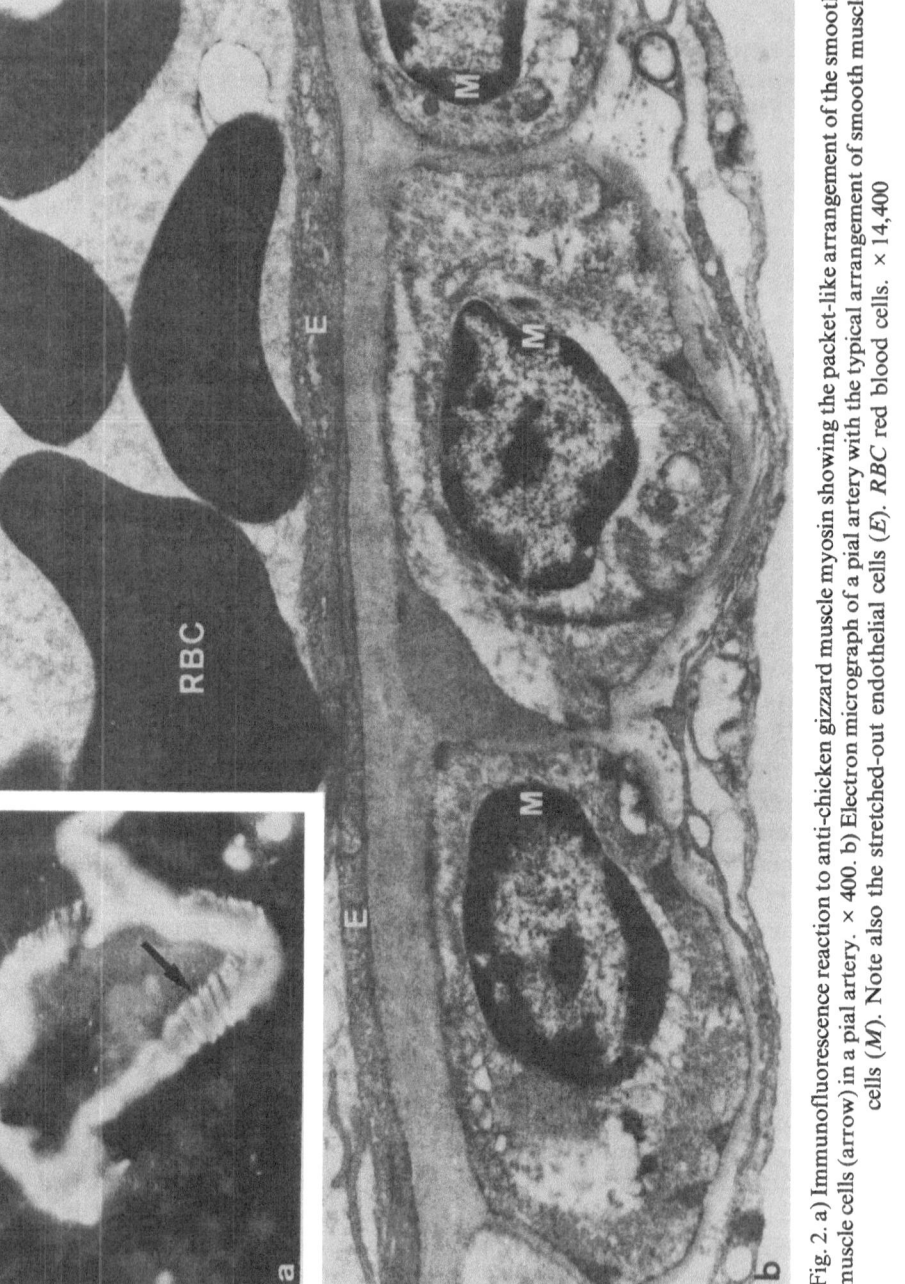

Fig. 2. a) Immunofluorescence reaction to anti-chicken gizzard muscle myosin showing the packet-like arrangement of the smooth muscle cells (arrow) in a pial artery. × 400. b) Electron micrograph of a pial artery with the typical arrangement of smooth muscle cells (*M*). Note also the stretched-out endothelial cells (*E*). *RBC* red blood cells. × 14,400

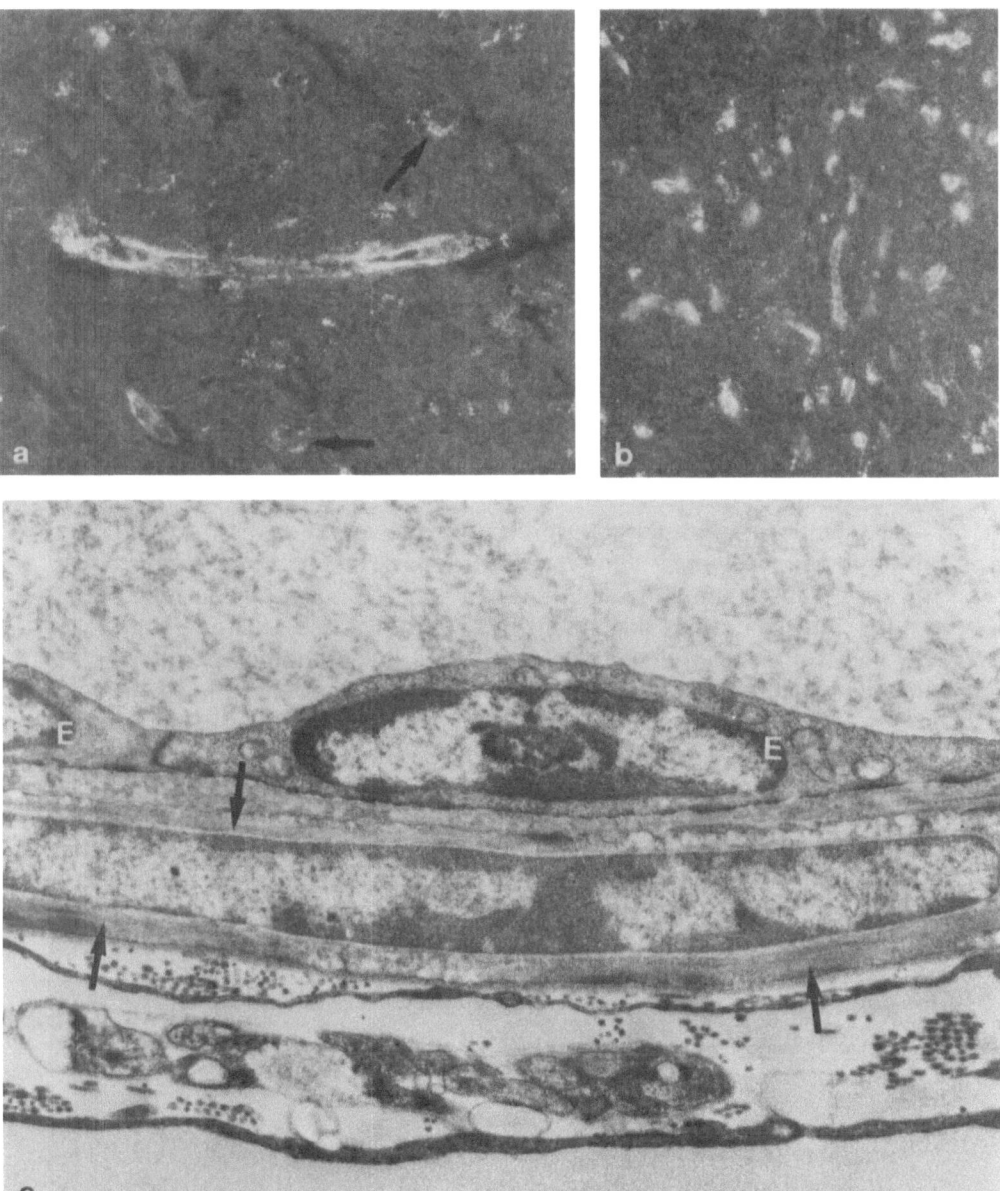

Fig. 3. a) Immunofluorescence reaction to anti-chicken gizzard muscle myosin marking the vessel wall of a cortical vein with specific greenish fluorescence. Some neuronal cells exhibit an unspecific yellow fluorescence of cytoplasmatic lipofuscin-granules (arrows). × 400. b) Immunofluorescence reaction to anti-thymus myosin marking also intracerebral capillaries. × 250. c) Electron micrograph of a pial vein with a pericytic cell containing filaments (arrows). Note the occurrence of some focally dense areas. *E* endothelial cell. × 14,400

Discussion

Perfusion-fixation led to a statistically significant dilatation of pial blood vessels which affected the arteries even more than the veins. The vessels did not collapse after the cessation of normal perfusion pressure. In the present experimental series, the shrinkage factor due to dehydration during the fixation procedure, or due to the use of different concentrations of fixatives seemed to have a minor effect, the measured differences were not statistically significant. These results underline that special care is required, when defining absolute pial or intraparenchymal venous diameters from histological sections only.

The immunohistochemistry revealed smooth muscle myosin in the vessel walls of pial and intraparenchymal veins and venules. The findings using anti-chicken gizzard muscle myosin may well allow the visualization of the correct distribution of smooth muscle myosin in the cerebral blood vessels. The results using anti-thymus myosin remained doubtful because of the strong background fluorescence, as was also observed in previous investigations[16].

The small morphological differences which exist between the different types of the so-called "intermediate" filaments, with a diameter of approximately 10 nm, are not constant or characteristic enough for a definite ultrastructural identification. The filaments in question may contain actin (4–7 nm diameter), myosin (11–16 nm diameter), the protein vimentin which may be found in vascular endothelial cells, and the protein desmin (also called "skeletin") which may among others also be found in smooth muscle cells[9]. The ultimate thickness of the filaments may also be affected by the chosen preparative procedure. Whether the filaments in endothelial and pericytic cells within the venous vessel walls consist of contractile elements giving the cells a sort of smooth mucle cell-like function, has still to be demonstrated by the application of immunocyto-chemical methods at an ultrastructural level. However, with contractile elements being absent in endothelial cells and pericytes, the observations of a marked venoconstriction in response to sympathetic stimulation[3,4] and the topical microapplication of noradrenalin[7,20] would be difficult to understand. It seems likely, therefore, to assume the observed filaments to be contractile elements.

Acknowledgement

The authors gratefully acknowledge the donation of the above mentioned antisera by Prof. U. Gröschel-Stewart, Darmstadt.

This work was supported by the Austrian *Fonds zur Förderung der wissenschaftlichen Forschung* (Project No. 4368).

References

1. Auer, L. M.: The pathogenesis of hypertensive encephalopathy. Acta neurochir. (Wien) Suppl. *27* (1978), 1—111.
2. Auer, L. M., Haydn, F.: Multichannel videoangiometry for continuous measurement of pial microvessels. Acta Neurol. Scand. Suppl. *72*, Vol. 60 (1979), 208—209.

3. Auer, L. M., Johansson, B. B.: Pial venous constriction during cervical sympathetic stimulation in the cat. Acta Physiol. Scand. *110* (1980), 203—205.
4. Auer, L. M., Johansson, B. B.: Extent and timecourse of pial venous and arterial constriction during cervical sympathetic stimulation in cats. In: The Cerebral Veins (Auer, L. M., Loew, F., eds.), pp. 131—136. Wien-New York: Springer. 1983.
5. Bubis, J. J., Luse, A. S.: An electron microscopic study of the cerebral blood vessels of the opossum. Z. Zellforsch. *62* (1963), 16—25.
6. Drenckhahn, D., Gröschel-Stewart, U.: Localization of myosin, actin, and tropomyosin in rat intestinal epithelium: Immunohistochemical studies at the light and electron microscope levels. J. Cell. Biol. *86* (1980), 475—482.
7. Edvinsson, L., McCulloch, J., Uddman, R.: Sympathetic and peptidergic mechanisms in cerebral veins. J. Cereb. Blood Flow Metabol. *1*, Suppl. 1 (1981), 327—328.
8. Frederickson, R. G., Low, F. N.: Blood vessels and tissue space associated with the brain of the rat. Amer. J. Anat. *125* (1969), 123—146.
9. Ghadially, F. N.: Ultrastructural Pathology of the Cell and Matrix (2nd ed.). London-Boston-Sydney-Wellington-Durban-Toronto: Butterworths. 1982.
10. Gröschel-Stewart, U., Schreiber, J., Mahlmeister, C., Weber, K.: Production of specific antibodies to contractile proteins, and their use in immunofluorescence microscopy. I. Antibodies to smooth and striated chicken muscle myosins. Histochemistry *46* (1976), 229—236.
11. Hager, H.: Elektronenmikroskopische Untersuchungen über die Feinstruktur der Blutgefäße und perivasculären Räume im Säugetiergehirn. Ein Beitrag zur Kenntnis der morphologischen Grundlagen der sogenannten Bluthirnschranke. Acta neuropathol. (Berl.) *1* (1961), 9—33.
12. Hammersen, F.: Zur Ultrastruktur der Venenwand. Zbl. Phlebol. *6* (1967), 221—237.
13. Maynard, E. A., Schultz, R. L., Pease, D. C.: Electron microscopy of the vascular bed of rat cerebral cortex. Amer. J. Anat. *100* (1957), 409—433.
14. Movat, H. Z., Fernando, V. P.: The fine structure of the terminal vascular bed. IV. The venules and their perivascular cells. Exp. mol. Pathol. *3* (1964), 98—114.
15. Nelson, E., Blinzinger, K., Hager, H.: Electron microscopic observations on the subarachnoidal and perivascular spaces of the Syrian hamster brain. Neurology (Minneap.) *11* (1961), 285—295.
16. Owman, C., Edvinsson, L., Hardebo, J. E., Gröschel-Stewart, U., Unsicker, K., Walles, B.: Immunohistochemical demonstration of actin and myosin in brain capillaries. Adv. Neurol. *20* (1978), 35—37.
17. Rhodin, J. A. G.: Ultrastructure of mammalian venous capillaries, venules, and small collecting veins. J. Ultrastruct. Res. *25* (1968), 452—500.
18. Roggendorf, W., Cervós-Navarro, J., Lazaro-Lacalle, M. D.: Ultrastructure of venules in the cat brain. Cell Tiss. Res. *192* (1978), 461—474.
19. Takahashi, M.: Fine structure of the rat intracranial veins. Acta Anat. Nippon *43* (1968), 238—254.
20. Ulrich, K., Auer, L. M., Kuschinsky, W.: Cat pial venoconstriction by topical microapplication of norepinephrine. J. Cereb. Blood Flow Metabol. *2* (1982), 109—111.
21. Wiedemann, M.: Dimensions of blood vessels from distributing artery to collecting vein. Circulat. Res. *12* (1963), 375—378.

Authors' addresses: Univ.-Doz. Dr. Dr. G. F. Walter (requests for reprints), Laboratory of Neuropathology, Institute of Pathological Anatomy, A-8036 Graz, Austria, Univ.-Prof. Dr. L. M. Auer, University Clinic of Neurosurgery, A-8036 Graz, Austria, Dr. Ichiro Sayama, Department of Surgical Neurology, Research Institute for Brain and Blood Vessels, Akita, Japan.

Max-Planck-Institut für Systemphysiologie, Dortmund, Federal Republic of Germany

The Vascular System of the Rat Cerebral Cortex—Basic Pattern of Leptomeningeal Vessels and Numerical Densities of Neocortical Arteries and Veins

T. Bär and T. Scheck

Summary

The vascular supply of the hemispheric cortex is similarly organized in the rat and in man. The arteries form polygonal nets on the brain surface which communicate in the pia-arachnoid. The cortex is supplied by branches from the pial net which penetrate the nervous tissue.

The leptomeningeal veins receive many branches which abruptly emerge from the cortex. These veins are not arranged in a pattern comparable to the arterial network in that the venous tributories are rather circumscribed.

The numerical density of the arteries penetrating the neocortex of the rat is 4 times greater than in man. The rat's leptomeningeal venous system receives more than 20 times as many cortical veins per mm² cortical surface than that of the human. Thus, there results a specific ratio between the penetrating arteries and the centrifugally draining veins of 0.8 for the rat and between 4–5 in man.

Keywords: Cerebral cortex; artery; vein; morphometry.

Introduction

The rat brain is frequently used as a model in experiments directed towards the blood-brain barrier and cerebral blood flow and metabolism. This seems justified because the basic pattern of the brain's vascular supply looks similar in different mammals. However, specific anatomical features may limit a direct comparison of results taken from different species[3].

The common principle in the organization of the cerebrovascular system can be demonstrated if the rat brain is compared with the foetal or newborn human brain[7, 2]. At this period of development the whole leptomeningeal vascular system is visible and not complicated by the development of gyri.

The similar basic pattern of the vascular system may be an expression of common principles which govern the development of the brain in mammals[6].

In the present study the basic arrangement of the leptomeningeal vessels is summarised. Apart from similarities, there are quantitative morphologic differences which may be important for an adequate interpretation of data taken from the rat.

Materials and Methods

30 brains of adult male and female Sprague Dawley rats were studied. The animals were fixed by perfusion-fixation via the ascending aorta (3% glutaraldehyde, 3% p-formaldehyde, $0.1 \, mol \cdot l^{-1}$ cacodylate buffer pH 7.4). After a short rinse with the buffer medium at body temperature, the vascular system was visualized by injection with India ink (the water soluble black ink was diluted 1 : 1 with 8% gelatine solution). The injected animals were stored at 4 °C for at least 3 hours. The brains were carefully removed from the skull and examined under a Leitz Elvar stereomicroscope equipped with a Polaroid camera. The numerical density of penetrating arteries and veins was estimated by direct counting under the microscope using a test area of $0.132 \, mm^2$. The positions of the measuring field are marked in Fig. 1. A total of 109 test fields were evaluated.

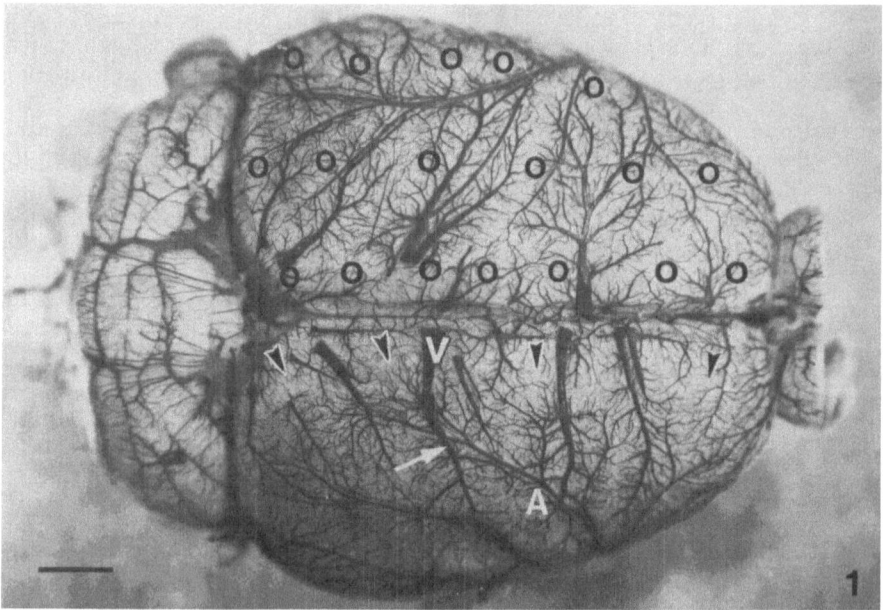

Fig. 1. Dorsal aspect of an adult rat brain. The vascular system is filled with an India ink/gelatine mixture. The superior sagittal sinus is removed. The small circles indicate the position of the test fields for counting the cortical arteries and veins. *A* artery, *V* vein, arrow heads: localization of the arterial boundary zone (note the fine arterial network), arrow: crossing between the main parietal branch of the middle cerebral artery and a superior cerebral vein. Bar represents 2 mm

Results

The main *arterial supply* reaches the rat brain from the ventral aspect. The major arteries show a fan-like, dichotomous branching pattern over the convexity of the hemispheres (Figs. 1 and 2). The distal branches result in a polygonal network of anastomosing small arteries (Figs. 1 and 3). Thus, the branches of the leptomeningeal arteries communicate on the pial surface of the brain. The watershed zone, between the areas of supply of the major arteries, extends in a parasagittal direction from frontal to occipital (Fig. 1).

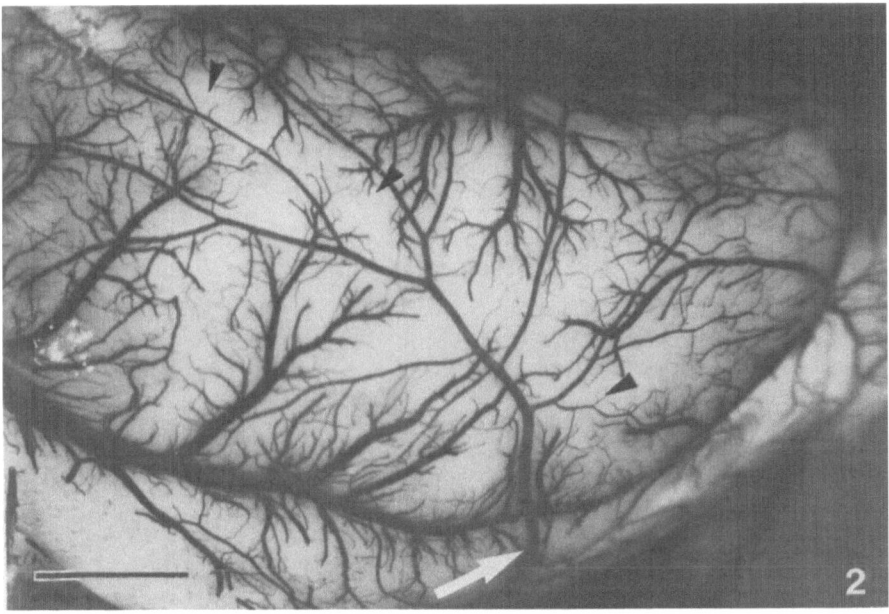

Fig. 2. Lateral aspect of a rat brain. The vascular system is visualized by the injection of black ink. Note the dichotomous branching pattern of the artery. Arrow heads: boundary zone between superior and inferior cerebral veins. Bar represents 2 mm

The *reflow of blood* from the neocortex to the leptomeningeal veins extends into the superior sagittal sinus (tributary of superior cerebral veins) or the inferior cerebral veins. A venous boundary zone develops between these main tributaries (Fig. 2). The proximal parts of the main branches of the middle cerebral artery are usually localized within the venous boundary zones.

Leptomeningeal arteries and veins show obvious differences in their branching patterns. The arteries develop a continuous network which is composed of nearly straight lined meshes (Fig. 3). An increase in curvature is observed in the boundary zones. Branches from pial arteries either penetrate the cortex adjacent to the parent vessel or they enter the nervous tissue at a certain distance after having formed almost rectangular loops (Fig. 3). The veins, which emerge from the cortex to enter the lepto-meninges, show a dendritic arborization resembling finely branched roots (Fig. 4). A complete anastomosing network, comparable to that formed by the arteries is lacking. The leptomeningeal veins collect the blood from rather circumscribed areas. The fields of drainage are interconnected by anastomoses of varying calibres. Venous capillaries can be found among these vessels (Figs. 3 and 4).

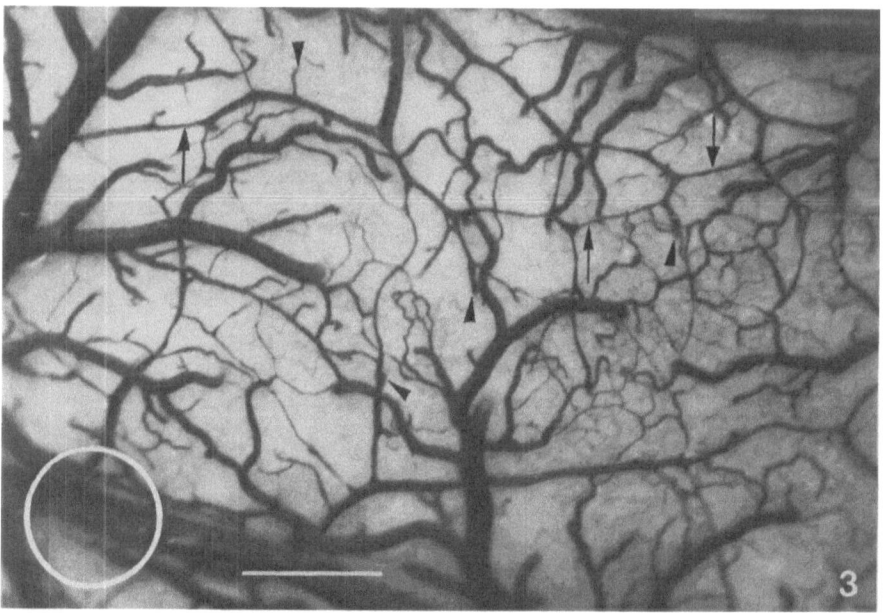

Fig. 3. Details of the leptomeningeal arteries and veins. The penetrating cortical arteries are connected by nearly straight lined segments building up a polygonal network in the pia arachnoid (arrows). The veins do not directly communicate at their points of emergence. Note the interconnected capillary-like venous vessels. Arrow heads: penetrating arteries, closed circle: size of the test field for counting. Bar indicates 0.5 mm

The number of arteries and veins which penetrate the neocortex was estimated in several regions of selected brains (Fig. 1). The number of penetrating arteries and veins per mm² ranged from 15—33 and 20—44, respectively. On average 24 ± 7 arteries and 31 ± 10 veins, penetrating into or emerging from the underlying cortex, were estimated in the rat brain (Tab. 1). The neocortex of the right hemisphere seems to be supplied by more (25 ± 7.7) arteries and drained by fewer veins (29 ± 9) than the left side (arteries: 22 ± 6.6; veins: 34 ± 10).

The differences in numerical density between arteries and veins, and between the right and left hemisphere were tested using the U-test. There is a

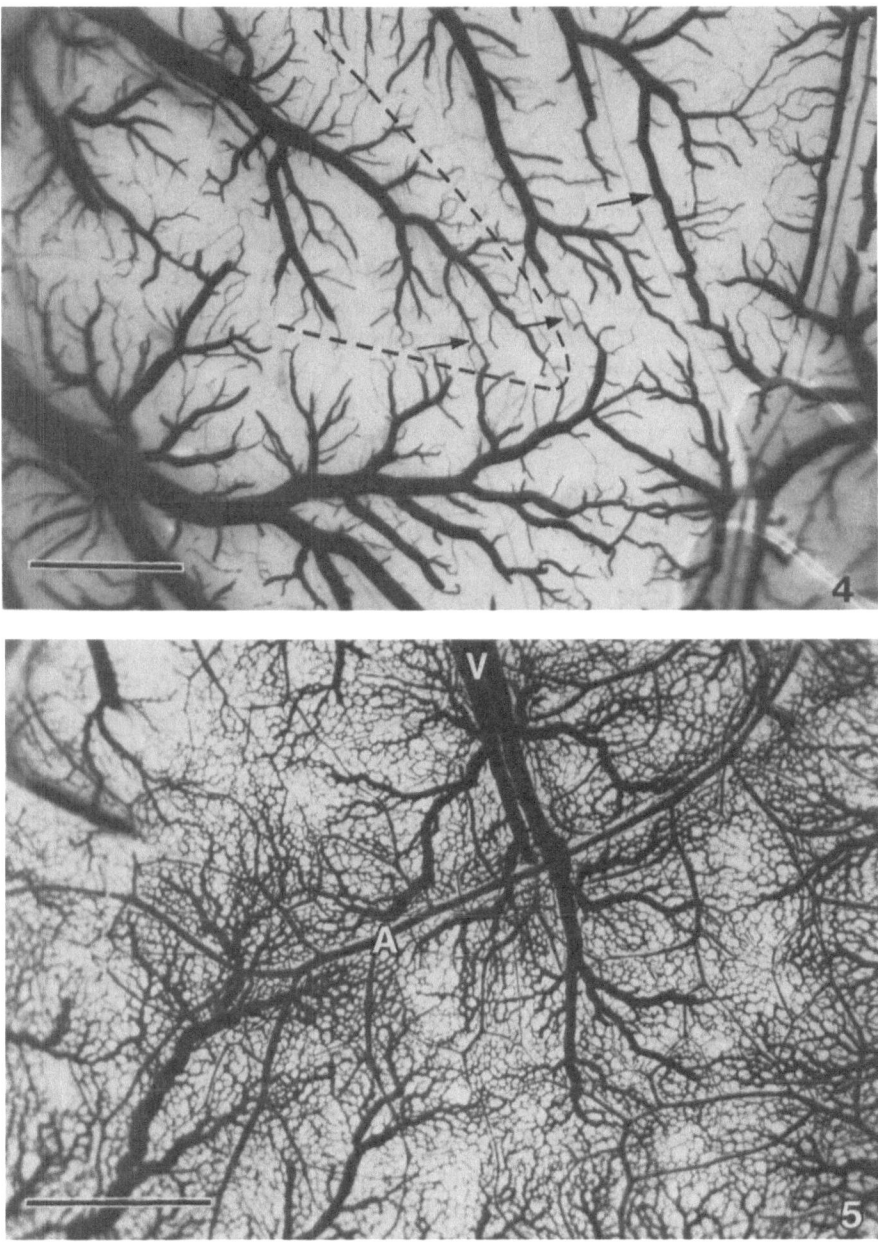

Fig. 4. Visualization of leptomeningeal veins from a lateral aspect of the hemisphere. Note the root-like branching pattern and the circumscribed fields of venous drainage. Arrows point to anastomoses of varying calibres. Bar represents 1 mm

Fig. 5. Leptomeningeal vascular system of the occipital part of the hemisphere of a rat brain one week after birth. Note the differences between the extremely densely meshed venous capillaries and the loosely woven arterial network. *A* artery, *V* vein. Bar represents 1 mm

significant difference between the medians of the right side and of the left side at the 99.5%—level for arteries and at the 99.0% level for veins (in two sided tests).

Discussion

The arrangement and the structural features of vessels are related to their circulatory function. The arterial blood flow is rather homogeneously distributed over the hemispheric surface. Local disturbances of the arterial supply can be compensated by anastomoses. The increasing curvature of the arterial segments, which bridge the boundary zones, may be associated with the special conditions of flow (varying pressure gradients) in the watershed area of the major arteries.

The venous tributories in the leptomeninges are connected by some larger superficial venous anastomoses. However, single venous trunks generally drain circumscribed areas which are usually bridged only by capillary-like segments. The capillary-like venous anastomoses in the pia-arachnoid are interpreted as being remnants of the dense venous capillary net which is present during the early postnatal period (Fig. 5). In adults these small vessels may represent a pre-formed reserve of anastomoses, which may open under conditions of disturbed venous reflow.

The numerical density of vessels entering or leaving the neocortex shows a remarkable difference between the rat and man. On average one mm² of the rat cortex is supplied by 24 arteries and drained by 31 veins which enter the pia-arachnoid. In the human brain 6 arteries and 1–2 veins were counted per mm² [5]. The diameters of the counted vessels varied within the same range in both species [1,5]. In the human brain an average of 4–5 times more arteries than veins are connected with the pial circulation. In the rat neocortex, however, there are approximately 20% more cortical veins than arteries. These favour the assumption that the reflow to the Galenic veins plays a more important role in the human than in the rat.

It remains to be checked by a dimensional analysis whether the difference between the rat and man in the numerical density of cortical arteries will fit in an allometric relation.

Table 1. *Numerical Density of Cortical Arteries and Veins on Area of Brain Surface* (Na; mm⁻²)

Species	Arteries	Veins	Art./Vein.
Human	6	1.2	5 *
Human	—	—	4 **
Rat	24	31	0.8

  * Data from Harnarine-Singh *et al.* 1972.
 ** Duvernoy *et al.* 1981.

References

1. Bär, T.: Distribution of radially penetrating arteries and veins in the neocortex of rat. In: Cerebral Microcirculation and Metabolism (Cervós-Navarro, J., Fritschka, E., eds.), pp. 1—8. New York: Raven Press. 1981.
2. van den Bergh, R., van der Eecken, H.: Anatomy and embryology of cerebral circulation. Progr. Brain Res. *30* (1968), 1—25.
3. Zeman, W., Innes, J. R. M. (eds.): Craigie's Neuroanatomy of the Rat. New York and London: Academic Press. 1963.
4. Duvernoy, H. M., Delon S., Vannson, J. L.: Cortical blood-vessels of the human brain. Brain Res. Bull. *7* (5) (1981), 519—579.
5. Harnarine-Singh, D., Geddes, G., Hyde, J. B.: Sizes and numbers of arteries and veins in normal human neopallium. J. Anat. *111* (1972), 171—179.
6. Kuhlenbeck, H.: The Central Nervous System of Vertebrates. Vol. 5, Part I: Derivatives of the Prosencephalon: Diencephalon and Telencephalon. Basel: Karger. 1977.
7. Ruckes, J.: Die arterielle Vascularisation der Pia mater des Neugeborenen. Z. Pathol. (Frankfurt) *76* (1967), 227—234.

Authors' address: Dr. T. Bär and T. Scheck, Max-Planck-Institut für Systemphysiologie, Rheinlanddamm 201, D-4600 Dortmund 1, Federal Republic of Germany.

[1] Department of Clinical Pharmacology, University Hospital of Lund, Sweden,
[2] Department of Oto-Rino-Laryngology, Malmö General Hospital, Malmö, Sweden,
[3] University Clinic of Neurosurgery, Karl-Franzens-Universität Graz, Graz, Austria

Autonomic Nerves and Morphological Organization of Cerebral Veins

L. Edvinsson[1], L. M. Auer[3], and R. Uddman[2]

Summary

Feline cerebral veins have been examined using histochemical and electron microscopic techniques. Numerous nerve fibres containing noradrenaline, substance P and vasoactive intestinal peptide were found around pial veins, the superior sagittal sinus and the cerebri magna vein, whereas the choroid plexus contained only few nerve fibres. Generally, large veins were surrounded by plexuses of nerve fibres whereas smaller veins only had single fibres.

Electron microscopy demonstrated that the walls of pial and choroid plexus veins consisted of endothelial cells, pericytes and collagenous material. The cerebri magna vein had a similar organization. Smooth muscle-like cells could, however, be observed in a few of these specimens.

Keywords: Feline cerebral veins; noradrenaline; substance P; vasoactive intestinal polypeptide.

Introduction

Recent studies have demonstrated that the microapplication of putative neurotransmitters may result in strong vasomotor responses in cerebral veins in situ[1,7,12]. Noradrenaline (NA) causes constriction, whereas substance P (SP) and vasoactive intestinal polypeptide (VIP) dilate pial arteries and veins. Sympathetic nerve stimulation also constricts the pial vessels[2]. It is thought that activation of perivascular nerves causes the release of NA which activates contractile cells in the vessels. However, the morphological prerequisites are poorly known. We have therefore examined cerebral veins (pial and choroid plexus veins, the cerebri magna vein and the superior sagittal sinus) by applying histochemical and electron microscopic techniques, showing the presence of perivascular nerves but with negligible amounts of smooth muscle-like cells.

Materials and Methods

Twelve cats of either sex, weighing between 2 and 4 kg, were used. The animals were killed by exsanguination under sodium barbiturate (Nembutal, Astra, Sweden) anaesthesia. The skull was opened and the dura mater removed. The superior sagittal sinus, the cerebri magna (CM), cortex pial veins, corpus callosal veins, choroid plexus veins from the third ventricle, and small arteries from the cerebral convexities were excised.

Histochemical Procedures

Specimens from the various veins were spread flat on microscope slides, desiccated over phosphorous pentoxide and exposed to paraformaldehyde gas at 80 °C for 1 h. All specimens were mounted in Entellan (Merck, FRG), and examined in a fluorescence microscope equipped with filters selected to give peak excitation at 405 nm[3].

In addition, specimens were fixed as whole mounts in a solution of picric acid and formaldehyde and processed for immunocytochemistry according to the protocol of Costa *et al.*[5]. Peptide-containing nerve fibres were demonstrated by an indirect immunofluorescence technique or by a peroxidase-antiperoxidase method using well characterized antisera[4]. Since the antibody is directed against only a small part of the antigenic molecule, cross reactivity with other peptides containing the same immunoreactive site cannot be excluded.

Electron Microscopy

Segments from cerebral veins were immersed for 2–4 hrs in a 0.15 M cacodylate buffer solution containing 3% glutaraldehyde and rinsed in cacodylate buffer. They were then postfixed in 1% OsO_4 in cacodylate buffer for 2 hrs, dehydrated in ethyl alcohol and embedded in Epon. Sections (1 μm in thickness) were stained with toluidine blue and examined using a light microscope. Ultra-thin sections were contrasted with 0.5% uranyl acetate and 1% lead citrate, and examined using a transmission electron microscope (Zeiss EM 10).

Results

Numerous NA-containing nerve fibres were revealed around the walls of all cerebral veins by the fluorescence method. The fibres formed plexuses around the larger veins, whereas only single fibres were seen around smaller veins (Fig. 1 a, b). On removal of the superior cervical ganglion, the NA-containing fibres disappeared. Nerve fibres containing SP and VIP were also seen around all cerebral veins (Fig. 1 c, d). Generally, the NA nerve fibres were more numerous than the fibres containing SP or VIP. Sympathectomy had no effect on the amount or distribution of the peptide-containing nerves.

Electron microscopy revealed that veins generally consisted of endothelial cells, pericytes and collagenous material. These findings clearly contrasted with the organization of pial arteries, which contained regular smooth muscle cells (Fig. 2). Numerous micropinocytic vesicles were observed in pericytes and at the abluminal side of endothelial cells in pial veins (Figs. 2–3), indicating a role in transport between the blood and cerebrospinal fluid circulatory systems. Smooth muscle-like cells and nerve bundles were observed in a few specimens from the cerebri magna vein (Fig. 4).

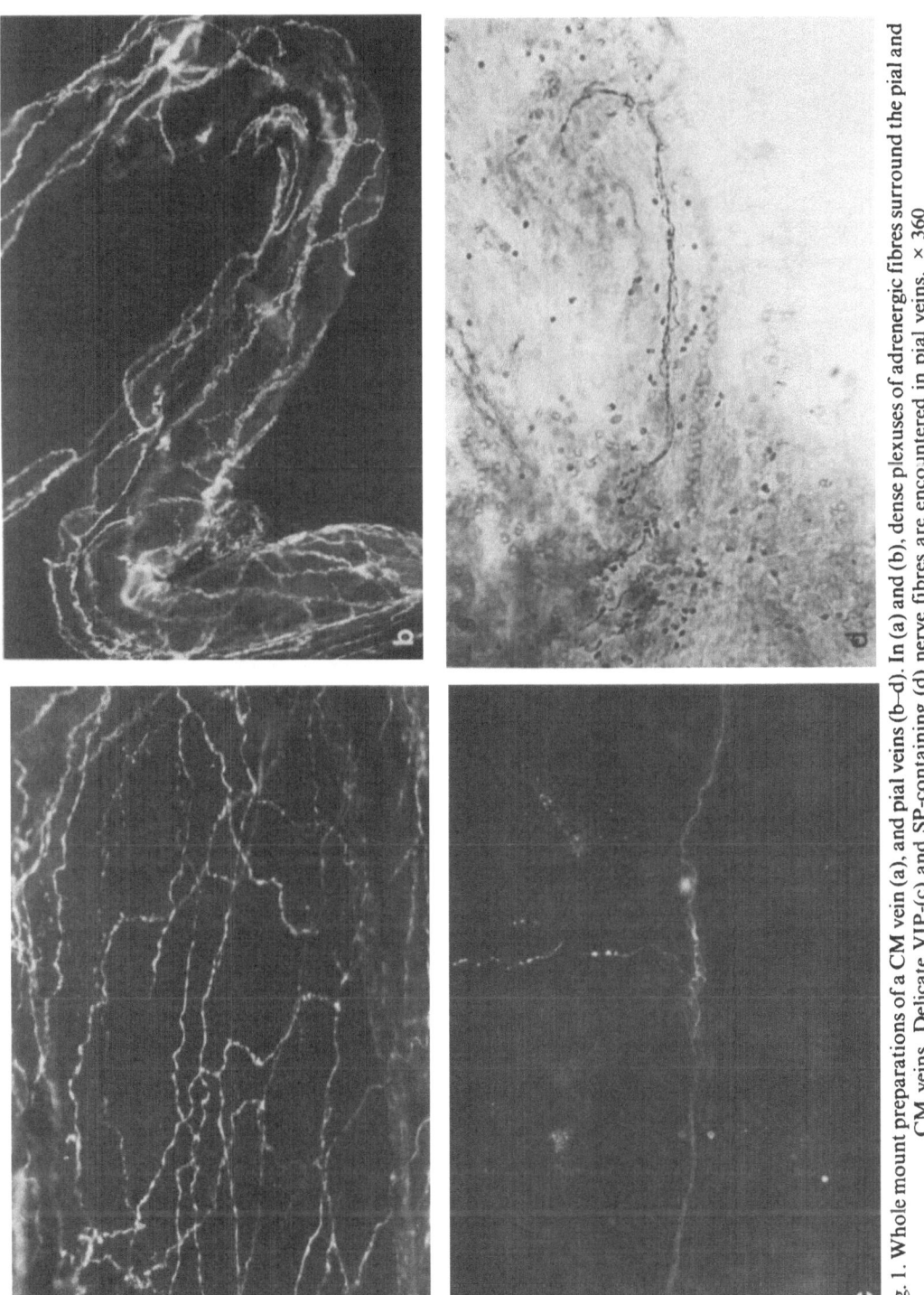

Fig. 1. Whole mount preparations of a CM vein (a), and pial veins (b–d). In (a) and (b), dense plexuses of adrenergic fibres surround the pial and CM veins. Delicate VIP-(c) and SP-containing (d) nerve fibres are encountered in pial veins. × 360

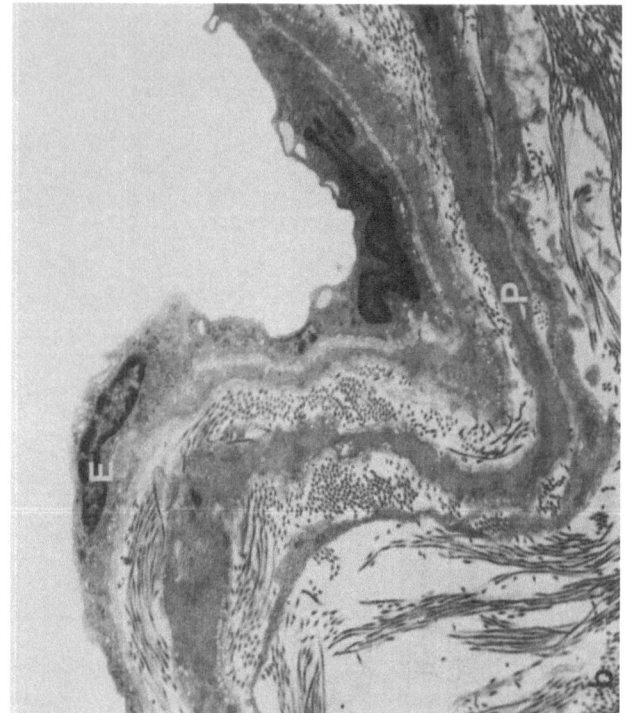

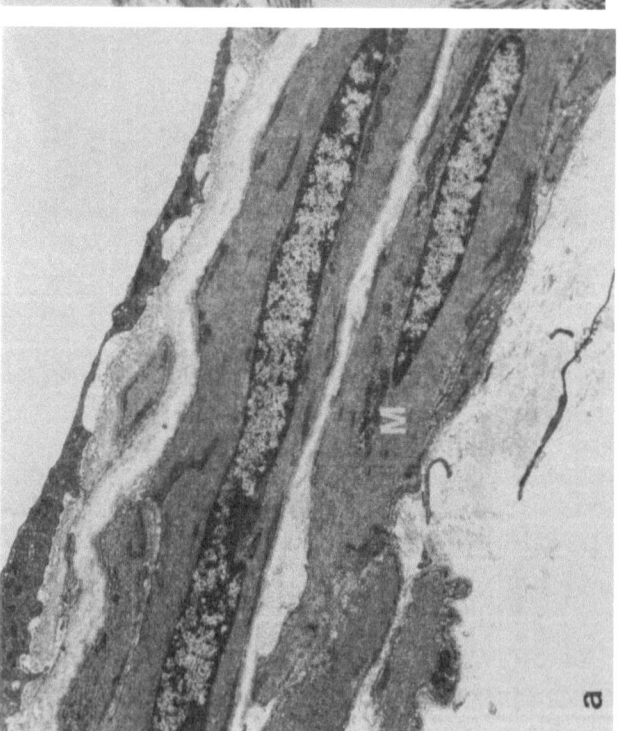

Fig. 2. Electron micrographs comparing the structural organization of a pial artery (a) and a pial vein (b). *E* endothelial cell, *P* pericyte, *M* smooth muscle cell. × 7,400

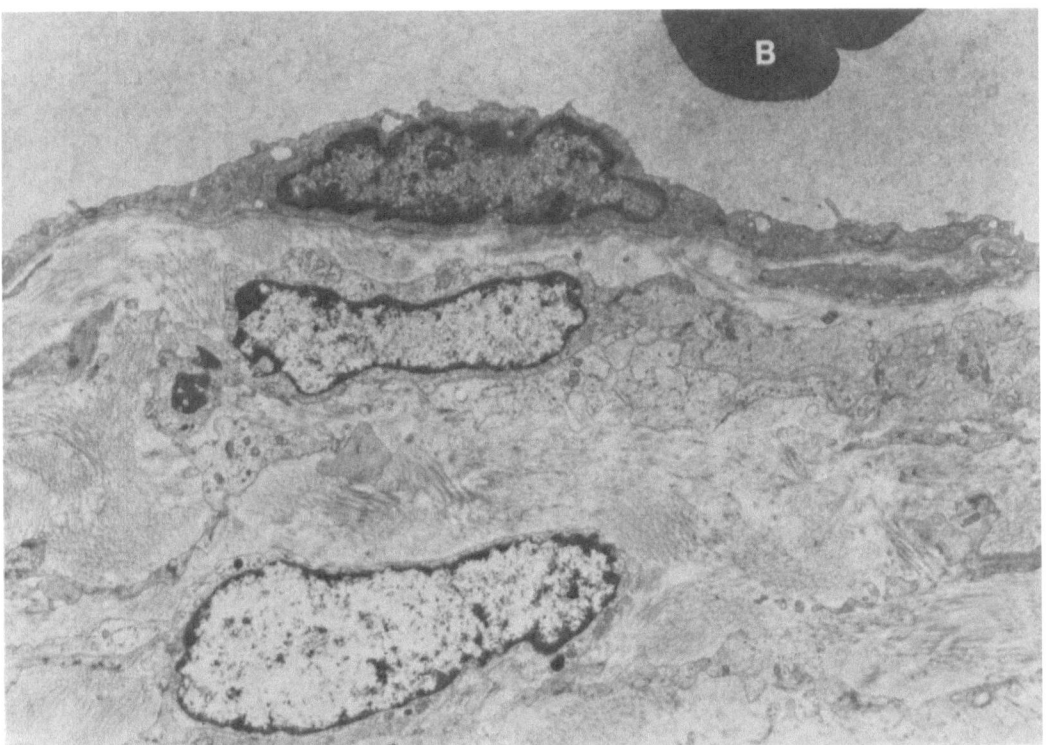

Fig. 3. Electron micrograph of a choroid plexus vein. Beneath the endothelial cell are two pericyte-like cells and collagenous material. *B* red blood cell. × 9,400

Discussion

It is now well established that cerebral veins as well as arteries are supplied with nerve fibres containing NA, SP and VIP[7]. The density of the nerve supply only varies slightly between pial arteries and veins[7,9]. Also large collecting veins, such as the cerebri magna and the superior sagittal sinus, are supplied with perivascular nerves. Perivascular microinjections in situ of NA resulted in dose-dependent reductions in the calibre of both arteries and veins. Pial veins displayed a more pronounced constriction than the arteries. Thus, microinjection of NA constricted the veins by 32% whereas the arteries constricted by 22%. The effects of noradrenaline were attenuated by the concomitant administration of the α-adrenoceptor antagonist phentolamine. Fine varicosed nerve fibres containing VIP and SP formed a wide-meshed network around the pial veins. Perivascular microinjection of VIP or SP resulted in increases in the calibre of pial veins, which were slightly smaller than those observed in pial arteries[7].

Distinct smooth muscle cells appear to be absent in the walls not only of intracerebral veins but also of pial veins in the rat and cat[8,10,11]. Only occasionally are smooth muscle-like cells observed in the cerebri magna

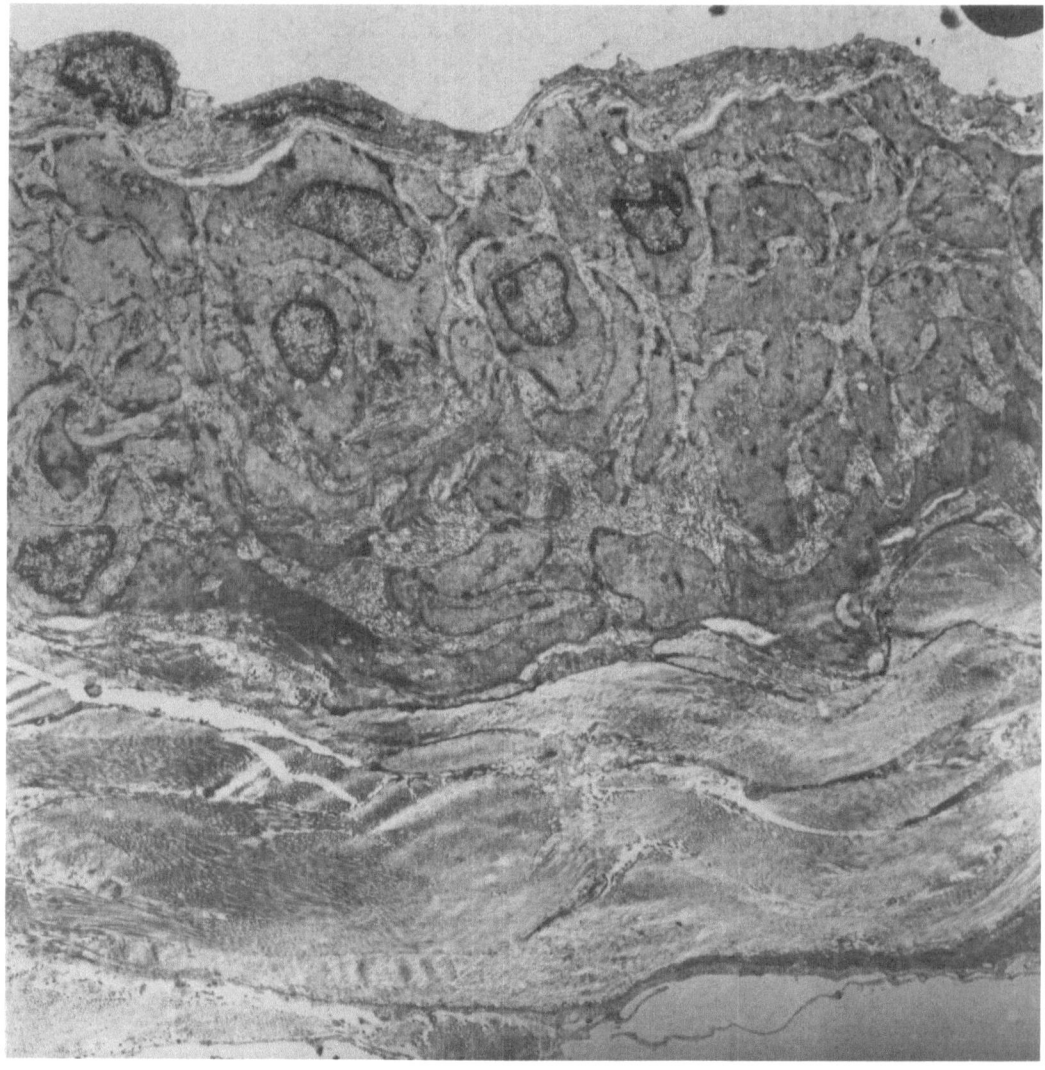

Fig. 4. Electron micrograph of a portion of the cerebri magna vein demonstrating smooth muscle-like cells in the vessel wall beneath the endothelium. × 7,400

vein. Exposure of isolated feline pial veins in vitro to 124 mM potassium or NA induce small contractions, whereas adjacent arteries show strong contractions[6]. Thus, the poor content of smooth muscle cells in veins as compared with arteries may well explain the difference observed in contraction (see Fig. 2). The amount of contractile force might, however, still be enough since the veins have to contract against a low pressure only.

In conclusion, cerebral veins and arteries receive numerous autonomic nerve fibres, the veins have a less well developed smooth muscle layer and respond poorly to most agents both in vitro and in situ. One exception is the strong contractions induced by sympathetic nerve stimulation or during microapplication of NA in situ. It is therefore conceivable that the sympathetic nerves may have an important role in the regulation of venomotor tone and of intracranial pressure.

Acknowledgements

Mrs. Marianne Palmegren is gratefully acknowledged for expert assistance with the electron microscopy examinations. This study was supported by the Swedish Medical Research Council (Grant No. 5958) and the Swedish National Association Against Heart and Chest Diseases.

References

1. Auer, L. M., Johansson, B. B., Kuschinsky, W., Edvinsson, L.: Sympathoadrenergic activity of cat pial vessels. J. Cerebr. Blood Flow Metabol. Suppl. *1* (1981), 311—312.
2. Auer, L. M., Johansson, B. B., Lund, S.: Reaction of pial arteries and veins to sympathetic stimulation in the cat. Stroke *12* (1981), 528—531.
3. Björklund, A., Falck, B., Owman, C.: Fluorescence microscopic and microspectrofluorometric techniques for the cellular localization and characterization of biogenic amines. In: Methods of Investigations and Diagnostic Endocrinology. Vol. 1: The Thyroid and Biogenic Amines (Rall, J. E., Kopin, I. J., eds.), pp. 318—368. Amsterdam: North Holland Publ. Co. 1972.
4. Coons, A. H., Leduc, E. H., Conolly, I. M.: Studies on antibody production. I. A method for the histochemical demonstration of specific antibody and its application to a study of the hyperimmune rabbit. J. exp. Med. *102* (1955), 49—60.
5. Costa, M., Buffa, R., Furness, J. B., Solcia, E.: Immunohistochemical localization of polypeptides in peripheral autonomic nerves using whole mount preparations. Histochemistry *65* (1980), 157—165.
6. Edvinsson, L., Högestätt, E. D., Uddman, R., Auer, L. M.: Cerebral Veins: Fluorescence histochemistry, electron microscopy and in vitro reactivity. J. Cerebr. Blood Flow Metabol. *3* (1983), 226—230.
7. Edvinsson, L., McCulloch, J., Uddman, R.: Feline cerebral veins and arteries: Comparison of autonomic innervation and vasomotor responses. J. Physiol. *325* (1982), 161—173.
8. Lange, W., Halata, Z.: Comparative studies on the pre- and postterminal blood vessels in the cerebellar cortex of Rhesus monkey, cat and rat. Anat. Embryol. *158* (1979), 51—62.
9. Nakakita, K., Imai, H., Kamei, I., Naka, Y., Nakai, K., Itakura, T., Komai, N.: Innervation of the cerebral veins as compared with the cerebral arteries: A histochemical and electron microscopic study. J. Cerebr. Blood Flow Metabol. *3* (1983), 127—132.
10. Roggendorf, W., Cervós-Navarro, J., Lazaro-Lacalle, M. D.: Ultrastructure of venules in the cat brain. Cell. Tissue Res. *192* (1978), 461—474.
11. Takahashi, M.: On the fine structure of the rat intracranial veins. Acta Anat. Nippon *43* (1968), 238—254.
12. Ulrich, K., Auer, L. M., Kuschinsky, W.: Cat pial venoconstriction by topical microapplication of norepinephrine. J. Cerebr. Blood Flow Metabol. *2* (1982), 109—111.

Author's address: L. Edvinsson, M.D., Department of Clinical Pharmacology, University Hospital of Lund, E-blocket, S-221 85 Lund, Sweden.

Department of Anatomy, University of Munich, Federal Republic of Germany

The Post-terminal Blood Vessels in the Cerebellar Cortex of the Rhesus Monkey, Cat, and Rat

W. LANGE

Summary

The postterminal vessels and veins in the rhesus monkey, cat and rat exhibit the same mural structure found in capillaries. The wall consists only of an endothelium and occasional pericytes embedded in the basal lamina. Even the large veins which run to the pial veins show this simple mural structure.

Keywords: Cerebellar cortex; veins; rhesus monkey; cat; rat.

The investigations of the vascularization of the central nervous system, especially of the angioarchitecture, which followed the pioneer work of Pfeifer (1928)[3] have contributed essentially to a better understanding of brain function.

The first detailed description of the cerebellar vascularization were made by Uchimura[5] and Fazzari[1] who investigated the angioarchitecture of the cerebellum after India ink injection.

According to these findings, as well as our own findings[2] in the rhesus monkey, cat and rat, the India-ink filled vessels penetrate the cerebellar cortex perpendicularly from the pia mater. Some of them only extend into the Purkinje cell layer and others extend into the central medullary ray of the folium. In their course through the cerebellar cortex these vessels give off small and large vessels perpendicular to their course and these branches give rise to capillary networks for each layer.

One of these vessels, which runs immediately underneath the Purkinje cells parallel to this layer, was interpreted by Uchimura as an artery, and was associated with the high degree of vulnerability of Purkinje cells.

In order to better understand the angioarchitecture and course of the veins we perfused human material retrogradely from the venous side with India ink, filling the veins incompletely. It can be seen quite clearly, that only some larger vessels are stained which originate deep in the medullary layer and become continuously wider in diameter towards the cortical surface.

It is noteworthy that the venous vessels of the molecular layer show only an endothelial layer and, in some divisions of the course of the vessel, possibly some pericytes are visible; however, other structural parts of the vascular wall apparently are missing. Thus, these vessels could be interpreted as extremely wide capillaries, since the structure of their walls completely resembles that of capillaries. The only possibility of obtaining a clear distinction between preterminal arterial vessels and postterminal vessels was provided by injecting microspheres which could not go through the capillaries. The diameter of the microspheres we used was $10 \pm 2 \mu m$. In semi-thin sections it can be shown that those vascular parts, through which the microspheres of this diameter do not pass, are vessels with extremely thin walls which resemble capillaries under light microscopy. Using the electron microscope, however, a significantly thicker vascular wall can be seen. The basal lamina contains pericytes and, occasionally, processes of smooth muscle cells which only rarely exhibit typically dark areas and intracellular caveolae.

The structure of the vessels with a diameter less than $10 \mu m$, in which no microspheres were found, resembles that of capillaries. The endothelium is sometimes seamless and Weibel-Palade bodies are only occasionally present. The basal lamina is thin. The number of pericytes contained in the basal lamina is low and they sometimes contain filaments. This type of vessel is found in all the layers of the cerebellar cortex, as well as in the central medullary rays of the various folia. The examination of a semi-thin section shows that capillaries are often transformed into vessels with a large luminal diameter, but with a mural structure which does not differ from that of capillaries. At the junction of two venules, or at the site of the opening of a capillary into a venule, a pericyte can often be found embedded as if in a niche. The structure of the wall of the veins undergoes no further changes before reaching the molecular layer. A vein can attain a diameter of 4 to 5 times that of an artery from the same layer, although its wall remains very thin. Such a large vein simply consists of an endothelium supported by a thin basal lamina, which then is split by enclosed pericytes. Usually a perivascular space develops before the vessel emerges from the molecular layer. The wall of the pia mater veins located just above the cerebellar cortex, which derive from the veins emerging from the molecular layer, have the same structure.

A very special point of the angioarchitecture of the cerebellar cortex has to be stressed. Since Uchimura[5] studied the angioarchitecture of the cerebellar cortex, Purkinje cells have been considered to be highly susceptible to a deficiency in oxygen, because a big arterial vessel might run parallel to this layer without capillaries branching off.

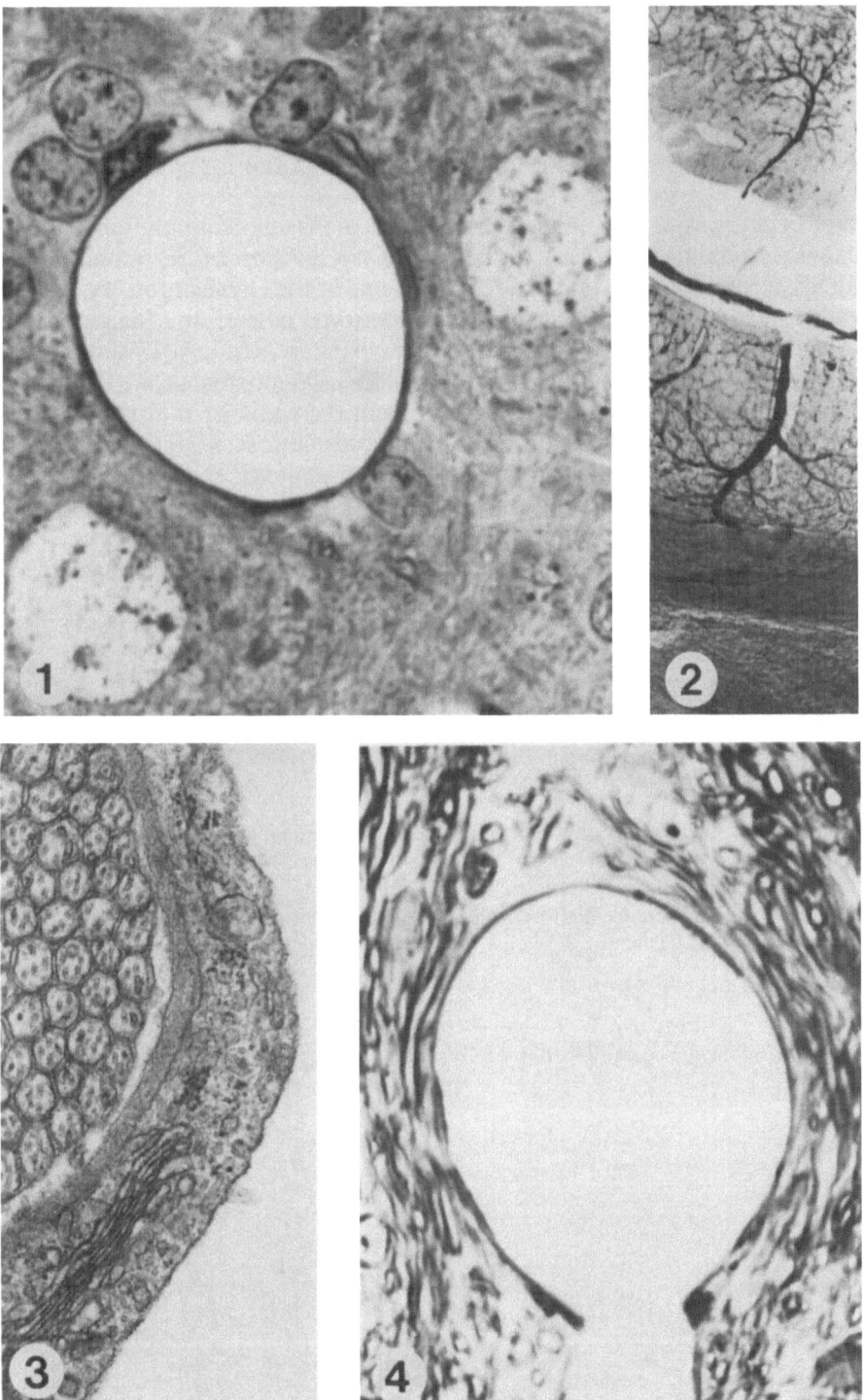

Fig. 1. Large venous vessel just underneath the Purkinje cell layer. × 3,000, cat

Fig. 2. India ink filled veins of the cerebellar cortex. × 30, man

Fig. 3. Large vein in the molecular layer of rhesus monkey. The wall of this vessel consists only of an endothelium and a basal lamina. × 29,400

Fig. 4. Opening of a capillary vessel into a vein in the central medullary ray. × 3,000, rat

In the location described by Uchimura, semi-thin sections reveal a large-calibre vessel with the diameter and the mural structure of a vein, and also capillaries in the Purkinje cell layer. The lack of a capillary network and the presence of large artery cannot be confirmed in the material investigated here.

Roggendorf et al.[4] found in the venules of the telencephalic cortex of the cat unequivocal myocytes, which provide this division of the vessel with the ability to contract. In the venules and veins of the rhesus monkey, cat and rat we never found myocytes, however pericytes containing filaments were found. The venous limb in the cerebellar cortex is considerably shorter than in the telencephalic cortex, since the cerebellar cortex is much thinner. Furthermore, we did not find myocytes in the walls of pial veins located directly above the cerebellum. We therefore suppose, that at least in this part of the brain, and according to the morphological findings obtained hitherto, a postcapillary regulation is lacking.

Additional investigations using immunohistochemical methods are planned to find out whether pericytes possibly contain contractile material.

References

1. Fazzari, I.: Le arterie del cerveletto. Studio anatomo-comparativo ed embriologico. Mem. R. Acad. naz. Lincei, Cl. Sci. fis. mat. et nat., Scr. VI, *4* (1931), 334—416.
2. Lange, W., Halata, Z.: Comparative Studies on the Pre- and Postterminal Blood Vessels in the Cerebellar Cortex of Rhesus monkey, Cat and Rat. Anat. Embryol. *158* (1979), 51—62.
3. Pfeifer, R. A.: Die Angioarchitektonik der Großhirnrinde. Springer: Berlin. 1928.
4. Roggendorf, W., Cervós-Navarro, J., Lazaro-Lacalle, M. D.: Ultrastructure of venules in the cat brain. Cell Tissue Res. *192* (1978), 461—474.
5. Uchimura, Y.: Über die Blutversorgung der Kleinhirnrinde und ihre Bedeutung für die Pathologie des Kleinhirns. Z. Neur. *120* (1929), 774—782.

Author's address: Dr. W. Lange, Department of Anatomy, University of Munich, D-8000 München, Federal Republic of Germany.

Laboratory for Surgical Research, Wilhelmina Gasthuis, Amsterdam, The Netherlands

Veins of the Rat Brain

K. C. HODDE

Summary

Venous patterns of the rat brain consistently differ in the various parts of the brain. This paper describes the relationship between pial arteries and veins of the rat cerebral and cerebellar cortices, as seen in Scanning Electron Microscopy of corrosion cast preparations. The cerebral pial veins display a number of places where they are crossed over by the supplying arterioles of the cortex. The cerebellar veins and arteries located in the fissures near the surface of the cerebellum run closely together for considerable distances whilst giving off or receiving branches.

It can also be seen that the cross-sectional area of the pial cortical veins is always larger than that of the accompanying arteries. As an extreme example of this relationship, the very large venous outflow of the pineal microcirculation is shown.

In the choroid plexus of the lateral ventricles the supplying arterioles are wrapped in a capillary sheath, in contrast to the large veins. The morphology of the long and short portal vessels of the hypophysis is shown to be different from that of the veins. It is concluded that venous morphology, pattern of distribution and their relation to accompanying arteries is consistently different for various parts of the brain. The possible functional implications are discussed.

Keywords: Vascular casts; SEM; venous morphology; rat brain.

Introduction

The veins of the rat brain and the venous drainage of the rat head have received much less attention than has the arterial supply. This might partly be due to its seemingly greater variability, or because of the certainly greater technical difficulty in obtaining useful anatomical specimens, as well as the lack of unambiguous criteria to recognize arteries from veins in preparations in which all vessels were injected with a contrast medium.

In principle, it is possible to make complete vessel system replications by making vascular casts, and, because of its large depth of focus and its high resolution the Scanning Electron Microscope (SEM) is a very suitable instrument with which to study the casts. Moreover, the replication of the vessel wall in resin casts is so accurate that the imprints of individual endothelial cells can be recognized. It has turned out that arterial and venous endothelial cells, both in fixed tissue and in a replicated form, consistently show a distinct and typical pattern. The arterial endothelial cells show ovoid contours, oriented along the length axis of the vessel, with bulging nuclei of similar shape. The venous cells have irregular contours and round, sharply delineated nuclei. These easily recognizable features in cast preparations now form an unambiguous criterion with which to recognize either vessel[2, 8]. See also Fig. 2.

The veins and arteries of several parts of the brain, notably the cerebral and cerebellar pial vessels, and some periventricular organs were studied in order to draw some comparative conclusions.

Materials and Methods

The cephalic blood vessel system of anaesthetized Wistar rats was filled with a low-viscosity, pre-polymerized resin to which, just before injection, a catalyst was added so that the mixture polymerized completely in situ, within 7–8 minutes. To this aim, a cannula was introduced through the left ventricle and secured in place just distally to the aortic valves. The aorta descendens was clamped, the right atrium opened and the blood washed out with heparinized saline. After injecting the resin and complete polymerization thereof, the head was removed and all tissue was corroded away in 15% KOH so that a resin cast of the complete blood vessel system remained. The specimens were dissected under an operation microscope and rendered suitable for SEM by freeze-drying and gold-sputtering.

Observations

The cerebral and cerebellar pial veins (Figs. 1–4). The cerebral pial vessels in the overview of Fig. 1 show the typical aspect of a dense network of arteries and veins which sometimes cross. The arteries are always found to overlie the veins when crossing occurs (Fig. 2) and that, in the cast preparations, the veins at these places always show deep depressions in their contours (Fig. 2). The cerebellar pial aspect is different in that the pial vessel network on the surface is much less dense. Usually the arteries and veins in the fissures run closely together for considerable distances (Fig. 4). No cross-overs occur as seen on the cerebral surface.

The cross-sectional area of veins compared to that of arteries always tends to be larger[9] and this is markedly so in the pineal organ. For 3 or 4 supplying arterioles of about 50 μm diameter there are 14 to 16 veins of 40–60 μm to be found (Fig. 5) which drain directly into v. cerebri magna[3]. Also, the choroid plexus shows very large draining veins. In the plexus of the lateral ventricles the arterioles display a peculiar capillary arrangement around them, incidentally draining into a large vein (Fig. 6). This constellation is seen only in the lateral plexus[4].

The neurohypophysis shows a very distinct venous drainage into the cavernous sinus (Fig. 7). This is a very consistent feature compared to the

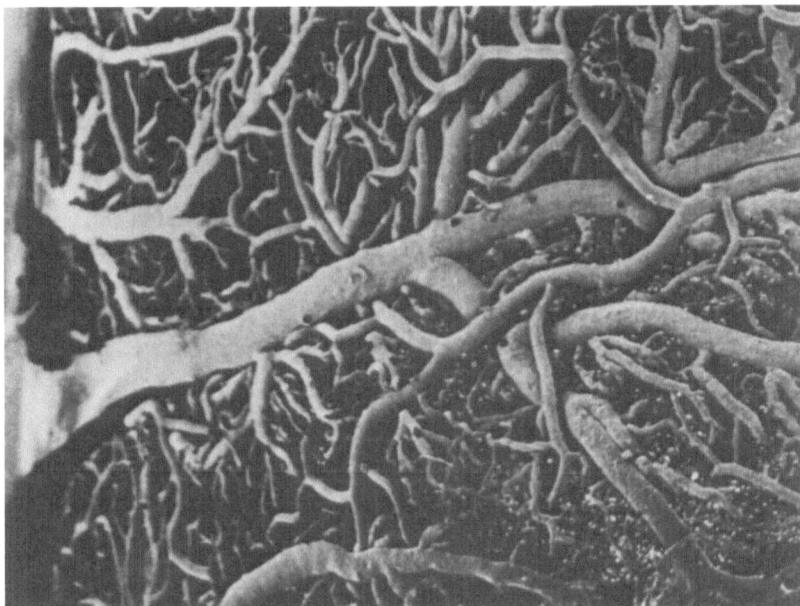

Fig. 1. Corrosion cast of rat cerebral pial vessels. To the left part of the superior sagittal sinus is just visible. Several arterio-venous cross-overs are noticeable. PW (Picture Width) is 3 mm

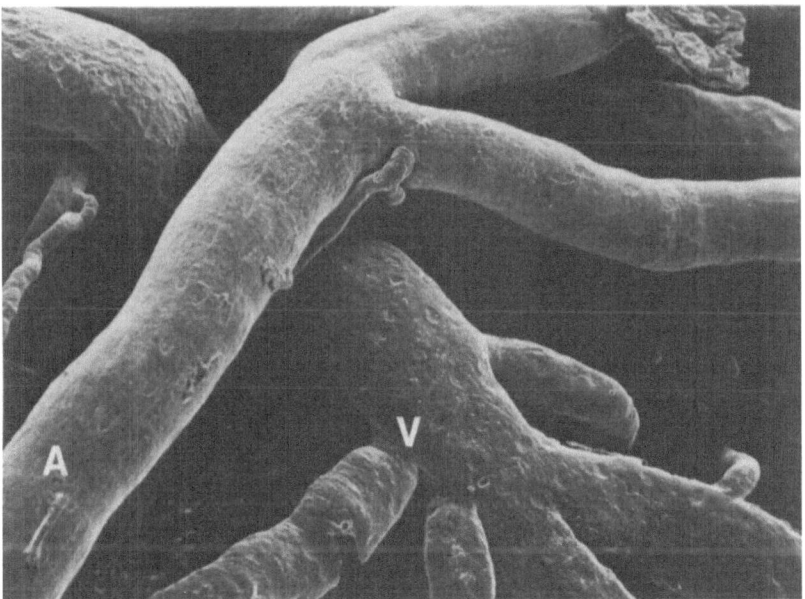

Fig. 2. Detail from same specimen. Artery (*A*) crossing over vein (*V*). Note the typical endothelial cell imprints, typical for arteries and veins respectively. PW is 300 μm

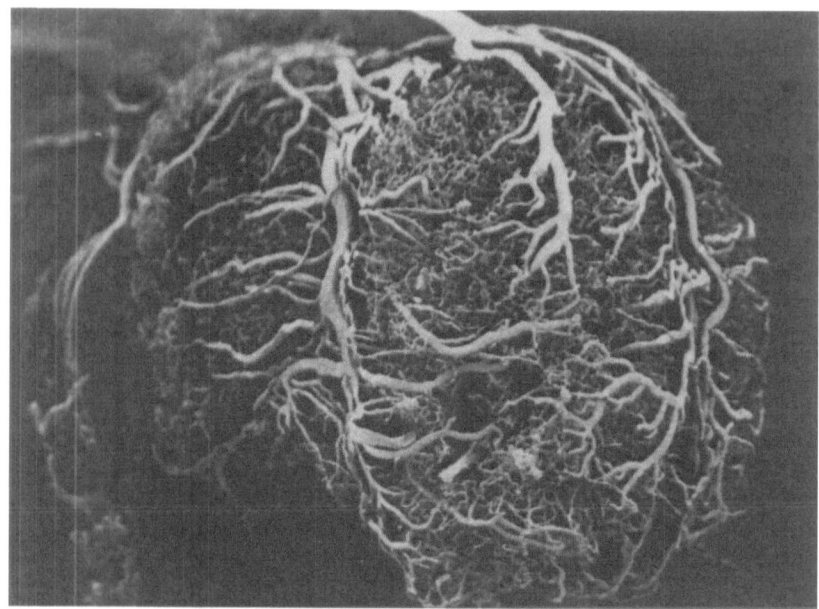

Fig. 3. Corrosion cast of rat paraflocculus vessels, right side, lateral view. PW is 3 mm

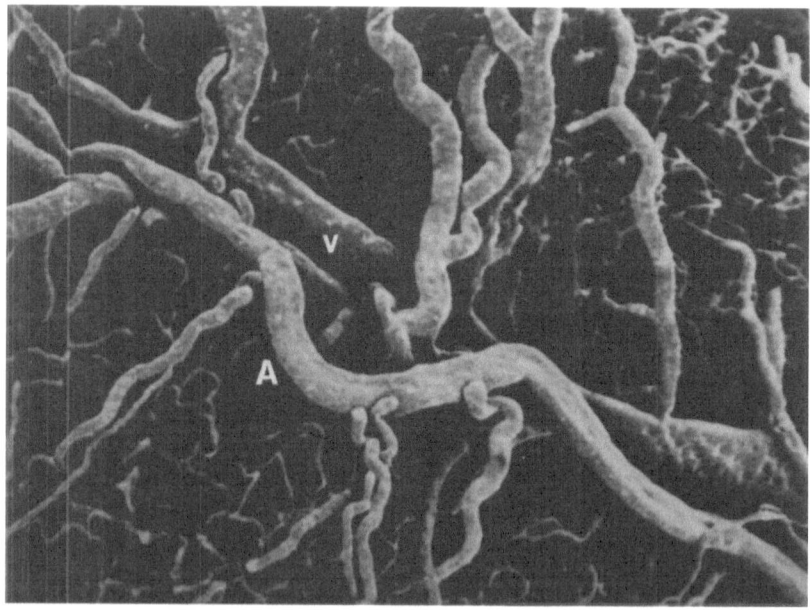

Fig. 4. Detail of same specimen as in previous Fig. Artery (*A*) and vein (*V*) run close together. PW is 1 mm

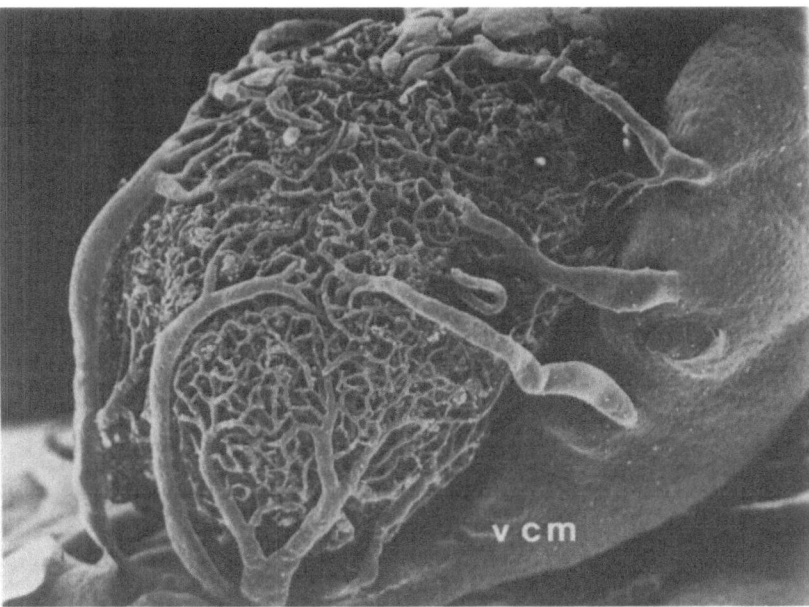

Fig. 5. Cast of the pineal gland vessels, mounted upside down. Note the many veins connected with the v. cerebri magna (*vcm*). PW is 1 mm

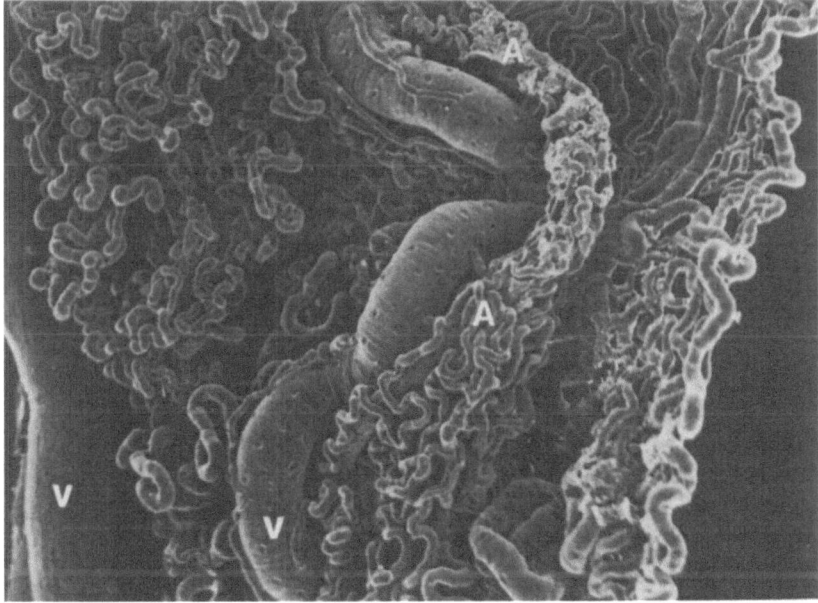

Fig. 6. Cast of the choroid plexus vessels of the lateral ventricle. Note the large veins (*V*) and their typical endothelial cell imprint pattern. *A–A* indicates an arteriole within the capillary wreath visible. PW is 500 μm

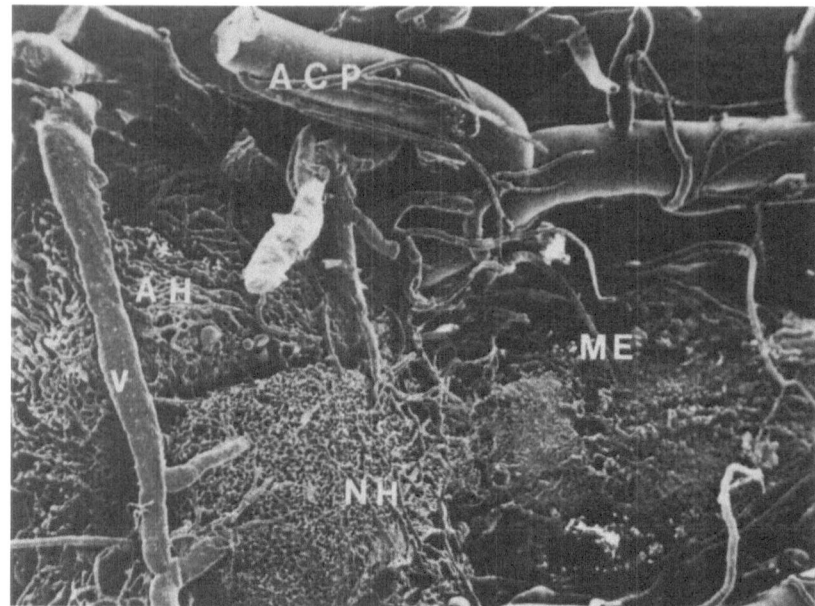

Fig. 7. Cast of the hypophyseal vessels, dorsal view. *ACP* a. comm. post.; *NH* neurohypophysis; *AH* adenohypophysis; *ME* median eminence; *V* draining vein of the neurohypophysis into the cavernous sinus. PW is 3 mm

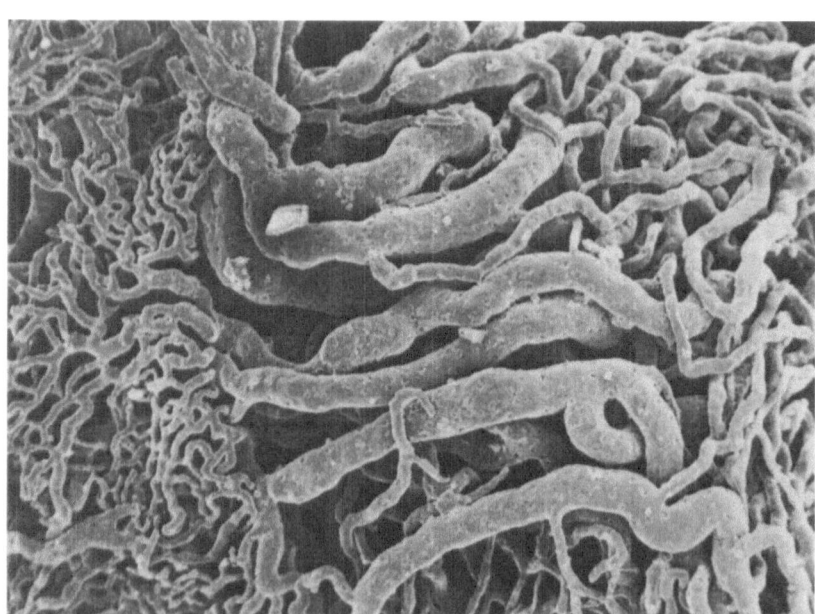

Fig. 8. Detail of similar preparation as in Fig. 7, rotated 90° anticlockwise. The middle part shows about 9 short portal vessels, to the left continuous with the capillaries of the neurohypophysis and to the right with the sinusoids of the adenohypophysis. PW is 1 mm

veins of the adenohypophysis, which vary in size and location when draining into the cavernous sinus.

For morphological comparison, the short portal vessels are depicted in Fig. 8. Like the long portal vessels they interconnect capillary networks from the neurohypophysis and median eminence respectively, with the sinusoids of the adenohypophysis. In shape and size they are very similar to one another, and the endothelial cell imprint pattern in the cast is similar to that of veins[5].

Discussion and Conclusion

The occurrence of veins being crossed-over by arteries on the pial surface of the cerebral cortex seems to indicate, at least in the cast preparations, an effect on venous flow which is not seen elsewhere in the areas studied. One might speculate that these spots function as valves in this valveless system, which in turn is connected through another valveless system in the musculo-skeletal wall with the visceral cavity where large pessure changes occur[1].

Also, the venous lengths between the arterial cross-overs then could be considered as functional units ("Pumpenstrecken", Kiss[6], p. 364) where arterial pulsations force the blood towards the sinuses. Contrary to previous reports arterio-venous anastomoses have not been seen in the areas studied.

The cerebellar pial surface typically displays a number of pairs of arteries and veins. Here one might think of the arterio-venous coupling described by v. Lanz (Kiss[6], Vollmerhaus[10]). The arterial pulsation has a noticable effect on venous flow in systems where arteries and veins run closely together within an enclosed space. So far, arterio-venous coupling with this effect in the brain has only been described in animals with an extensive carotid *rete mirabile*[7].

In the hypophysis, the neurohypophyseal drainage is a consistent feature in terms of size and location, just like in the pineal gland. This is in contrast to the adenohypophyseal drainage, which might stem from a different embryological origin.

The long and short portal vessels appear identical in shape except for their respective lengths; both groups display a similar asymmetry: capillaries on the one side and sinusoids on the other.

Although the actual flow direction is still a matter of argument rather than of demonstration, it seems that, if form follows function, the flow direction in both groups of portal vessels must be the same.

In conclusion it can be said that the pial veins of the various parts of the rat brain consistently show distinctly different patterns, both *per se* as well as in relation to the accompanying arteries.

References

1. Batson, O. V.: More About Veins. Proc. Xth Int. Cong. Anat., Tokyo 1975. Science Council of Japan (Yamada, E., ed.), pp. 551, p. 1.
2. Hodde, K. C., Miodoński, A., Bakker, C., Veltman, W. A. M.: Scanning Electron Microscopy of micro-corrosion casts with special attention on arterio-venous

differences and application to the rat's cochlea. SEM/1977/part II, IIT Research Institute, Chicago, IL 60616, pp. 477—484.

3. Hodde, K. C.: The vascularization of the rat pineal organ. In: The Pineal Gland of Vertebrates Including Man (Ariëns Kappers, J., Pévet, P., eds.), Progr. Brain Res. (Elsevier) *52* (1979 a), 39—44.

4. Hodde, K. C.: The vascularization of the choroid plexus of the lateral ventricle of the rat. Beitr. elektronenmikr. Direktabb. Oberfl. (BEDO) *12* (1979 b), 395—400.

5. Hodde, K. C.: Vascular casts of the rat pituitary gland. BEDO *14* (1981), 593—602.

6. Kiss, F.: Funktionell-morphologische Angaben zum venösen Kreislauf. Acta anat. *30* (1957), 358—370.

7. König, H. E.: Arteriovenöse Koppelungen in der Dura mater des Rindes. Berl. Münch. Tierärztl. Wchschr. *92* (1979), 1—3.

8. Miodoński, A., Hodde, K. C., Bakker, C.: Rasterelektronenmikroskopie von Plastik-Korrosions-Präparaten: morphologische Unterschiede zwischen Arterien und Venen. Direktabb. Oberfl. (BEDO) *9* (1976), 435—442.

9. Noordergraaf, A.: Circulatory System Dynamics. Biophysics and Bioengineering Series, Vol. 1. New York: Academic Press. 1978.

10. Vollmerhaus, B.: Die Arteria und Vena ovarica des Hausrindes als Beispiel einer funktionellen Kopplung viszeraler Gefässe. Anat. Anz. (Erg. H.) *112* (1963), 258—264.

Author's address: K. C. Hodde, M.D., Laboratory for Surgical Research, Wilhelmina Gasthuis, 1st Helmersstr. 104, NL-1054 EG Amsterdam, The Netherlands.

Department of Zoology (Head: Prof. Dr. Hans Adam), University of Salzburg, Austria

Scanning Electron Microscopy of Corrosion Casts of Cerebral Vessels. Advantages and Limitations[1]

A. Lametschwandtner and U. Lametschwandtner-Albrecht

Summary

The investigation of vascular corrosion casts (injection replicae) with the scanning electron microscope (SEM) enables one to study the vascular patterns of tissues and organs from large supplying arteries and draining veins to the finest capillaries. The present paper reviews the advantages and limitations of this method, and especially discusses the work done on the vertebrate brain.

Keywords: Cerebral vessels; scanning electron microscopy.

Introduction

The scanning electron microscopy of vascular corrosion casts[24,27] has become a widely used method to study microanatomical aspects of the vasculature of tissues and organs[14]. This is shown by the increasing number of papers dealing either with the methodical procedure of how to prepare suitable vascular casts[5,11,16,32] or with the microangioarchitecture of various tissues and organs under normal (for references see[35]) or pathological conditions[15,22,23,34]. There are, however, surprisingly few studies done with this method on the cerebral vascular system of vertebrates (for references see[2,9,11,12,20]).

The present paper aims to present the advantages and disadvantages of the method and consider some future trends.

Material and Methods

The data available from the literature concern the brain vascular system of the hagfish[1,20], the common toad (for references see[20]), the crested newt[18], the rat (for references see[2,12,25]), the cat[21], the rabbit, the dog, the sheep, the monkey[2,30] and the human[4]. Methodological details are described in some additional papers[3,6−8,11,19,26,29].

[1] This work was supported by the Jubiläums-Fonds der Oesterreichischen National-bank (Project No. 1868).

Results

Summarizing the present knowledge about the scanning electron microscopy of vascular corrosion casts these advantages are named by most of the authors:

1. A "Quasi three-dimensional" image of the vascular bed is possible. This is due to the high depth of focus given by the specific way the scanning electron microscope works.

2. A large specimen can be studied. The modern generation of scanning electron microscopes has a very large specimen chamber.

2 a. Too large specimens can be dissected under the binocular microscope. The resin mixtures used nowadays provide a good stability of the casts.

3. Mounted casts can be manipulated extensively under the electron beam. Full rotation, tilting from 0–90° and shifting of the specimen in x-, y- and z-direction enable one to clarify very complex branching patterns or vascular courses.

4. In the casts arteries and veins can be discerned by their specific surface impressions (arteries: elongated impressions of endothelial cell nuclei, veins: oval to round impressions[10]).

5. Vascular structures, like circular constrictions[10], sphincters or changes of the luminal endothelial cell surface in general[31] can be shown. The replication quality of the resin mixtures is very high[35] and the replication itself is very good[35].

The advantage listed above result in good facilities when studying such complex vascular patterns as occur in the vertebrate brain[2,4,12,16,20]. Thus

1. a good spatial impression of the vascular architecture of even the smallest capillary bed of every cerebral region can be perceived.

2. Vascular connections between areas far apart from each other but functionally belonging together can be demonstrated without errors[17].

3. Every vessel may be studied individually with respect to its origin, course, branching (number of branches, branching angles, directions), supplying and draining areas.

4. Interesting cerebral regions may be exposed and prepared under the dissecting microscope. A suitable microinstrumentarium (glass needles and glass hooks, fine scissors and a micromanipulator) enables us to do this.

5. Intimal lesions, like arteriosclerotic plaques, can be visualized at very early stages.

Together with the standard light microscopical methods (India-ink injected specimens treated according the method of Spalteholz[33] and latex-injection for gross dissection[11]) these advantages have brought new insights into the brain vascular structure and have led to altered concepts of the functions of some areas of the cerebral nervous system[2,20].

But there are still shortcomings the technique suffers from. In summary they are:

1. No information about the direction of blood flow within the casted area.

2. No information about the actually "open" vessels (especially at the capillary level) since physiologically "closed" vessel might be opened artificially by the injection itself.

3. Unsatisfactory control of the actual injection pressure (i.e. pressure with which the resin enters the brain or other organs of interest). The viscosity of the injected resin, injection cannulas (diameter), injection volume per unit time, physical properties of the vessel cannulated and others factors can all create poor pressure conditions. Partial filling or extravasation may result.

4. Unknown effects of the injections themselves (chemically) on the wall of the blood vessel or its luminal surface structures during the injection procedure.

5. Too little knowledge about physical and chemical properties (polymerization; resistance against heat, dying and decalcification solutions, cleaning agents, spattering; behaviour under the electron beam; shrinkage).

To minimize these disadvantages several modifications are being undertaken. While some of the disadvantages can probably be elimitated in the near future (points 3–5), the others still remain a subject of speculation.

References

1. Adam, H., Lametschwandtner, A., Wimmer, C.: Hypothalamohypophyseal relations in cyclostomes and amphibia. Proc. Int. Symp. Neuroendocr. Regul. Mechan. Serbian Acad. Sci., Belgrad 1979, pp. 53—62.
2. Bergland, R. M., Page, R. B.: Pituitary-brain vascular relations: A new paradigm. Science 204 (1979), 18—24.
3. Castenholz, A., Zöltzer, H., Erhardt, H.: Structures imitating myocytes and pericytes in corrosion casts of terminal blood vessels. A methodical approach to the phenomenon of "plastic strips" in SEM. Mikroskopie 39 (1982), 95—106.
4. Duvernoy, H. M., Delon, S., Vannson, J. L.: Cortical blood vessels of the human brain. Brain Res. Bull. 7 (1981), 519—579.
5. Gannon, B. J.: Vascular casting. In: Principles and Techniques of Scanning Electron Microscopy. Biological Applications, Vol. 6 (Hayat, M. A., ed.). New York: Van Nostrand-Reinhold. 1978.
6. Gannon, B., J.: Prepolymerization of methacrylate corrosion casting media for microvascular replication using ultraviolet light. J. Anat. 128 (1979), 665.
7. Gannon, B., J.: Preparation of microvascular corrosion casting media: Procedure for partial polymerization of methyl methacrylate using ultraviolet light. Biomed. Res. 2 (1981), Suppl. 227—233.
8. Grunt, T.: Normale und pathologische Angioarchitektur der Haut unter besonderer Berücksichtigung der Blutgefäßversorgung des humanen Basalioms und der Blutgefäßversorgung eines experimentell erzeugten Tumors bei einer Labormaus. Wissenschaftliche Hausarbeit am Zoologischen Institut der Universität Salzburg, Salzburg 1981.

9. Heinz, E.: Möglichkeiten und Grenzen der Darstellung des Blutgefäßsystems des zentralen Nervensystems von Wirbeltieren an Hand rasterelektronenmikroskopischer Gefäßausgußpräparate. Wissenschaftliche Hausarbeit am Zoologischen Institut der Universität Salzburg, Salzburg 1981.

10. Hodde, K. C., Miodonski, A., Bakker, C., Veltman, W. A. M.: Scanning electron microscopy of microcorrosion casts with special attention on arterio-venous differences and application to the rats cochlea. Scan. Electr. Micr. 2 (1977), 477—484.

11. Hodde, K. C., Nowell, J. A.: SEM of micro-corrosion casts. Scan. Electr. Micr. 2 (1980), 89—106.

12. Hodde, K. C.: Cephalic vascular patterns in the rat. Thesis. University of Amsterdam. 1981.

13. Hodde, K. C.: Veins of the rat brain. In: The Cerebral Veins (Auer, L. M., Loew, F., eds.), pp. 85—92. Wien-New York: Springer. 1983.

14. Kessel, R. G., Kardon, R. H.: Tissues and organs: A text-atlas of scanning electron microscopy. San Francisco: Freeman and Co. 1979.

15. Kus, J., Miodonski, A., Olszewski, E., Tyrankiewicz, R.: Morphology of arteries, veins, and capillaries in cancer of the larynx. Scanning electron microscopical study on microcorrosion casts. J. Cancer Res. Clin. Oncol. 100 (1981), 271—283.

16. Lametschwandtner, A., Simonsberger, P., Adam, H.: Scanning electron microscopical studies of corrosion casts. The vascularization of the paraventricular organ (Organon vasculosum hypothalami) of the toad, Bufo bufo L. Mikroskopie 32 (1976), 195—203.

17. Lametschwandtner, A., Simonsberger, P., Adam, H.: Vascularization of the pars distalis of the hypophysis in the toad, Bufo bufo (L.) (Amphibia, Anura). A comparative light microscopical and scanning electron microscopical study I. Cell Tiss. Res. 179 (1977), 1—10.

18. Lametschwandtner, A., Miodonski, A., Adam, H.: The choroid plexus of the lateral ventricle of Triturus cristatus (L.). A SEM-study of vascular corrosion casts. 1980 (unpublished).

19. Lametschwandtner, A., Miodonski, A., Simonsberger, P.: On the prevention of specimen charging in scanning electron microscopy of vascular corrosion casts by attaching conductive bridges. Mikroskopie 36 (1980), 270—273.

20. Lametschwandtner, A.: Das Gefäßbett des Gehirns und der Hypophyse von Myxine glutinosa (L.) (Cyclostomata, Myxinoidea) and Bufo bufo (L.) (Amphibia, Anura). Eine rasterelektronenmikroskopische Untersuchung an Gefäßausgußpräparaten (Korrosionspräparaten). Mikroskopie 39 (1982), 35—42.

21. Miodonski, A., Poborowska, J., Grubenthal, A. F.: SEM study of the choroid plexus of the lateral ventricle in the cat. Anat. Embryol. 155 (1979), 323—331.

22. Miodonski, A., Kus, J., Olszewski, E., Tyrankiewicz, R.: Scanning electron microscopic studies on blood vessels in cancer of the larynx. Arch. Otolaryngol. 106 (1980), 321—332.

23. Miodonski, A., Kus, J., Tyrankiewics, R.: SEM blood vessel casts analysis. In: Three Dimensional Microanatomy of Cells and Tissue Surfaces (DiDio, L. J. A., Motta, P. M., Allen, D. J., eds.), pp. 71—87. Amsterdam: Elsevier North Holland Inc. 1981.

24. Murakami, T.: Application of the scanning electron microscope to the study of the fine distribution of blood vessels. Arch. histol. jap. 32 (1971), 445—454.

25. Murakami, T.: Pliable methacrylate casts of blood vessels: use in a SEM-study of the microcirculation in rat hypophysis. Arch. histol. jap. 38 (1975), 151—168.

26. Murakami, T.: Methyl-methacrylate injection replica method. In: Principles and Techniques of Scanning Electron Microscopy. Biological Applications, Vol. 2 (Hayat, M. A., ed.), pp. 159—169. New York: Van Nostrand-Reinhold. 1978.

27. Nowell, J. A., Pangborn, J., Tyler, W. S.: SEM study of the avian lung. Scan. electr. microsc. 1970, pp. 249—256.

28. Nowell, J. A., Lohse, C. L.: Injection replication of the microvasculature for SEM. Scan. electr. microsc. 1974, pp. 267—274.

29. Ohtani, O.: Microcirculation studies by injection replica method. Biomed. Res. *2* (1981), 219—226.

30. Page, R. M., Bergland, R. M.: The neurohypophyseal capillary bed. I. Anatomy and arterial supply. Amer. J. Anat. *148* (1977), 345—358.

31. Reidy, M. A., Levesque, M. J.: A scanning electron microscopic study of arterial endothelial cells using vascular casts. Atherosclerosis *28* (1977), 463—470.

32. Rosenbauer, K. A., Notermans, H. P., Jansen, B.: Techniken zur Anfertigung von Gefäßausgußpräparaten für rasterelektronenmikroskopische Untersuchungen. Präparator *26* (1980), 291—298.

33. Spalteholtz, W.: Über das Durchsichtigmachen von menschlichen und tierischen Präparaten, 2. Aufl. Leipzig 1914.

34. Staindl, O., Lametschwandtner, A.: Die Angioarchitektur solid-zystischer Basaliome. Eine rasterelektronenmikroskopische Untersuchung an Korrosionspräparaten. HNO *29* (1981), 112—117.

35. Weiger, T.: Die Verwendbarkeit der polymerisierenden Kunststoffe Methylmethacrylat und Mercox CL zur Herstellung von Korrosionspräparaten zur rasterelektronen-mikroskopischen Untersuchung der Gefäßstruktur tierischer Gewebe und Organe. Wissenschaftliche Hausarbeit am Zoologischen Institut der Universität Salzburg, Salzburg 1981.

Author's address: Doz. Dr. Mag. A. Lametschwandtner, Zoologisches Institut der Universität Salzburg, Akademiestrasse 26, A-5020 Salzburg, Austria.

Venous Blood Brain Barrier

Department of Neurology, University of Lund, Lund Hospital, Lund, Sweden

The Venous Blood-Brain Barrier

B. B. Johansson

Summary

There is no morphological evidence that the permeability of cerebral veins essentially differs from that of capillaries and arterial vessels under normal conditions. No quantitative data are available regarding the leakage of substances through the blood-brain barrier in various sections of the vascular tree under pathological conditions. A survey on previous publications of the author and on related literature is given.

Keywords: Cerebral veins; permeability; tight junctions; pinocytosis.

Introduction

Few studies have concentrated on the venous side of the blood-brain barrier and information on this topic is therefore scarce. In contrast to many other vascular beds there is no evidence that the post-capillary venous section is more permeable than the pre-capillary or capillary part of the vascular tree.

The Blood-Brain Barrier Under Normal Conditions

Wolff was the first to point out that cerebral vessels contain very few pinocytotic vesicles[26]. Using the electron dense tracer horse-radish peroxidase Reese and Karnowsky demonstrated that the tracer did not pass the endothelial cell membrane of cerebral arteries, capillaries or veins[23]. They confirmed the observation of relatively scarce pinocytotic vesicles and also demonstrated that the endothelial cells were joined together by tight junctions. The luminal site of the endothelial cell membrane seemed to constitute the barrier to protein since protein tracers could fairly easily pass through the basement membrane once they had entered the endothelial cells. An old observation that substances injected into the ventricular system freely entered the brain was also confirmed. Later studies have indicated that there might be a slight passage of protein in some arterial sections in the cortex but not in veins under physiological conditions[24].

The brain vessels are unique in many other aspects, e.g. the presence of enzymatic barriers[2] and specific transport mechanisms[5,18]. The endothelial cells contain several times more mitochondria than other vascular beds, indicating a high metabolic activity[19]. The luminal and abluminal surfaces of the cell membrane are not identical. Thus, the enzymes alkaline phosphatase and γ-glutamyl transpeptidase are found on both sides whereas Na^+, K^+-ATPase and 5'-nucleotidase are found only on the abluminal side[3]. Whether these characteristics are confined to a particular segment of the vascular tree is not known, but from a physiological point of view it seems most likely that these mechanisms are most important in the microvessels. In fact, many of the above characteristics are typical of high resistance epithelium rather than of endothelial cells. Furthermore, the electrical resistance of microvessels at the brain surface is similar to that of a tight epithelium[6].

Another interesting aspect of the blood-brain barrier is that the endothelial cells are not freely permeable to water as previously thought[22]. It has been pointed out that the blood-brain barrier is functionally similar to membranes known to regulate water permeability. The water permeability seems to be under neurogenic and endocrine influences[21,22]. What significance these observations may have for the net uptake of water and the regulation of brain water volume is a matter of controversy.

The Blood-Brain Barrier Under Pathological Conditions

Extensive damage to the blood-brain barrier is seen under conditions such as brain tumours, trauma, embolism (gas, fat or solid), meningitis and cerebrovascular disorders. Further, it can be altered by toxic and chemical influences and by x-ray treatment. Various pathological changes have been observed, e.g. gross disruption of the vascular wall, opening of the tight junctions, widening of the basal lamina and an enhanced number of vesicles (for a recent review see Lee[15]). Information about veins in particular is scarce. A specific damage to the blood-brain barrier in *veins* occurs in multiple sclerosis where perivenous leakage is an early and often long-lasting phenomenon[4].

An interesting type of blood-brain barrier dysfunction is the transitory "opening" that occurs in connection with acute hypertension and epileptic seizures. It seems to be directly related to the tension in the vessel wall. Thus dilated vessels are particularly prone to increase their permeability during acute hypertension or a local increase in perfusion pressure[11]. Since the disturbance is functional rather than structural and is rapidly reversible, this experimental model has been extensively studied. Whether the passage of protein tracers is through the endothelial cell membrane or between the endothelial cells (through "opened" tight junctions) is still controversial[9,16,17,20,25]. A markedly enhanced pinocytotic activity has been observed by electron microscopy but the significance of this finding has been questioned. The tracer horseradish peroxidase has occasionally been observed between tight junctions[17] but this in itself is no proof that the junctions have been opened since it has been shown that pinocytotic vesicles can empty their contents into such junctions. Whereas many studies stress that the tracers are seen particularly on the precapillary side, leakage around veins does occur[9,17] and has been nicely shown under in vivo conditions by Auer[1].

By labelling the luminal side of the endothelial cell membrane with a choleragenoid-peroxidase conjugate it has definitely been shown that vesicles can be formed from the luminal side[8] (one argument has been that the vessels might transport the tracer out from the brain to the blood after leakage through the tight junction).

The mechanism behind pinocytosis is still being debated. However, membrane fusion is essential for all kinds of endo- and exocytosis. Several drugs can alter membranes and prevent endocytotic mechanisms such as pinocytosis. Such drugs are also efficient in preventing the blood-brain barrier from opening during acute hypertension[7,10,12,14]. On the other hand, ethanol and nitrous oxide—both increasing the fluidity of membranes—increase the passage of albumin into the brain during acute blood pressure elevation[13]. These results would seem to favour the possibility that passage is transendothelial rather than interendothelial. However, since it is not known whether the drugs used could also influence tight junctions, the results are not conclusive.

References

1. Auer, L. M.: The pathogenesis of hypertensive encephalopathy. Acta neurochir. (Wien), Suppl. 27 (1978), 1—111.
2. Bertler, Å., Falck, B., Owman, Ch., Rosengren, E.: The localization of mono-aminergic blood-brain barrier mechanisms. Pharmacol. Rev. 18 (1966), 369—385.
3. Betz, A. L., Firth, J. A., Goldstein, G. W.: Polarity of the blood-brain barrier: distribution of enzymes between the luminal and antoluminal membranes of brain capillary endothelial cells. Brain Res. 192 (1980), 17—28.
4. Broman, T.: The permeability of the cerebrospinal vessels in normal and pathological conditions. Copenhagen: Munksgaard. 1949.
5. Crone, C.: Permeability of capillaries in various organs as determined by use of indicator diffusion method. Acta Physiol. Scand. 58 (1963), 292—305.
6. Crone, C., Olesen, S. P.: Electrical resistance of brain microvascular endothelium. Brain Res. 242 (1982), 49—55.
7. Hansson, H. A., Johansson, B. B.: Prevention of protein extravasation in the rat brain by an anion transport inhibitor in acute experimental hypertension in rats. Acta Physiol. Scand. 105 (1979), 513—517.
8. Hansson, H.-A., Johansson, B. B.: Induction of pinocytosis in cerebral vessels by acute hypertension and by hyperosmolar solutions. J. Neurosci. Res. 5 (1980), 183—190.
9. Hansson, H.-A., Johansson, B. B., Blomstrand, C.: Ultrastructural studies on cerebrovascular permeability in acute hypertension. Acta neuropathol. (Berl.) 32 (1975), 187—198.

10. Hardebo, J. E., Johansson, B. B.: Effect of an anion transport inhibitor on blood-brain barrier lesions in acute hypertension. Possible prevention of transendothelial vesicular transport. Acta neuropathol. (Berl.) *51* (1980), 33—38.
11. Johansson, B. B.: The blood-brain barrier in acute and chronic hypertension. In: The Cerebral Microvasculature. Advances in Experimental Medicine and Biology 131 (Eisenberg, H. M., Suddith, R. L., eds.), pp. 211—226. New York: Plenum Press. 1980.
12. Johansson, B. B.: Pharmacological modification of hypertensive blood-brain barrier opening. Acta Pharm. Toxicol. *48* (1981), 242—247.
13. Johansson, B. B., Linder, L.-E.: Do nitrous oxide and lidocain modify the blood-brain barrier dysfunction in acute hypertension in the rat? Acta Anaesthesiol. Scand. *24* (1980), 65—68.
14. Johansson, B. B., Auer, L. M., Linder, L.-E.: Phenothiazine protection of the blood-brain barrier during acute hypertension: evidence for a modification of the endothelial cell membrane. Stroke *13* (1982), 220—225.
15. Lee, J. C.: Anatomy of the blood-brain barrier under normal and pathological conditions. In: Histology and Histopathology of the Nervous System (Haymaker, W., Adams, R. D., eds.), pp. 798—869. Springfield, Ill., U.S.A.: Ch. C Thomas. 1982.
16. Nag, S., Robertson, D., Dinsdale, H. B.: Cerebral cortical changes in acute hypertension. An ultrastructural study. Lab. Invest. *36* (1977), 150—161.
17. Nagy, Z., Mathieson, G., Huttner, I.: Blood-brain barrier opening to horseradish peroxidase in acute arterial hypertension. Acta neuropathol. (Berl.) *48* (1979), 45—53.
18. Oldendorf, W. H.: The blood-brain barrier. Exp. Eye Res. (Suppl.) *25* (1977), 177—190.
19. Oldendorf, W. H., Cornford, M. E., Brown, W. J.: The large apparent work capability of the blood-brain barrier: a study of the mitochondrial content of capillary endothelial cells in brain and other tissues of the rat. Ann. Neurol. *1* (1977), 409—417.
20. Petito, C., Schaefer, J. A., Plum, F.: Ultrastructural characteristics of the brain and blood-brain barrier in experimental seizures. Brain Res. *127* (1977), 251—267.
21. Reichle, M. E., Grubb, R. L.: Regulation of brain water permeability by centrally-released vasopressin. Brain Res. *143* (1978), 191—194.
22. Raichle, M. E., Harman, B. K., Eichling, J. O., Sharpe, L. G.: Central noradrenergic regulation of cerebral blood flow and vascular permeability. Proc. Nat. Acad. Sci. (Wash.) *72* (1975), 3726—3730.
23. Reese, T. S., Karnovsky, M. J.: Fine structural localization of a blood-brain barrier to exogenous peroxidase. J. Cell. Biol. *34* (1967), 207—217.
24. Westergaard, E., Brightman, M. W.: Transport of proteins across normal cerebral arterioles. J. Comp. Neurol. *152* (1973), 17—44.
25. Westergaard, E., van Deurs, B., Brønsted, H. E.: Increased vesicular transfer of horseradish peroxidase across cerebral endothelium, evoked by acute hypertension. Acta neuropath. (Berl.) *37* (1977), 141—152.
26. Wolff, J.: Beiträge zur Ultrastruktur der Kapillären der normalen Großhirnrinde. Z. Zellforsch. *60* (1963), 409—431.

Author's address: B. B. Johansson, M.D., Department of Neurology, University of Lund, Lund Hospital, S-221 85 Lund, Sweden.

Institute of Medical Physiology, Department A, University of Copenhagen, The Panum Institute, Copenhagen N, Denmark

Electrical Resistance of Brain Postcapillary Venular Endothelium of Known Morphology

S.-P. Olesen and M. Bundgaard

Summary

The permeability of pial venules and venous capillaries in the brain of the frog (*Rana temporaria*) has been examined. The electrical resistance of the vascular wall was determined from the intravascular potential profile resulting from a current introduced into the lumen of the vessel. The average resistance was $1{,}870\,\Omega\text{cm}^2$, equivalent to a permeability as low as $5 \times 10^{-7}\,\text{cm}\,\text{sec}^{-1}$, which is similar to values obtained for the blood-brain barrier in mammals. Electron microscopical examinations of the same microvessels identified the endothelial lining as the barrier layer. The vessels were devoid of glial investment and the interendothelial clefts were obliterated by elaborate tight junctions impermeable to lanthanum ions. Further, it was demonstrated that the venous microvessels of the brain parenchyma were similar to the pial blood vessels with respect to endothelial ultrastructure and tightness to lanthanum ions. It is concluded that the frog has an endothelial blood-brain barrier of a tightness comparable to that of the mammalian barrier.

Keywords: Frog-blood-brain barrier; electrical resistance; electron microscopy.

Introduction

Blood vessels of the frog brain drain into a plexus of venous capillaries and venules located between the dorsal brain surface and the arachnoid membrane[7]. These vessels are easily accessible for single capillary measurements. Further, assuming that the blood-brain barrier extends into the subarachnoid space, such recordings would provide us with a unique opportunity to obtain direct information on the blood-brain barrier.

Well-defined segments of single pial microvessels in the frog were examined. The overall ion permeability of the vascular wall, expressed by its electrical resistance, was determined. The electrical resistance was found to be as high as that of a tight epithelium, thus supporting the assumption that pial microvessels can be used as a model of the blood-brain barrier. In an attempt to explain the low permeability in terms of structure, the same vessels were examined by electron microscopy. The pial venules and capillaries appeared without glial investment, implying that the observed high resistance can be localized to the endothelial lining. This view was confirmed by the observation that ionic lanthanum did not permeate the endothelium.

Materials and Methods

The theory behind the electrical resistance measurements is based on the analogy between a leaky ion-conducting cable and a blood vessel embedded in a conducting medium. This is illustrated in Fig. 1. Current, introduced into the lumen of a vessel through a microelectrode, creates a potential V_0 at the electrode tip and a potential displacement inside the vessel that decays exponentially away from the current source, due to leakage through the capillary wall[6]. From the length constant λ (cm) of the exponential potential decay (the distance in which the potential has fallen to $1/e$), the radius of the vessel a (cm) and the resistivity of the blood ρ_i (Ω cm), the membrane resistance R_m can be determined as,

$$R_m = 2\rho_i\lambda^2/a \quad (\Omega\,cm^2)$$

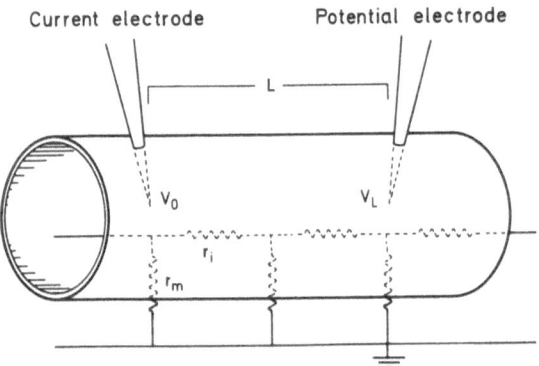

Fig. 1. Theoretical model for the electrical resistance measurements. r_i (Ω cm^{-1}) is the internal resistance per unit length vessel and r_m (Ω cm) the membrane resistance per unit length vessel. Current introduced into the lumen through the left microelectrode creates a potential V_0 (V) at the electrode tip and a potential $V_L = V_0 e^{-L/\lambda}$ (V) at a distance L (cm) from the current source where it is being measured. $\lambda = \sqrt{r_m/r_i}$ is the length constant of potential decay

The experiments were carried out on adult frogs (*Rana temporaria*) as previously described[4]. In brief: Following anaesthesia by immersion in 5% urethane, the dorsal surface of the telencephalic part of the brain was exposed by removing the frontoparietal bone by a dental drill. After removal of the arachnoid membrane, the network of pial venules and capillaries was directly accessible for microelectrode impalement. The brain surface was kept moist by constant superfusion with frog Ringer's solution. Two reference electrodes were placed at the rim of the skull and the microelectrodes, used for introduction of the current and intravascular potential measurement, were mounted on micromanipulators. Current from a constant current generator was introduced into the lumen of a vessel and the measured intravascular potentials were amplified and recorded. Three to four potential recordings at different distances from the current source were necessary to make a reliable determination of λ. The vessel radius was determined by means of a calibrated grid placed in the eyepieces of the microscope. The specific resistivity of frog blood was 130 Ωcm.

Brains, exposed as mentioned above, were fixed by in situ immersion for electron microscopy. The preparative procedures have been described in detail in Ref.[2]. The normal ultrastructure of pial and cerebral microvessels was studied in three animals. In two frogs the telencephalon was superfused with a solution containing lanthanum ions for 30 minutes prior to fixation. By using a phosphate-buffered fixative, the lanthanum ions were precipitated as lanthanum phosphate. The electron dense precipitates reflected the distribution of lanthanum ions in the tissue.

Results

The electrical resistance of the vascular wall was determined in 40 pial venous capillaries and venules (average diameter = 33.4 μm; range 12.8–57.6 μm). A mean value of 1,870 Ωcm² (SD 639 Ωcm²; n = 40) was obtained. The average length constant was 1,026 μm (SD 257 μm; n = 40).

Electron microscopical examinations of pial blood vessels within the same range of diameters showed that the vascular wall was made up of a continuous endothelium, a basal lamina (surrounding scattered pericytes) and a layer of loose connective tissue. The characteristic glial investment of

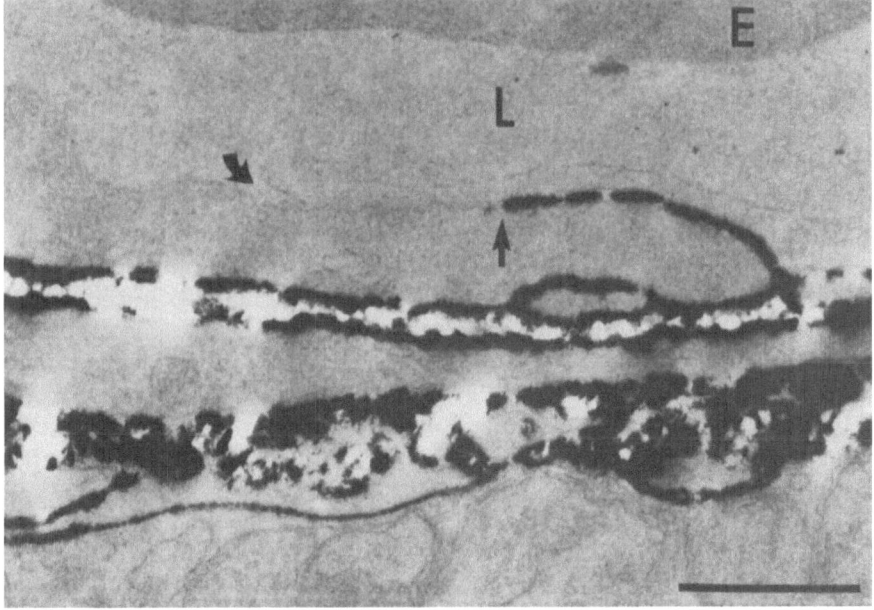

Fig. 2. Electron micrograph of an unstained section of a venule near the dorsal surface of the frog brain. The brain was superfused with ionic lanthanum prior to fixation. The diffusion of lanthanum ions appears to be stopped abruptly in the interendothelial cleft by the tight junction (straight arrow). The luminal opening of the cleft is indicated by the curved arrow. L vascular lumen. E erythrocyte. Bar, 0.5 μm

parenchymal blood vessels did not extend beyond the brain surface. The endothelial cells were joined by elaborate tight junctions appearing as 4–10 punctate contacts or fusions between neighbouring membranes. These junctions were impermeable to lanthanum ions administered in the superfusate. The endothelium of parenchymal blood vessels near the dorsal surface of the telencephalon was structurally similar to that of the pial vessels. Particularly, the junctions had the same appearance. After superfusion with ionic lanthanum, the tracer permeated the perivascular sheath of glial end-feet and diffused into the interendothelial clefts where further penetration was stopped abruptly by the junctions (Fig. 2).

Discussion

We have demonstrated that the brain vascular endothelium has a very high electrical resistance, nearly 100 times higher than that of the vascular endothelium of striated muscle, and 1,000 times higher than that of the mesenteric vessels[8, 3]. On the other hand, the observed resistance is comparable to values obtained on a typical tight epithelium such as the frog skin $(2,000 \, \Omega \, cm^2)$[10].

The high electrical resistance of the wall of frog pial blood vessels reflects a very low permeability to the dominating univalent ions in plasma (Na^+, K^+, Cl^-). The relation between the partial electrical conductance $G_j (\Omega^{-1} \, cm^{-2})$ and the ion permeability $P_j (cm \, sec^{-1})$ is given by

$$G_j = z_j^2 \cdot c_j \cdot P_j \cdot \frac{F^2}{RT}$$

where z_j is valency, c_j ionic concentration, and F, R and T have their usual significance. Using this expression, one arrives at a permeability of about $5 \times 10^{-7} \, cm \, sec^{-1}$. This is similar to the ion permeability of the blood-brain barrier obtained by tracer methods in mammals[1, 5, 9].

The electron microscopical examinations have shown that the very low permeability to ions can be attributed to the endothelium of both the parenchymal and the pial microvessels. No other elements in the vascular wall seem to offer appreciable restriction to diffusion. These findings corroborate our electrical measurements and strongly indicate that the endothelial blood-brain barrier extends to the pial vessels. Consequently, the easily accessible plexus of microvessels on the dorsal surface of the frog brain represents a useful model for direct recordings of blood-brain barrier characteristics.

References

1. Bradbury, M. W. B.: The concept of a blood-brain barrier. New York: J. Wiley. 1979.
2. Bundgaard, M.: Ultrastructure of frog cerebral and pial microvessels and their impermeability to lanthanum ions. Brain Res. *241* (1982), 57—65.
3. Crone, C., Christensen, O.: Electrical resistance of a capillary endothelium. J. Gen. Physiol. *77* (1981), 349—371.
4. Crone, C., Olesen, S. P.: Electrical resistance of brain microvascular endothelium. Brain Res. *241* (1982), 49—55.

5. Hansen, A. J., Lund-Andersen, H., Crone, C.: K$^+$-permeability of the blood-brain barrier, investigated by aid of a K$^+$-sensitive microelectrode. Acta physiol. scand. *101* (1977), 438—445.

6. Hodgkin, A. L.: The ionic basis of electrical activity in nerve and muscle. Biol. Rev. *26* (1951), 339—409.

7. Lametschwandtner, A., Albrecht, U., Adam, H.: The vascularization of the anuran brain. Olfactory bulb and telecephalon. A scanning microscopic study of vascular corrosion casts. Acta zool. (Stockholm) *61* (1980), 225—238.

8. Olesen, S. P., Crone, C.: Electrical resistance of muscle capillary endothelium. Biophys. J. *42* (1983), 31—41.

9. Smith, Q. R., Johanson, C. E., Woodbury, D. M.: Uptake of ^{36}Cl and ^{22}Na by the brain-cerebrospinal fluid system: Comparison of the permeability of the blood-brain and blood-cerebrospinal fluid barriers. J. Neurochem. *37* (1981), 117—124.

10. Ussing, H. H., Windhager, E. E.: Nature of shunt path and active sodium transport path through frog skin epithelium. Acta physiol. scand. *61* (1964), 484—504.

Authors' address: S.-P. Olesen, M. D., Institute of Medical Physiology, Department A, The Panum Institute, Blegdamsvej 3, DK-2200 Copenhagen N, Denmark.

Department of Neurosurgery, School of Medicine, Keio University, Shinjuku-ku, Tokyo, Japan

Cerebral Microcirculation in Experimental Sagittal Sinus Occlusion in Dogs

S. Sato, Y. Miyahara, Y. Dohmoto, T. Kawase, and S. Toya

Summary

The epicerebral microcirculation was investigated by fluorescein microangiography before and after sinus occlusion. The observations were performed through a glass-covered skull window in dogs. The diameter of the pial vein was relatively increased within a short interval after sinus occlusion. Diapedesis of fluorescein dye from the pial vein was observed thirty minutes after the sinus occlusion prior to petechial haemorrhage. This indicates that the blood-brain-barrier destruction progresses gradually and that the small particles of sodium fluorescein easily pass through the blood-brain-barrier. Petechial haemorrhages were observed one hour after the sinus occlusion, prior to intracerebral haemorrhage. Sinus occlusion leads to an increase in the permeability of the pial veins, resulting in petechial haemorrhage, which is responsible for subarachnoid haemorrhage.

From observations made for more than one hour, two types of microcirculatory change in the pial veins were observed. One change is of the "no-filling" type and the other is of the intravascular stasis type. The "no-filling" type is thought to lead to an extensive haemorrhagic infarction thereafter.

Keywords: Sagittal sinus occlusion; microcirculation; blood-brain-barrier.

Introduction

In 1825, Ribes first observed sinus thrombosis at autopsy[11]. With the introduction and the prevalence of cerebral angiography[8] and computed tomography[2,4], clinical reports of sinus thrombosis have increased. Sacrifice of cortical veins or dual sinus by operative procedures has often been experienced[5]. Even if the clinical analysis of a sinovenous circulatory diturbance is more significant, there are however, no detailed studies of microcirculatory changes following cerebral venous occlusion. The purpose of this study is to examine the microcirculation after experimental sinus occlusion using serial fluorescein cerebral microangiography.

Methods and Materials

Ten mongrel dogs weighing 10 to 15 kg were used in this study. They were anaesthetized with 10 mg/kg of sodium pentobarbital. An endotracheal tube was inserted after tracheostomy and connected to an artificial respirator. The experiments were conducted under a control respirator using room air. Preparatory to undertaking fluorescein microangiography, a polyethelene tube (inner diameter of 0.5 mm) was inserted into the right subclavian artery: the tube was fixed in this position both for continuous blood pressure monitoring, and as an injection channel to carry the fluorescein dye (0.5%). A 10 × 10 mm glass-covered skull window was made on the right parietal cortex before the occlusion of the sagittal sinus. A 5 mm burr hole was made at the point of vertex, and the superior sagittal sinus was exposed. The sinus was coagulated using bipolar forceps.

Since full details of the methods for fluorescein microangiography have been reported previously[7], only an outline is presented here. A Nikon motor-driven camera equipped with a Konan microscope K-280 and a rechargeable repeating flash was used for consecutive shooting of 3.8 exposures per second or 1.3 exposures per second. An Asahi Bunko Kogakusha filter (420–480 nm) was mounted in front of the flash as an exciting filter, and a Kodak-Wratten No. 12 filter was added in front of the camera as a barrier filter. After the photographic equipment had been assembled, 1 ml/kg fluorescein sodium solution (0.5%) was injected manually at a rapid rate via the polyethelene tube placed into the subclavian artery. Consecutive exposures were made as soon as the injection began, and a total of 36 photographs was taken.

After the completion of the initial fluorescein microangiography, the superior sagittal sinus was occluded without delay. Series of fluorescein microangiography were taken once or twice every hour until six hours after the occlusion of the superior sagittal sinus. The films were developed with double sensitivity, and the photographs were enlarged to 13 × 18 cm when detailed analysis was required. In order to investigate the microcirculation, measurements were made of the regional circulation time (RCT). To determine the RCT, we measured the time taken for the fluorescein dye from its entry into the arterioles to its entry into the venules. Arterioles and venules, with an internal diameter of about 60 μm located in the field of the glass-covered skull window, were observed on enlarged angiograms.

Results

The initial RCT was between 1.6 seconds and 3.8 seconds. In all the experiments RCT was prolonged after sinus occlusion (Fig. 1). In four dogs, the RCT was shorter than twice the time of the initial RCT, six hours after the sinus occlusion. The RCT of one dog was prolonged 4.6 seconds four hours after the sinus occlusion, and remained unchanged until six hours after sinus occlusion. Four dogs showed no filling of the fluorescein dye into the pial veins thirty minutes after the sinus occlusion. Another dog showed no filling of the fluorescein dye up to six hours after sinus occlusion.

Following sinus occlusion, the relative diameter of the pial vein was compared with that in the initial fluorescein microangiograph (Fig. 2a) in all experiments. In all experiments fluorescein microangiography, performed thirty minutes after sinus occlusion, showed diapedesis of the fluorescein dye from the pial veins, prior to the observation of petechial haemorrhage (Fig. 2b). Other experiments showed no filling of the fluorescein dye into the pial veins and extravasation of blood around the pial veins was observed (Fig. 2c). Five experiments showed intravascular

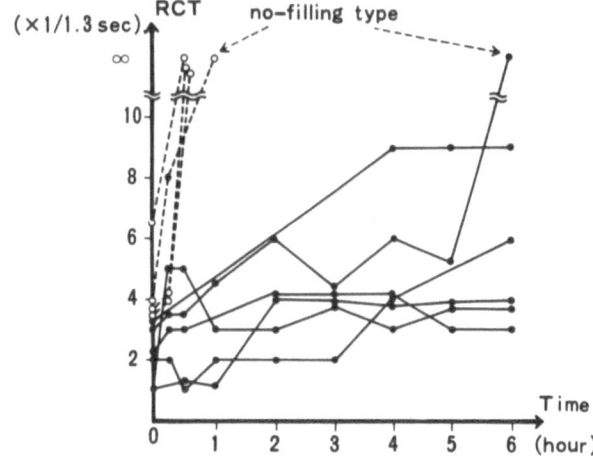

Fig. 1. The changes in regional circulation time (*RCT*) after superior sagittal sinus occlusion; RCT was prolonged in all experiments after the occlusion, and four experiments showed no filling of the fluorescein dye into the venous circulation 30 minutes after the occlusion

stasis, diapedesis of the fluorescein dye and petechial haemorrhage from the pial veins (Fig. 2 d). Petechial haemorrhage was observed prior to the haemorrhagic infarction demonstrated by pathology.

Discussion

The fact that diapedesis of the fluorescein dye from the pial veins was observed first, means that the destruction of the blood-brain-barrier occurs first in the pial veins. This explains why the pial veins which are located in the subarachnoid space, are so easily damaged by haemodynamic mechanical forces, compared with the intracerebral vessels. Diapedesis of the fluorescein dye from the pial veins was observed prior to the petechial haemorrhage. This indicates that the destruction of blood-brain-barrier progresses gradually and that the small particles of sodium fluorescein easily pass through the blood-brain-barrier of the pial veins in a congestive state. These experimental results support the clinical fact that the cerebrospinal fluid remains clear directly after sinus occlusion[3]. Petechial haemorrhage from the pial veins following diapedesis of the fluorescein dye brings about eventual subarachnoid haemorrhage. The same phenomenon was reported in an article on experimental brain trauma[12].

In this study, two types of cerebro-venous microcirculatory change were observed; one was "no-filling" of the fluorescein dye into the pial veins, and the other was an intravascular stasis and eventual diapedesis of the fluorescein dye. It is assumed that the differences in the venous microcirculation depend upon the differences in the collateral circulation. It is thought that, in "no-filling" cases, the thrombus in the sinus accelerates

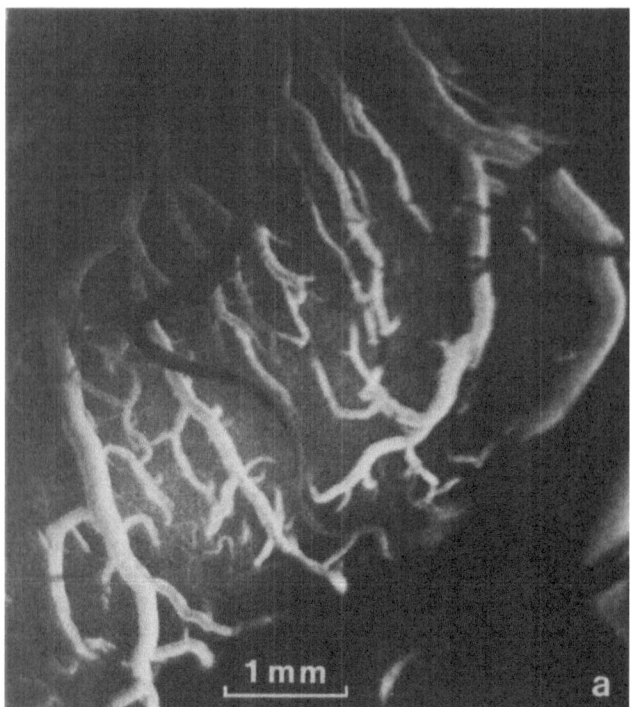

Fig. 2. Serial fluorescein venous microangiograms taken through the glasscovered dog's skull window before and after the superior sagittal sinus occlusion. a) Initial fluorescein microangiogram. b) Fluorescein microangiogram taken thirty minutes after the occlusion showing the diapedesis of the fluorescein dye around the pial vein. c) Fluorescein microangiogram taken six hours after the occlusion showing a filling defect in the pial vein of the fluorescein dye. d) Fluorescein microangiogram taken two hours after the occlusion showing the petechial haemorrhage around the pial vein and intravascular stasis of the fluorescein dye

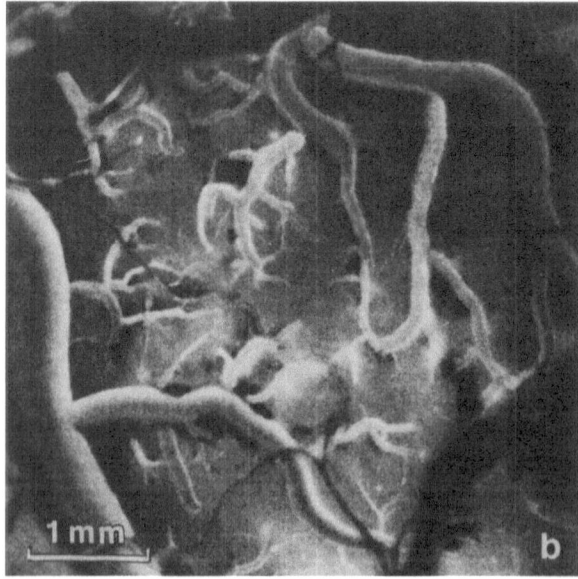

Fig. 2b

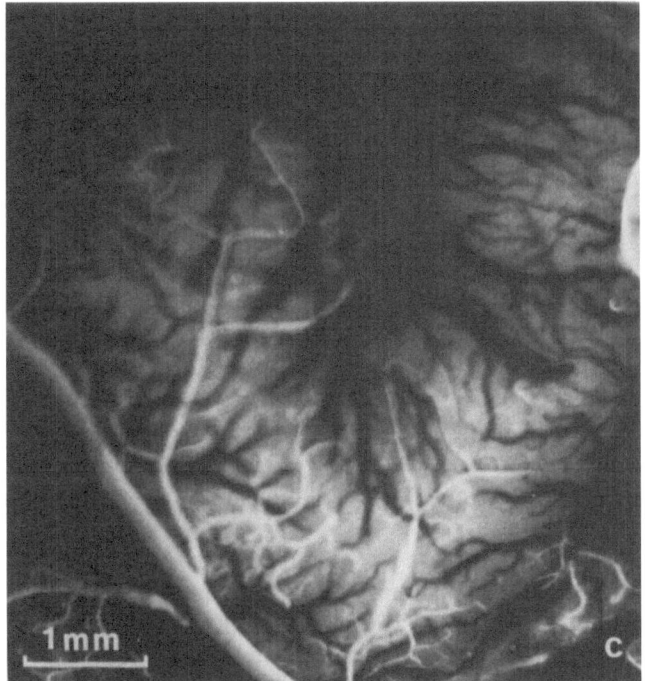

Fig. 2c

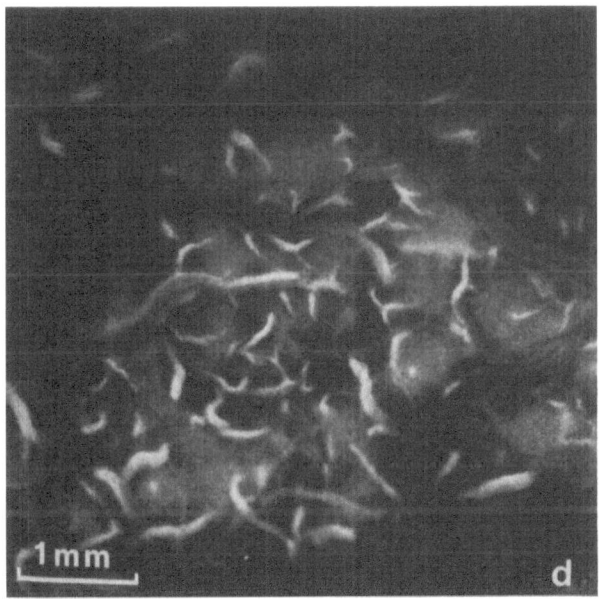

Fig. 2d

intravascular coagulation in the cerebral cortical vein, which leads to haemorrhagic infarction[1]. The same phenomenon was mentioned by Krayenbühl in 1954[9]. There are two categories of angiographical findings in venous thrombosis; the filling defect of the venous phase and the venous shunt type category. In these two categories, the prognosis was different, and the experimental venous microcirculatory changes seem to correspond with the results reported by Krayenbühl.

Some reports suggest that an increased intracranial pressure caused by the absorption disturbance of the cerebrospinal fluid through the Pacchionian bodies, is responsible for the thrombosis of the affected vein[6, 10]. But the finding that the "no-filling" phenomenon of the pial veins occurs soon after sinus occlusion, suggests that venous stagnation itself, rather than the intracranial hypertension, may be mainly responsible for the intravascular coagulation in the cortical cerebral veins.

In this experiment, observations were made for only six hours after sinus occlusion, but long-term observations of more than six hours may make it possible to observe that the intravascular stasis situation might develop into a "no-filling" situation.

It is concluded that in the early stages of superior sagittal occlusion, increased permeability of the pial veins and the destruction of the blood-brain-barrier preceeds subarachnoid haemorrhage. In the case of the "no-filling" situation , the thrombus in the superior sagittal sinus accelerates intravascular coagulation in the cortical veins, which then leads to haemorrhagic infarction.

Acknowledgement

Special thanks are given to Dr. Shigeo Toya, Takeshi Kawase and other neurosurgeons in Keio University, for providing the necessary time and encouragement.

References

1. Bailey, O. T.: Dural thrombosis in early life. J. pediat. *11* (1937), 755—771.
2. Barnes, B. D., Winestoch, D. F.: Dynamic radionuclide scanning in the diagnosis of thrombosis of the superior sagittal sinus. Neurology *27* (1977), 656—661.
3. Biemond, A.: Brains disease, pp. 562—565. Amsterdam: Elsevier Pub. 1970.
4. Bunanno, F. S., Moody, D. M., Ball, M. R., Lastor, D. W.: Computed cranial tomographic findings in cerebral sinovenous occlusion. J. comput. assist. Tomogr. *2* (1978), 281—290.
5. Dandy, W. E.: Removal of longitudinal sinus involved in tumor. Arch. surg. *41* (1940), 244—256.
6. Hakmatpamar, J.: Cerebral Circulation and Perfusion in Experimental Increased Intracranial Pressure. J. Neurosurg. *32* (1970), 21.
7. Kawase, T., Iisaka, Y., Toya, S., *et al.*: Experimental study in cerebral microcirculation. I. It's methods and result. Keio Igaku. (Tokyo) *51* (1974), 351—360.
8. Kinal, M. E.: Traumatic thrombosis of dural venous sinuses in closed head injuries. J. Neurosurg. *27* (1967), 142—145.

9. Krayenbühl, H. A.: Cerebral venous thrombosis diagnostic value of cerebral angiography. Schweiz. Arch. Neurol. Psychiat. *74* (1954), 261—287.
10. Krayenbühl, H. A.: Cerebral venous and sinus thrombosis. Clin. Neurosurg. *14* (1967), 1—23.
11. Ribes, F.: Des recherches faites sur la phlébite. Rev. Med. *3* (1825), 5.
12. Smith, D. M., Ducker, T. B., Kemp, L. G.: Experimental in vivo microcirculation dynamics brain trauma. J. Neurosurg. *30* (1969), 664—672.

Authors' address: S. Sato, M.D., Department of Neurosurgery, School of Medicine, Keio University, Shinanomachi-35, Shinjuku-ku, Tokyo, Japan 160.

Department of Physiology and [1]Institute Surgery Research, University of Munich, Federal Republic of Germany

Effects of Bradykinin on Permeability and Diameter of Cerebral Vessels

M. Wahl, A. Unterberg[1], and A. Baethmann[1]

Summary

The effect of bradykinin on cerebral vessels was investigated using superfusion of the cortical surface. Na^+-fluorescein or fluorescein-labelled dextran (FITC-dextran) were intravenously administered to indicate a change of vascular permeability. Using intravital fluorescence microscopy and microphotography the following results were obtained: bradykinin (10^{-9} to 10^{-2} M) did not increase the permeability to FITC-dextran of different molecular weights (20,000–70,000). Similarly, 10^{-9} to 10^{-7} M bradykinin did not induce leakage of Na^+-fluorescein (MW 376). With 10^{-6} M bradykinin and higher concentrations, an increasing extravasation of Na^+-fluorescein was detected. After 1 min of superfusion the pial arteries (60–250 μ) dilated dose dependently with bradykinin concentrations above 10^{-7} M. However, the dilatation decreased within 30 min of superfusion. In contrast, constriction of the veins (60–270 μ) was observed, increasing with the duration of the superfusion period. Considering these effects, one can say that bradykinin may play a part in the formation and spread of vasogenic brain oedema.

Keywords: Bradykinin; pial veins; pial arteries; vasogenic brain oedema.

Introduction

Bradykinin has been reported to change the diameter of resistance and capacitance vessels as well as vascular permeability in many vascular beds. Since components of an intracerebral kinin system have been found, it is conceivable that bradykinin influences the above mentioned parameters in the brain under physiological conditions. In addition, bradykinin may be involved in pathological events. Damage of the blood-brain-barrier (BBB) by cold injury of the brain induces an uptake of oedema fluid and of kininogen from the intravascular space into the brain parenchyma. The formation of kinin has been found in damaged brain tissue [1] and ventriculo-cisternal perfusion with bradykinin increased the cerebral water content [3].

Therefore, it is possible, that bradykinin enhances the dysfunction of the BBB and the spread of oedema. The aim of the present study was to test whether bradykinin changes the permeability and diameter of cerebral vessels.

Methods

Experiments were performed in cats of both sexes which were anaesthetized with α-chloralose (40–50 mg/kg i.v.) and immobilized during artificial ventilation. Arterial pH, P_{co_2} and P_{o_2} were controlled. Endtidal CO_2, arterial blood pressure and body temperature were measured continuously. After trepanation using a cooled dental drill, the brain and dura were covered with a 1–3 cm layer of paraffin oil. The dura was then removed under oil. Extraparenchymal vessels of the parietal lobe were investigated by fluorescence microscopy. Microphotographs of the cortex were taken with Kodak Ektachrome 400 film. The cortical surface was superfused (5 ml within 30 min) via plastic tubes perforating the wall of a reservoir around the trephined area. The walls were made from rapid polymerizing dental cement. Inert mock cerebrospinal fluid (CSF) containing increasing concentrations of bradykinin-triacetate (Serva) was used for superfusion. Na^+-fluorescein or fluorescein-labelled dextran (FITC-dextran) were infused i.v. as BBB indicators. Penetration of a marker from the intravascular space into the parenchyma indicated an increased barrier permeability. The suitability of the method to study changes in vascular permeability has already been demonstrated by Svensjö et al.[2] on the vessels of the hamster cheek pouch.

Results and Discussion

During control superfusion with mock CSF a penetration of the markers from the intravasal space into the brain parenchyma was not observed. The effects of bradykinin on BBB permeability are summarized in Table 1. Bradykinin (10^{-9} to 10^{-2} M) did not increase the permeability to FITC-dextran of different molecular weights (20,000 to 70,000). Similarly 10^{-9} to 10^{-7} M bradykinin did not induce leakage of Na^+-fluorescein (MW: 376). At 10^{-6} M bradykinin and higher concentrations an increasing extravasation of Na^+-fluorescein was detected. It started with single fluorescent spots concentrated at leakage sites in the venules and became more diffuse and generalized at higher concentrations.

During cortical superfusion bradykinin induced dose dependent dilatations of the pial arteries (60–250 μ) starting at 4×10^{-7} M. After 1' superfusion a maximal dilatation of 37% was measured with 4×10^{-5} M bradykinin. After 30 min of continuous superfusion the dilatation was reduced. This might be explained by tachyphylaxis. However, it might also be due to an outward diffusion of bradykinin from the superfusion fluid into the natural CSF in the cortical subarachnoid space adjacent to the trephined area. A release of constrictory factors from the brain parenchyma might also explain the finding.

In contrast to the arteries, 1 min superfusion with bradykinin did not induce a significant change of pial venous diameter (60–270 μ) over the whole concentration range tested. With increasing duration of the superfusion period, a dose dependent constriction of the veins developed.

Table 1. *Effects of Bradykinin on Blood-Brain-Barrier-Permeability*

Bradykinin (M)	substance	BBB Marker MW	Stokes' Radius (Å)	Effect
10^{-9}—10^{-7}	Na$^+$-fluor-escein	376	5.5	no increase
10^{-6}—10^{-3}				stepwise increase
10^{-9}—10^{-2}	FITC-Dextran	20,000	30	
		40,000	45	no increase
		70,000	60	

The constriction reached a maximum of about 14% with 4×10^{-4} M after 30' superfusion. It appears likely that the slowly developing venous constriction is due to an indirect effect of bradykinin. Bradykinin might release constrictory prostaglandins as discussed by Wilkens and Back[5] in veins of the isolated perfused canine hind limb.

Conclusions

Bradykinin, acting from the parenchymal side, induces an apparently selective permeability of the BBB. Considering the Stokes' radius of Na$^+$-fluorescein, the results could be explained by an opening of functional pores of at least 11 Å in diameter. Besides this effect on permeability, bradykinin may facilitate the passage of small molecules through the BBB by arteriolar dilatation, which has already been found in a previous study[4], and venous constriction. Both these effects of bradykinin may play a part in the formation and spread of vasogenic brain oedema.

References

1. Baethmann, A., Maier-Hauff, K., Lange, M., Kempski, O.: Activation of the kallikrein-kinin-system in brain tissue secondary to cerebral injury and ischemia. J. Cerebr. Blood Flow Metabol. *1*, Suppl. 1 (1981), 218—219.
2. Svensjö, E., Arfors, K.-E., Raymond, R. M., Grega, G. J.: Morphological and physiologically correlation of bradykinin-induced macromolecular efflux. Amer. J. Physiol. *236* (1979), 600—606.
3. Unterberg, A., Baethmann, A., Geiger, R.: The kallikrein-kinin-system as mediator in cerebral oedema. Agents and Actions, Suppl. 9 (1982), 402—407.
4. Wahl, M.: The effect of morphine, enkephalin and bradykinin on cerebral vascular resistance in cats. J. Cerebr. Blood Flow Metabol. *1*, Suppl. 1 (1981), 323—324.

5. Wilkens, H. J., Back, N.: Reversal of bradykinin action: Possible involvement of prostaglandins or dual receptors. Pharmacol. Res. Comm. *12* (1980), 411—422.

Authors' address: Prof. Dr. M. Wahl, Department of Physiology, University of Munich, Pettenkoferstrasse 12, D-8000 München 2, Federal Republic of Germany.

Institute for Surgical Research, Klinikum Grosshadern, University of Munich, Federal Republic of Germany

The Kallikrein-Kinin-System as Mediator in Vasogenic Brain Oedema *

A. Unterberg and A. Baethmann

Summary

Ventriculo-cisternal perfusion in dogs with bradykinin causes brain oedema in the grey and white matter, whereas perfusion with plasma leads to oedema only in the periventricular white matter. Formation of kinins, indicated by consumption of kininogens, was observed in more than 50% of the experiments with ventricular perfusion using plasma. In this subgroup, a correlation was found between the amount of kinins released and the degree of white matter oedema. Activation of the kallikrein-kinin-system in cerebral tissue could enhance damage to the blood-brain barrier, lead to disturbances of the microcirculation or induce cytotoxic effects. The current findings provide further support for a role of the KK-system as mediator of secondary brain damage. Methods which interfere specifically with the formation or function of kinins should be expected to benefit the treatment of brain oedema.

Keywords: Kallikrein-Kinin-system; vasogenic brain oedema.

Introduction

In vasogenic brain oedema a protein-rich, plasma-like fluid enters the cerebral parenchyma through a defect in the blood-brain barrier in areas of tissue damage. Activation, or release of brain oedema factors derived from plasma, or from focal brain tissue necrosis is considered the basis of the secondary damage[1, 5, 9].

The kallikrein-kinin-(KK)-system is a highly attractive mediator candidate since kinins were found to dilate arteries and arterioles and enhance the permeability of the peripheral blood vessels[3, 8]. Formation of kinins has been found secondary to dilution of plasma with cerebrospinal

* Supported by Deutsche Forschungsgemeinschaft Ba 452/3,5.

fluid in vitro[7]. A similar process might occur in vasogenic brain oedema, if plasma entering the brain tissue is diluted by extracellular fluid. The release of kinins into the central nervous system was recently observed in focal and perifocal brain tissue subjected to cold injury[5].

In this study it was analyzed, whether the direct exposure of cerebral parenchyma to plasma causes brain oedema, and, moreover, whether the kallikrein-kinin-system is involved in this process. Ventriculo-cisternal perfusion of the dog brain was performed with homologous plasma as a carrier of kininogens, or with bradykinin dissolved in mock cerebrospinal fluid to circumvent the blood-brain barrier.

Method

The experiments were performed in mongrel dogs of 11–13 kg body weight during pentobarbital anaesthesia and immobilization. The blood pressure, tidal CO_2 and the acid-base status were all controlled. Both lateral ventricles of the brain and the cisterna magna were punctured and plastic tubes inserted for ventriculo-cisternal perfusion. The perfusion rate was 1 ml/min. The perfusion pressure was adjusted to 10 mm Hg and continuously controlled.

30 minutes of perfusion with mock CSF was used during control conditions, which was followed by a 3 hrs perfusion with homologous plasma (n = 13), or bradykinin (n = 10), respectively. One group of animals was perfused for 3 hrs with mock CSF only (controls). At termination of perfusion, the brain was removed for the determination of electrolytes and water in samples of the cortex, nucleus caudatus and periventricular white matter. Kininogen- and bradykinin-concentrations were measured in the perfusate prior to and after ventricular passage. 14-C-Inulin was added to the perfusate as a volume marker. Bradykinin was determined by rat uterus bioassay according to Mann, Geiger and Werle[6], kininogens in the plasma perfusate were measured after conversion to kinins according to the methods of Diniz and Carvalho[2].

Results and Discussion

The perfusion of the cerebral ventricles with bradykinin (inflow-concentration: about 3,000 ng/ml) led to a marked increase in tissue water content in the periventricular white matter, nucleus caudatus and cerebral cortex (Table 1). The Na^+-content was markedly increased in the nucleus caudatus, whilst only marginally raised in the white matter or cerebral cortex.

Contrary to experiments with bradykinin, the perfusion with homologous plasma led to oedema and an increase in Na^+-content in the white matter only. The basal ganglia and cerebral cortex were not affected. This observation might be explained by a difference in the penetration between the bradykinin and the plasma-perfusate due to different viscosities.

The uptake of water into the oedematous white matter occurred in relation to the uptake of Na^+-ions, which was demonstrated by the regression:

$$y = -232.6 + 5.91 \, x, \ r = 0.53, \ n = 22, \ p < 0.01,$$

where x = water content (g/100 g FW) and y = Na^+ (mM/kg DW).

Table 1. *Cerebral Water Content (mean ± SEM) After Ventricular Perfusion with Mock CSF, Plasma, or Bradykinin in mV100 g F.W.*

	Cortex	Nuc. caudatus	White matter
Mock CSF (controls)	80.9 ± 0.6	80.3 ± 0.3	66.2 ± 1.0
Plasma	80.9 ± 0.8	80.3 ± 0.7	68.5 ± 1.5*
Bradykinin	81.6 ± 0.4*	81.9 ± 0.4*	68.0 ± 1.7*

* Significantly different from controls (p < 0.05 or less).
From: Unterberg, *et al.*[9].

Table 2. *Bradykinin (ng/ml; mean ± SEM) of In- and Outflowing Perfusate*

Inflow	Outflow		
	60 min	120 min	180 min
2,476 ± 322	1,019 ± 217*	672 ± 292*	713 ± 257*

* Significantly different from inflow-concentrations (p < 0.01 or less).
From: Unterberg, *et al.*[9].

A relationship between the uptake of water and sodium into oedematous tissue supports the contention that perfusion with bradykinin, or plasma respectively, led to a vasogenic type of oedema associated with an increase in vascular permeability. During the passage of the bradykinin-perfusate through the ventricular system, clearance within the periventricular brain tissue, or choroid plexus led to a decrease in bradykinin of 60% or more, of the inflowing perfusate concentration (Table 2). The decrease in concentration cannot be attributed to dilution of the perfusate by freshly formed CSF, as can be assumed from the concentration profile of the 14-C-inulin volume marker. A clearance of approximately 1,500–1,900 ng bradykinin/min from the perfusate suggests the presence of highly active kininases in the periventricular brain tissue[4].

Kininogen-concentrations in the plasma perfusate were between 1,500–4,000 ng/ml (kinin-equivalents). In 5 experiments, kininogen-concentrations decreased markedly in the perfusate during ventricular passage, suggesting their consumption was due to the formation of kinins. Dilution by endogeneously formed CSF was ruled out as a cause of the decrease. In a few additional experiments, the kininogen-consumption was not found during ventricular perfusion with plasma.

Our results show undoubtedly that, exposure of brain tissue to bradykinin, or plasma respectively causes brain oedema. The question

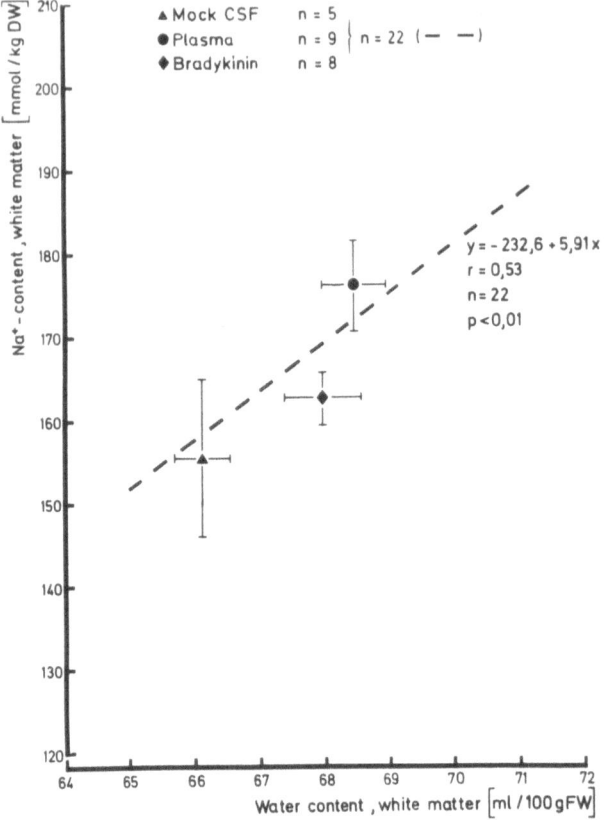

Fig. 1. Regression analysis between cerebral water- and a sodium-content in periventricular white matter of experiments where the ventricular system was perfused with mock CSF, plasma, or bradykinin in mock CSF. — — — symbolizes the regression curve where all experiments were taken together. There is a close correlation between the influx of water and uptake of sodium in the white matter

arises whether or not the formation of brain oedema secondary to the cerebral exposure to plasma resulted from an activation of the KK-system. To answer that question a regression analysis between the formation of kinins and the water content in the oedematous white matter was carried out. The analysis was conducted only in experiments where the formation of kinins was observed. The following regression function was obtained:

$y = 67.48 + 0.019 x$, $r = 0.98$, $p < 0.05$,

where y = water content (ml/100 g FW), x = formation of kinins.

The relationship found between both parameters supports the conclusion that, formation of brain oedema in the white matter in animals with ventriculo-cisternal perfusion with plasma resulted from an activation of the kallikrein-kinin-system.

Acknowledgement

The technical and secretarial help of Miss Isolde Juna, Ulrike Goerke, Angelika Konrad and Sylvia Schneider is gratefully appreciated.

References

1. Baethmann, A., Oettinger, W., Rothenfußer, W., Kempski, O., Unterberg, A., Geiger, R.: Brain edema factors: Current state with particular reference to plasma constituents and glutamate. In: Brain Edema, Advanc. Neurol. *28* (Cervós-Navarro, J., Ferszt, R., eds.), pp. 171—195. New York: Raven Press. 1980.
2. Diniz, C. R., Carvalho, J. F.: A micromethod for determination of bradykininogen under several conditions. In: Structure and Function of Biologically Active Peptides: Bradykinin, Kallikrein and Congeners (Erdös, E. G., ed.), Ann. N. Y. Acad. Sci. *104*, pp. 77—89, 1963.
3. Frey, E. K., Kraut, H., Werle, E.: Das Kallikrein-Kinin-System und seine Inhibitoren. Stuttgart: F. Enke. 1968.
4. Fuxe, K., Ganten, D., Köhler, C., Schüll, B., Speck, G.: Evidence for differential localization of angiotensin-I-converting enzyme and renin in the corpus striatum of rat. Acta Physiol. Scand. *110* (1980), 321—332.
5. Maier-Hauff, K., Lange, M., Schürer, L., Kempski, O., Baethmann, A.: Cerebral uptake and consumption of plasma-kininogens in brain trauma. Agents and Actions, Suppl. *9* (1982), 408—412.
6. Mann, K., Geiger, R., Werle, E.: A sensitive kinin-liberating assay of kininogenase in rat uteri, isolated glomeruli and tubules of rat kidney. In: Kinins: Pharmacodynamics and Biological Roles (Sicuteri, F., Back, N., Haberland, G. L., eds.), pp. 65—73. New York: Plenum Publ. 1976.
7. Sicuteri, F., Fanciulacci, M., Bavazzano, A., Franchi, G., Del Bianco, P. L.: Kinins and intracranial haemorrhages. Angiology *21* (1970), 193—210.
8. Svensjö, E., Arfors, K. E., Raymond, R., Grega, G. J.: Morphological and physiological correlation of bradykinin-induced macromolecular efflux. Amer. J. Physiol. *236* (1979), 600—606.
9. Unterberg, A., Baethmann, A., Geiger, R.: The kallikrein-kinin-system as mediator in cerebral edema. Agents and Actions, Suppl. *9* (1982), 402—407.

Authors' address: Dr. A. Unterberg, Institute for Surgical Research, Klinikum Grosshadern, University of Munich, D-8000 München 70, Federal Republic of Germany.

Neurogenic Regulation of Cerebral Veins

[1] Department of Neurosurgery, Graz University, Austria and [2] Department of Neurology, University of Lund, Sweden

Extent and Timecourse of Pial Venous and Arterial Constriction During Cervical Sympathetic Stimulation in Cats

L. M. AUER[1] and B. B. JOHANSSON[2]

Summary

Cervical sympathetic stimulation (90 s, 15 Hz, 10 ms) was performed in 38 normoxic, normocapnic and normotensive cats under barbiturate and nitrous-oxide anaesthesia. Pial venous and arterial reactions were observed through closed cranial windows. Diameter variations of 261 venous and 166 arterial vessel portions were continuously analysed using a multichannel videoangiometer with an in vivo accuracy around 2%. Pial veins constricted by 12.7% and pial arteries by 11.9%. Veins > 150 μm resting diameter constricted significantly more than vessels ⩽ 150 μm. Veins started to constrict synchronously with arteries whereas arteries reached maximal constriction significantly earlier than veins. The venoconstriction, however, lasted longer than the arterial constriction and continued for about 20 seconds after the end of the stimulation.

The time course strongly indicates that the reduction in the venous diameter is not a passive decrease secondary to arterial constriction and reduced blood flow. The observations support the hypothesis that sympathetic stimulation under resting conditions is able to decrease both cerebral blood flow and cerebral volume.

Keywords: Sympathetic stimulation; pial veins; pial arteries; cat.

Introduction

In a recent report[6] it was shown that pial veins constrict more than pial arteries during cervical sympathetic stimulation. The question asked in the present study was whether the venoconstriction is an active phenomenon or a consequence of the arterial constriction and the resulting decrease in cerebral blood flow. The dynamic behaviour of the pial veins and arteries was therefore studied with the angiometer technique[2], paying special attention to eventual differences between veins and arteries as to the time of onset of constriction during sympathetic stimulation.

9*

Materials and Methods

Experiments were performed in 38 adult cats with a body weight of 1.5 to 3 kg under sodium-pentobarbital anaesthesia (30 mg/kg) and subsequent respiration with a 3:1 mixture $N_2O:O_2$ following endotracheal intubation. Mean arterial pressure (MAP) was continuously recorded via a femoral catheter. Frequent blood gas samples were taken from a contralateral femoral catheter for analysis of PaO_2 and $PaCO_2$ using an AVL gas check. After preparing a cervical sympathetic trunk and mounting it on bipolar silver electrodes, an ipsilateral parietal cranial window was made in a closed version as described earlier in detail[1]. Pial vessels were observed using a Leitz intravital microscope with Ultrapak objectives, and diameter changes recorded with an attached TV-multichannel videoangiometer[2]. Analysis of grey value changes on single TV-lines crossing a pial vessel as shown on a TV monitor gives continuous information on diameter variations in the respective blood vessel. With this 3-channel device, 3 vessel portions can simultaneously and continuously be analysed, the data being calibrated in micrometers and recorded on a penwriter. Multiple replay of single experiments stored on videotape allow the analysis of the whole vasculature visible at fifty-fold magnification through the microscope. Sympathetic stimulation was performed during a 90 s period using square wave pulses of 10 ms duration at 15 Hz so as to induce maximal dilatation of the ipsilateral pupil. Blood gases were frequently determined and maintained at normal levels (PaO_2 103.3 ± 1.4 mmHg mean ± SEM; $PaCO_2$ 30.0 ± 0.3 mmHg).

The interval from the onset of sympathetic stimulation to the beginning of vascular constriction as well as the interval to maximum constriction was determined in 156 pial veins and 92 pial arteries and statistically compared using variance analysis (F-Test, Duncan-Test). The magnitude of the maximal response of 261 pial veins and 166 pial arteries was also analysed. In 7 animals, the time-course differences between arteries and veins were calculated in every individual animal and statistically evaluated.

Results

Throughout the experiments, MAP showed no significant variation. The percentage constriction was uniform among vessels of similar resting diameters in single experiments but varied from animal to animal. Three percent of vessels reacted within a range of ± 2% as compared with the resting level, *i.e.* to an extent below the resolution of the technique under in vivo conditions and were excluded from further evaluation[2]. Mean percent-constriction of pial veins with resting diameters of 30–486 μm and arteries with resting diameters 30–283 μm are presented in Fig. 1 and Table 1. Vessels were separated according to their resting calibres, *i.e.* ≤ 150 μm and > 150 μm; with this chosen threshold, a significantly different extent of venoconstriction was observed. Veins with a resting diameter > 150 μm (165–486 μm; mean ± SEM 235.7 ± 11.4) constricted by 15.5%, veins ≤ 150 μm (30–150 μm), mean 100.4 ± 3.4 μm) by 11.3%. By contrast, the constriction of the larger arteries (153–283 μm; mean 205.7 ± 5.5 μm) and smaller arteries (30–150 μm; mean 80.1 ± 3.2 μm) did not significantly differ. The constriction of large veins however, was significantly stronger than that of arteries.

The time-interval from the onset of stimulation to the onset of constriction was, however, identical in both veins and arteries of

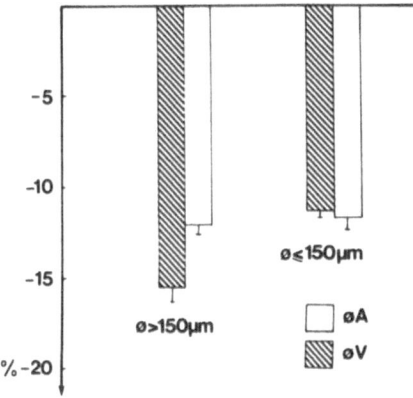

Fig. 1. Pial venous (black bars) and arterial (open bars) constriction during cervical sympathetic stimulation in percent of resting diameters. $> 150\,\mu m$ = large vessels, $\leqslant 150\,\mu m$ = small vessels. Bars indicate mean values $\pm$ SEM

Table 1. *Pial Vessel Constriction During Sympathetic Stimulation*

Ø	Arteries % mean ± SEM	Statistical comparison	Veins % mean ± SEM
Total	11.9 ± 0.4 n = 246	n. s.	12.7 ± 0.6 n = 331
$\leqslant 150\,\mu m$	11.7 ± 0.9 n = 140	n. s.	11.3 ± 0.4 n = 217
Statistical comparison	n. s.	—	p < 0.01
$> 150\,\mu m$	12.1 ± 0.5 n = 106	p < 0.01	15.5 ± 0.8 n = 114

Table 2. *Interval Between Start of Sympathetic Stimulation and Onset of Constriction*

Ø	Arteries Seconds mean ± SEM	Veins Seconds mean ± SEM
Total	11.1 ± 0.7 n = 92	11.3 ± 1.2 n = 156
$\leqslant 150\,\mu m$	11.9 ± 0.9 n = 59	11.9 ± 1.6 n = 115
$> 150\,\mu m$	9.7 ± 1.4 n = 33	9.6 ± 1.01 n = 41

Table 3. *Interval Between Start of Sympathetic Stimulation and Maximal Constriction*

∅	Arteries Seconds mean ± SEM	Veins Seconds mean ± SEM
Total	92.6 ± 5 n = 104	108.6 ± 3.9 n = 124
⩽ 150 μm	89.21 ± 7.07 n = 58	104.6 ± 5.03 n = 69
> 150 μm	96.78 ± 7 n = 46	113.64 ± 6.07 n = 55

corresponding size (Table 2). While arteries stopped constricting as soon as sympathetic stimulation ceased, veins continued to constrict for about 20 seconds (Table 3). Arteries returned to resting levels within 4.4 ± 0.3 min and veins within 5.3 ± 0.3 min.

Discussion

Although the action of the sympathetic nervous system on the cerebral vasculature has been a matter of considerable debate in recent years[11,14,18] functional studies on veins are scarce: Forbes and Cobb[13] noted in 1938, that pial veins constricted in response to sympathetic stimulation. Reichl and Walland[19], however, without giving any details, stated that veins did not respond to sympathetic stimulation using an open cranial window technique. The present data confirm our earlier observation that veins do constrict in response to sympathetic stimulation[3] and that their percentage decrease in diameter is even more marked than in arteries[6]; this has also been confirmed by Uddman et al.[20]. The time-course, veins starting to constrict synchronously with the arteries, indicates that the reduction of venous diameter is an active constriction and not a passive collapse due to a decrease in blood flow. Arteries would have to constrict earlier than veins in individual animals if the venous diameter decrease were passive. Furthermore, the time-difference between arteries and veins should in that case be shorter for small blood vessels than for large ones. The contrary was the case in the present series, *i.e.* the time interval was longer in small vessels than in large ones. Recent studies, including a topical microapplication technique, have confirmed that veins can constrict by an active sympatho-adrenergic mechanism probably mediated via alpha-receptors[5,7,16,21]. Assuming that about 70% of the total cerebral blood volume is located in the venous bed as in other body regions[17], these data should be of special interest with regard to the regulation of cerebral blood volume. In fact, Edvinsson et al.[12] reported that electrical stimulation of the nerve trunks to the superior cervical sympathetic ganglia produced an 18% decrease in cerebral blood volume and that the reduction was prevented by pretreatment with an alpha-adrenoreceptor antagonist.

The reduction of intracranial pressure noted during sympathetic stimulation[4,10,19] supports the hypothesis that cerebral blood volume can decrease under these circumstances. The observation that cervical sympathetic stimulation can reduce intracranial pressure suggests an effect not only on pial but also on intraparenchymal vessels. Possible physiological roles of the sympathetic nerves in the regulation of blood volume will be discussed in the following chapters of this volume[8,9]. The reason why maximum constriction is reached later in pial veins than in arteries is unknown but might be related to differences in the smooth muscle layer thickness, vascular nerve supply or intraluminal pressure in arteries and veins. Moreover, recent data suggest a different noradrenalin-induced neuromuscular coupling in pial veins than in arteries, possibly causing slower contraction[16,21].

Acknowledgement

This work was supported by the Austrian *Fonds zur Förderung der wissenschaftlichen Forschung* (Project No. 4368) and the Swedish Medical Research Council (Project No. 14X-4968).

We are grateful for the technical assistance of Miss Sylvia Schreiner. We also acknowledge the statistical calculations by Miss S. Auer.

References

1. Auer, L. M.: The pathogenesis of hypertensive encephalopathy. Acta neurochir. (Wien), Suppl. *27* (1978), 1—111.
2. Auer, L. M., Haydn, F.: Multichannel videoangiometry for continuous measurement of pial microvessels. Acta Neurol. Scand. *60*, Suppl. 72 (1979), 208—209.
3. Auer, L. M., Johansson, B. B.: Pial venous constriction during cervical sympathetic stimulation in the cat. Acta Physiol. Scand. *110* (1980), 203—205.
4. Auer, L. M., Johansson, B. B.: Cervical sympathetic nerve stimulation decreases intracranial pressure in the cat. Acta Physiol. Scand. *113* (1981), 565—566.
5. Auer, L. M., Johansson, B. B., Kuschinsky, W., Edvinsson, L.: Sympathoadrenergic activity of cat pial veins. J. Cerebr. Blood Flow Metabol. *1*, Suppl. 1 (1981), 311—312.
6. Auer, L. M., Johansson, B. B., Lund, St.: Reactions of pial arteries and veins to sympathetic stimulation in the cat. Stroke *12* (1981), 528—531.
7. Auer, L. M., Kuschinsky, W., Johansson, B. B., Edvinsson, L.: Sympathoadrenergic influence on pial veins and arteries in the cat. In: Cerebral Blood Flow: Effect of Nerves and Neurotransmitters (Heistad, D. D., Marcus, M. L., eds.), pp. 291—300. New York: Elsevier/North Holland Inc. 1982.
8. Auer, L. M., Sayama, I., Johansson, B. B., Leber, K.: Pial venous reaction to sympathetic stimulation during elevated ICP. In: The Cerebral Veins (Auer, L. M., Loew, F., eds.), pp. 143—153. Wien-New York: Springer. 1983.
9. Auer, L. M., Sayama, I., Johansson, B. B., Leber, K., Haselsberger, K.: Sympathoadrenergic influence on cerebral blood volume during increased intracranial pressure. In preparation.
10. Dorigotti, L., Glässer, A. H.: Decrease in cerebrospinal fluid-pressure induced by cervical sympathetic stimulation. Pharmacol. Res. Communications *4* (1972), 151—156.

11. Edvinsson, L., MacKenzie, E. T.: Amine mechanisms in the cerebral circulation. Pharmacol. Rev. *28* (1977), 275—348.
12. Edvinsson, L., Nielsen, K. C., Owman, Ch., West, K. A.: Evidence of vasoconstrictor sympathetic nerves in brain vessels of mice. Neurology *23* (1973), 73—77.
13. Forbes, H., Cobb, St.: Vasomotor control of cerebral vessels. Brain *61* (1938), 221—233.
14. Heistad, D., Marcus, M.: Response to "Do vasomotor nerves significantly regulate cerebral blood flow?" Circ. Res. *43* (1978), 494—495.
15. Kuschinsky, W., Wahl, M.: Alpha-receptor stimulation by endogenous and exogenous norepinephrine and blockade by phentolamine in pial arteries of cats. Circ. Res. *37* (1975), 168—174.
16. Kuschinsky, W., Ulrich, K., Auer, L. M.: In vivo study of pial venous reactions to norepinephrine. In: The Cerebral Veins (Auer, L. M., Loew, F., eds.), pp. 155—159. Wien-New York: Springer. 1983.
17. Mellander, S., Johansson, B.: Control of resistance, exchange and capacitance vessels in the peripheral circulation. Pharmacol. Rev. *20* (1968), 117—196.
18. Purves, M. J.: Do vasomotor nerves significantly regulate cerebral blood flow? Circ. Res. *43* (1978), 485—493.
19. Reichl, R., Walland, A.: Inhibition of neurosympathetic cerebroarterial constriction by clonidine in cats. Europ. J. Pharmacol. *68* (1980), 349—357.
20. Uddman, R., Edvinsson, L., McCulloch, J., Watt, P.: Differences in reactivity of pial arteries and veins to local chemical factors and putative neurotransmitters. In: The Cerebral Veins (Auer, L. M., Loew, F., eds.), pp. 161—166. Wien-New York: Springer. 1983.
21. Ulrich, K., Auer, L. M., Kuschinsky, W.: Cat pial venoconstriction by topical microapplication of norepinephrine. J. Cerebr. Blood Flow Metabol. *2* (1982), 109—111.

Authors' address: Univ.-Prof. Dr. L. M. Auer, Department of Neurosurgery, Graz University, A-8036 Graz, Austria.

[1] Department of Neurosurgery, Graz University, Austria, [2] Division of Surgical Neurology, Research Institute of Brain and Blood Vessels, Akita, Japan, and [3] Institute for Applied Statistics, Technical University, Graz, Austria

Response of Pial Veins and Intracranial Pressure to Cervical Sympathetic Stimulation During Hypercapnia

L. M. Auer, L. Iliff[1], I. Sayama[2], and J. Gölles[3]

Summary

In a series of 26 anaesthetized cats equipped with closed cranial windows, pial venous reactions to cervical sympathetic stimulation during normocapnia and various steps of hypercapnia up to $PaCO_2$ of 60 mm Hg were analysed. Whilst during normocapnia the large pial veins constricted by a mean of $15 \pm 1\%$, which was significantly more than the mean response of the small veins (up to 100 μm resting calibre $10.8 \pm 0.9\%$), their behaviour was inversed with rising CO_2, until, at a $PaCO_2$ of 60 mm Hg, the small pial veins constricted significantly more than the large ones. The regression lines for single CO_2-steps, resting calibres plotted against percent constriction during stimulation, were non-linear, and demonstrated a significant fall in sympathoadrenergic responsiveness from large to small veins during normocapnia. However, during hypercapnia ($PaCO_2 = 60$ mm Hg) there was a significant rise in responsiveness from large to small veins. During sympathetic stimulation, the ICP decreased by 8–10% with all CO_2-steps.

In 6 additional animals, hypercapnia with a $PaCO_2$ of 50 mm Hg was induced and pial venous dilatation was observed. Subsequent injection of the alpha-blocker yohimbine caused a 4% further dilatation, though statistically not significant.

It is concluded that cerebral blood volume is neurogenically controlled during normocapnia and hypercapnia mainly through the cerebral veins.

Keywords: Hypercapnia; pial veins; cerebral blood volume; intracranial pressure.

Introduction

Besides myogenic and metabolic regulation, neurogenic mechanisms have been suggested to influence the cerebral circulation. The physiological importance of the sympathoadrenergic system in regulating cerebral blood flow (CBF) under normal circumstances has been questioned because cerebral blood flow remained stable during sympathetic stimulation in normocapnic animals[1,9,10,16,]. Recent studies of the cerebral venous circulation revealed, however, a stronger sympathoadrenergic constriction as compared to that in arteries induced by cervical sympathetic stimulation[5]. The observation of a significant concomitant reduction in intracranial pressure (ICP)[7] suggested a control function of the sympathetic system on cerebral blood volume mainly via the cerebral veins, because 70–80% of peripheral blood volume is located in the veins[15]. Studies of the pial arterial reaction to sympathetic stimulation during hypercapnia revealed controversial data in different series, however, the data seem to indicate a different behaviour of small and large vessels at different levels of CO_2[8,13,14]. Pial veins have so far not been studied under hypercapnic conditions. Therefore, the present study has been undertaken to test the hypothesis that the sympathoadrenergic tone is elevated during hypercapnia in order to control cerebral blood volume.

Materials and Methods

Experiments were performed in 2 groups of cats totalling 32 animals. Anaesthesia was induced with $30\,mg\,kg^{-1}$ sodium-pentobarbital (Nembutal®), immobilization with $60\,\mu g\,kg^{-1}$ pancuronium-bromide (Pavulon®). After endotracheal intubation, a Loosco baby respirator was used for artificial ventilation using a 3:1 mixture of $N_2O:O_2$. One femoral artery was cannulated for continuous recording of the mean arterial pressure (MAP) and the ipsilateral femoral vein for drug administration. The contralateral femoral artery was cannulated to withdraw blood samples for blood gas analysis using an AVL-analyzer. In the 26 animals of group 1, one cervical sympathetic nerve was isolated, cut and mounted on bipolar silver electrodes. A parietal cranial window was made ipsilateral to the sympathetic preparation, covered with a glass shield and sealed with acrylic glue[2,5]. Pial venous calibres were continuously recorded by a penwriter attached to a Leitz intravital microscope[4]. ICP was measured by inserting a PVC-cannula into the cisterna magna. MAP and ICP were recorded using Hellige 1214 electromanometers and Statham P23dB pressure transducers. Body temperature was maintained between 37° and 38°C using a rectal thermosensor and a heating pad. After recording a normal steady state condition, the following experiments were performed:

Group 1: In 26 cats, the cervical sympathetic nerve was stimulated with an HSE type P stimulator for 90 seconds with square wave pulses of 10 milliseconds at 15 Hz so as to achieve maximal mydriasis and complete retraction of the membrana nictitans. When parameters had again reached their resting state, arterial carbon dioxide tension was stepwise increased from 30 mm Hg (29.9 ± 0.3 mm Hg) to 40 (40.2 ± 0.3 mm Hg), 50 (50.9 ± 0.4 mm Hg) and 60 (63.0 ± 0.9 mm Hg) by adding CO_2 to the respired gas mixture. As soon as vessel calibres had obtained a steady state condition, sympathetic stimulation was repeated at each step of $PaCO_2$. Following stimulation, the decrease in venous calibres and volume per unit of vessel length, calculated from diameter variations, was compared between single steps of $PaCO_2$, and regression analysis and analysis of variance were performed[12].

Group 2: In 6 animals, $PaCO_2$ was elevated to 50 mm Hg (52.3 ± 0.9 mm Hg) by adding CO_2 to the respired gas mixture. When a new steady state condition was achieved during hypercapnia, $0.25\,mg\,kg^{-1}$ of the alpha-adrenergic blocking substance yohimbine was injected intravenously, and further calibre changes in pial veins were observed. For statistical evaluation, analysis of variance was used.

Results

Group 1: Variations in 203 pial venous portions with resting calibres between 60 μm and 343 μm were investigated. The mean percent constriction of veins up to 100 μm and above 100 μm resting diameter at each step of $PaCO_2$ is given in Table 1: Under normocapnic conditions, veins constricted significantly during cervical sympathetic stimulation and large

Table 1. *Pial Venous Constriction During Sympathetic Stimulation and Step-Hypercapnia*

$PaCO_2$ mm Hg	$\emptyset V \leqslant 100$ μm mean percent ± SEM		$\emptyset V > 100$ μm mean percent ± SEM
30	—10.8 ± 0.9	p < 0.025	—15.0 ± 0.9
40	—6.9 ± 2.5	n.s.	—6.0 ± 0.8
50	—4.1 ± 1.1	n.s.	—3.1 ± 0.4
60	—5.3 ± 1.6	p < 0.0005	—1.6 ± 0.3

veins significantly more than small veins. As $PaCO_2$ was elevated, the percent constriction diminished and became equal in all ranges of calibre. At a $PaCO_2$ of 60 mm Hg, however, the constriction of the small veins increased again and became significantly stronger than the constriction of the large veins. Regression analysis (Fig. 1) revealed a non-linear relationship between the resting calibre and the percent venous reaction to stimulation at $PaCO_2$ 30, 50 and 60 mm Hg, and an almost linear relationship at a $PaCO_2$ of 40 mm Hg. Analysis of the regression curves showed a significantly rising venous constriction with rising resting calibre

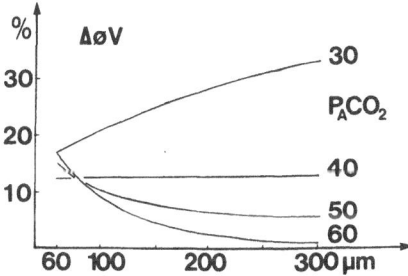

Fig. 1. Regression lines of percent pial venous reactions (ordinate) to sympathetic stimulation at various levels of $PaCO_2$ versus resting calibres (abscissa). $\emptyset V$ = percent reduction in pial venous volume per unit of vessel length

at a PaCO$_2$ of 30 mm Hg. At PaCO$_2$ 60 mm Hg, the situation became inversed; there was a significant fall in venous constriction with rising calibre. Investigations in a few pial veins with resting calibres below 50 mm Hg showed a tendency of vascular volume to decrease by 20–30% during stimulation at a PaCO$_2$ of 60 mm Hg, compared with 6% in veins with diameters around 300 μm. In veins with resting diameters above 100 μm, constriction diminishes significantly with rising PaCO$_2$ (p < 0.001). In 60 μm veins, constriction remains unchanged between PaCO$_2$ 30 and PaCO$_2$ 60 mm Hg.

Table 2. *Mean Percent ICP Changes (SEM) During Step-Hypercapnia and Sympathetic Stimulation*

Mean PaCO$_2$ (mm Hg)	Hypercapnia	Stimulation
30		—9.0 (1.6)
40	+ 33.0 (6.3)	—8.0 (2.7)
50	+ 94.7 (13.5)	—10.3 (4.7)
60	+ 132.7 (18.1)	—9.4 (2.3)

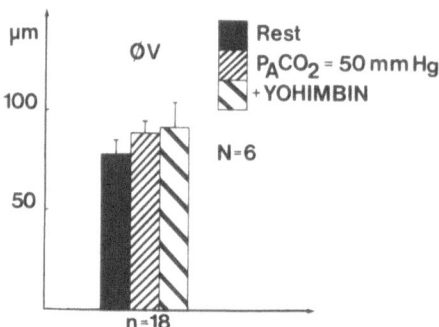

Fig. 2. Absolute pial venous calibre changes (∅V) from the resting condition (Rest) to hypercapnia (PaCO$_2$ = 50 mm Hg) and following the injection of yohimbine. N = number of animals, n = number of vessel portions

The influence of sympathetic stimulation and vasoconstriction on ICP in 18 animals is indicated for each step of PaCO$_2$ in Table 2. The mean resting ICP was 4.7 mm Hg. Between PaCO$_2$ 30 mm Hg and 60 mm Hg, a stimulation induced fall in ICP remained unchanged with 8–10%.

Group 2: Eighteen pial veins with a mean resting calibre of 76 ± 7 μm dilated to 82 ± 6 μm during hypercapnia with a PaCO$_2$ of 50 mm Hg (Fig. 2). The further 4.0 ± 1.7% dilatation after the injection of yohimbine was not significant.

Discussion

The observation that pial veins do not significantly dilate after pharmacological blockade of the alpha-receptors in group 2 is in agreement with earlier observations[6], and indicates that the sympathetic system does not effectively increase vascular tone during hypercapnia in anaesthetized cats. Therefore, the extent of venous constrictions induced by sympathetic stimulation in group 1 cannot have been masked by preincreased sympathetic tone. This result from anaesthetized animals does, however, not rule out a spontaneously rising sympathetic tone with hypercapnia in unanaesthetized individuals. The sensitivity of large pial veins to stimulation of the sympathetic nerve is not raised during hypercapnia.

By contrast, small pial veins become increasingly responsive with rising $PaCO_2$ and show even a tendency to constrict more during hypercapnia than during normocapnia. This observation suggests that, with increasing carbon dioxide tension, sympathoadrenergic venous constriction is shifted from large to small veins, where the large proportion of venous blood volume is located under resting conditions, and where the hypercapnic blood volume increase is more pronounced[2,3]. As a result, ICP is reduced by the same percentage during normocapnia and hypercapnia, and absolute ICP-reduction is even more pronounced during hypercapnia.

Results of this study support the hypothesis that cerebral blood volume can be controlled by the sympathetic system not only during normocapnia, but also during hypercapnia up to a $PaCO_2$ of 60 mm Hg.

The functional importance of this regulatory mechanism is underlined by observations in unanaesthetized animals which had an increased cerebral blood volume after sectioning of the sympathetic nerve[11].

Acknowledgement

This work was supported by the Austrian *Fonds zur Förderung der wissenschaftlichen Forschung* (Project No. 4368).

References

1. Alm, A., Bill, A.: The effect of stimulation of the cervical sympathetic chain on retinal oxygen tension and on uveal, retinal and cerebral blood flow in cats. Acta Physiol. Scand. *88* (1973), 84—94.
2. Auer, L. M.: The pathogenesis of hypertensive encephalopathy. Acta neurochir. (Wien), Suppl. *27* (1978), 1—111.
3. Auer, L. M.: The sausage-string phenomenon in acutely induced hypertension— arguments against the vasospasm-theory in the pathogenesis of acute hypertensive encephalopathy. Europ. Neurol. *17* (1978), 166—173.
4. Auer, L. M., Haydn, F.: Multichannel videoangiometry for continuous measurement of pial microvessels. Acta Neurol. Scand. *60*, Suppl. *72* (1979), 208—209.
5. Auer, L. M., Johansson, B. B.: Pial venous constriction during cervical sympathetic stimulation in the cat. Acta Physiol. Scand. *110* (1980), 203—205.
6. Auer, L. M., Trummer, U. G., Johansson, B. B.: Alpha-adrenoreceptor antagonists and pial vessel diameter during hypercapnia and hemorrhagic hypotension in the cat. Stroke *12* (1981), 847—851.

7. Auer, L. M., Johansson, B. B.: Cervical sympathetic nerve stimulation decreases intracranial pressure in the cat. Acta Physiol. Scand. *113* (1981), 565—566.
8. Auer, L. M., Kuschinsky, W., Johansson, B. B., Edvinsson, L.: Sympatho-adrenergic influence on pial veins and arteries in the cat. In: Cerebral Blood Flow: Effects of Nerves and Neurotransmitters (Heistad, D. D., Marcus, M. L., eds.), pp. 291—300. New York: Elsevier North Holland, Inc. 1982.
9. Busija, D. W., Marcus, M. L., Heistad, D. D.: Effects of sympathetic nerves on the cerebral circulation in cats. In: Cerebral Blood Flow: Effects of Nerves and Neurotransmitters (Heistad, D. D., Marcus, M. L., eds.), pp. 301—308. New York: Elsevier North Holland, Inc. 1982.
10. D'Alecy, L. G., Rose, J. C., Sellers, S. A.: Sympathetic modulation of hypercapnic cerebral vasodilatation in dogs. Circ. Res. *45* (1979), 771—785.
11. Edvinsson, L., Owman, Ch., West, K. A.: Changes in cerebral blood volume of mice at various time periods after superior cervical sympathectomy. Acta Physiol. Scand. *82* (1971), 521—526.
12. Gölles, J., Auer, L. M.: Regression and multiple test analysis for the physiological investigation of pial vessels in vivo. In preparation.
13. Kobayashi, S., Waltz, A. G., Rhoton, A. L.: Effects of stimulation of cervical sympathetic nerves on cortical blood flow and vascular reactivity. Neurology *21* (1971), 297—302.
14. Kontos, H. A., Wei, E. P.: Effects of sympathetic nerves on pial arterioles during hypercapnia and hypertension. In: Cerebral Blood Flow: Effects of Nerves and Neurotransmitters (Heistad, D. D., Marcus, M. L., eds.), pp. 275—280. New York: Elsevier North Holland, Inc. 1982.
15. Mellander, S., Johansson, B.: Control of resistance, exchange and capacitance vessels in the peripheral circulation. Pharmacol. Rev. *20* (1968), 117—196.
16. Meyer, M. W., Smith, K. A., Klassen, A. C.: Sympathetic regulation of cephalic blood flow. Stroke *8* (1977), 197—201.

Author's address: Univ.-Prof. Dr. Auer, Department of Neurosurgery, Graz University, A-8036 Graz, Austria.

[1]Department of Neurosurgery, Graz University, Austria, [2]Department of Surgical Neurology, Research Institute for Brain and Blood Vessels, Akita, Japan, and [3]Department of Neurology, University of Lund, Sweden

Pial Venous Reaction to Sympathetic Stimulation During Elevated Intracranial Pressure

L. M. Auer[1], I. Sayama[2], B. B. Johansson[3], and K. Leber[1]

Summary

Earlier experiments have shown that a sympathoadrenergic mechanism is able to decrease intracranial pressure by venous constriction and a consequent reduction in cerebral blood volume. In the present study using 16 cats, the intracranial pressure was elevated by two different models (cisternal infusion of mock CSF in 8 cats and brain oedema induced by water-intoxication in 8 cats) to test the hypothesis that cervical sympathetic stimulation is also effective under the circumstance of an elevated ICP. A cranial window technique and multichannel-videoangiometry were used to measure variations in pial venous calibres. The elevation of ICP up to 30 mm Hg with brain oedema and 50 mm Hg with CSF-infusion per se resulted in marked venous and arterial dilatation in the first model and arterial dilatation only in the second model. Pial venous constriction induced by sympathetic stimulation was significantly less with an elevated ICP. ICP reduction as a consequence of vascular constriction during stimulation remained around 15% between the resting condition and the situation of ICP elevated to 30 mm Hg by CSF infusion; at higher ICP levels up to 50 mm Hg, the percent ICP reduction with sympathetic stimulation fell to 10%.

The results indicate that the sympathoadrenergic system exerts a regulatory influence on cerebral blood volume via alterations in venous calibre both at a normal and increased ICP. The clinical importance of this mechanism for the control of an elevated ICP may, however, be limited.

Keywords: Sympathetic stimulation; pial venous constriction; elevated ICP; cerebral blood volume.

Introduction

Under resting conditions the cerebral blood volume constitutes 3–4% of the intracranial volume. Nevertheless, an increase in blood volume contributes to an important extent to the development of intracranial hypertension. Thus, hypercapnic cerebral vasodilatation causes a rise in a previously normal intracranial pressure up to 25–30 mm Hg[5,10].

Relaxation of the smooth muscle cells of the cerebral vessels has been assumed to cause an increase in cerebral blood volume at increased levels of intracranial pressure (ICP)[11]. Although the control of cerebral blood volume (CBV) is of great clinical interest, data regarding the regulation of CBV under physiological and pathological circumstances are scarce. Research in the field of the cerebral circulation has mainly focussed on cerebral blood flow and its regulation whilst the veins, which contain the larger part of the CBV, have been neglected.

We have shown earlier that cervical sympathetic stimulation induces a significant constriction of pial veins[3] as well as a decrease in intracranial pressure[6]. The observation that pial veins and arteries constricted simultaneously after the onset of cervical sympathetic stimulation indicated an active venoconstriction[8]. Moreover, pial veins constricted following the topical administration of norepinephrine[12]. On this basis we postulated that the sympathoadrenergic system may exert an influence on intracranial pressure via CBV under normal levels of intracranial pressure. In the present study we have investigated whether sympathetic stimulation has an effect also on an elevated intracranial pressure. Since the behaviour of cerebral venous vessels may be different in normal and oedematous tissue, experiments were performed in two series, *i.e.* ICP was increased either by the infusion of cerebral spinal fluid (CSF) or by acutely inducing brain oedema.

Materials and Methods

Experiments were performed in 16 cats of either sex with a body weight of 1.5–3 kg. After induction of anaesthesia with $30 \, mg \cdot kg^{-1}$ sodium-pentobarbital (Nembutal®), the animals were intubated endotracheally, immobilized with $60 \, \mu g \cdot kg^{-1}$ pancuroniumbromide (Pavulon®) and ventilated with a 3:1 mixture of $N_2O:O_2$. The aorta and vena cava inferior were cannulated with PVC catheters via a femoral artery and vein. Arterial blood samples were collected for blood gas analysis (AVL type 937C gas check). The mean arterial pressure (MAP) was recorded using a Statham P23dB pressure transducer and a Hellige type 1214 electromanometer. Body temperature was continuously monitored with a Philips rectal-thermosensor unit and maintained between 37° and 38 °C using a heating pad. The right sympathetic nerve was prepared, cut and mounted on bipolar silver electrodes, isolated in a plastic tube and connected to a HSE type P stimulation device. With the animal in the sphinx position, the head was fixed into a stereotaxic frame, and a parietal cranial window was made ipsilateral to the sympathetic preparation. After opening of the dura mater with microsurgical techniques, the cranial bone-window was allowed to fill up with the animal's own CSF, thereafter covered with a glass shield and sealed with acrylic tissue glue. In cats belonging to group B (see below), the glass window was fixed at the level of the lamina interna of the skull bone in order to prevent herniation of oedematous brain into the skull defect. ICP was registered via a plastic cannula in the cisterna magna, connected to a Statham P23dB transducer and a HSE electromanometer. Via a y-shaped tube, the plastic cannula was connected to a syringe filled with mock CSF, driven by a Harvard infusion-pump for CSF-infusion (group A). Following a steady state period under resting conditions, the cervical sympathetic nerve was stimulated for 90 seconds with square wave impulses of 10 msec at 15 Hz to achieve retraction of the membrana nictitans and maximal pupil dilatation (1–10 V).

Throughout the experiments, pial arteries and veins under the cranial window were observed using a Leitz intravital microscope with Ultropak objectives[1]. Vessel diameter variations were continuously monitored using a TV multichannel videoangiometer[2] and registered on a Rikadenki multichannel penwriter together with MAP and ICP. The TV picture of the pial surface was monitored on videotape for multiple replay of experiments and evaluation of a large number of pial vessel portions. Blood gases were maintained at normal levels in both series.

Group A. In 8 cats, ICP was increased stepwise to 20, 30, 40 and 50 mm Hg by cisternal infusion of 0.6 ml to 1.2 ml · min⁻¹ of mock CSF. Sympathetic stimulation was repeated at each level of ICP during a steady state period.

Group B. In 8 cats, ICP was increased by the intravenous infusion of 3.0–5.5 ml · min⁻¹ of distilled water with a low volume infusion pump under resting conditions and after elevation of ICP to 20 and 30 mm Hg. Sympathetic stimulation was repeated at each level of ICP.

Statistical evaluation was performed using variance-analysis (Duncan-Test).

The results from the experiments with cisternal infusion (group A) have been briefly presented elsewhere[9].

Results

Group A. PaO_2 was 98 ± 1.7 mm Hg before and 102.3 ± 3 mm Hg during the experiments; the corresponding values for $PaCO_2$ were 29.7 ± 0.3 mm Hg and 29.0 ± 0.73 mm Hg.

MAP was 93.6 ± 8 mm Hg under resting conditions and varied between 88 and 105 mm Hg during the experiments.

Pial arteries. Measurements were performed on a total of 50 pial arterial portions with resting diameters ranging from 24 to 198 μm. During the stepwise increase in ICP by cisternal infusion of CSF the arteries dilated from a mean resting diameter of 112.2 ± 6.6 μm (SEM) to 165.1 ± 16.1 μm at an level of ICP of 50 mm Hg (Fig. 1). An example of an individual experiment is shown in Fig. 2. During sympathetic stimulation the arterial

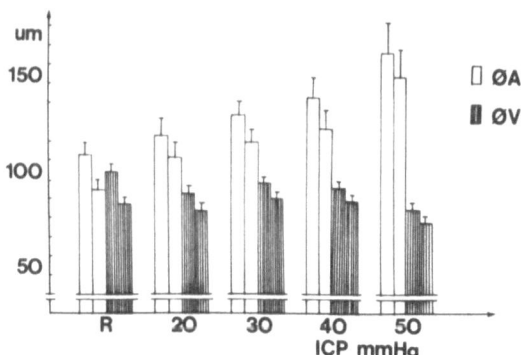

Fig. 1. Absolute mean calibres ± SEM of pial veins ($\emptyset V$) and pial arteries ($\emptyset A$) at various levels of ICP from resting conditions (R) in steps of 10 mm Hg up to 50 mm Hg. Elevation of ICP is achieved by cisternal infusion of mock CSF. The left bar of each pair represents the calibre before sympathetic stimulation, the right bar represents the calibre during stimulation

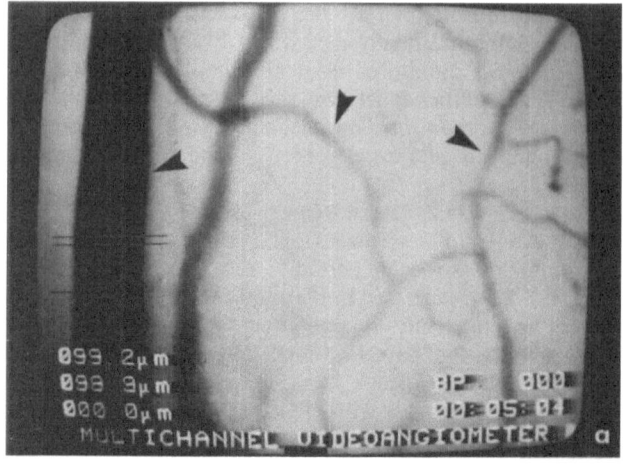

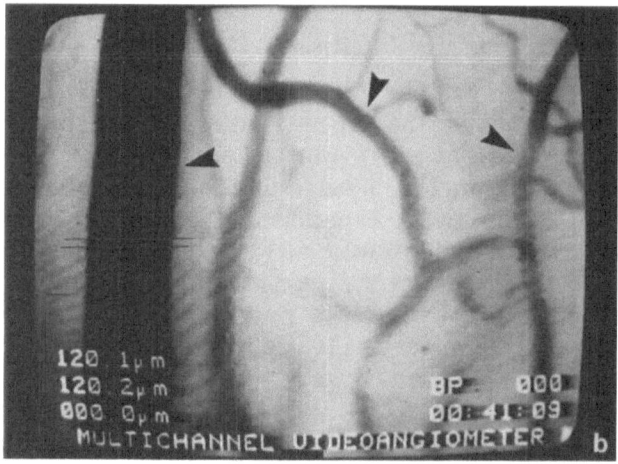

Fig. 2. Videopictures from the pial surface of a cat a) under resting conditions at normal ICP, b) at an ICP of 50 mmHg elevated by the cisternal infusion of mock CSF. Note marked arterial dilatation, big vessels as well as small ones (arrows)

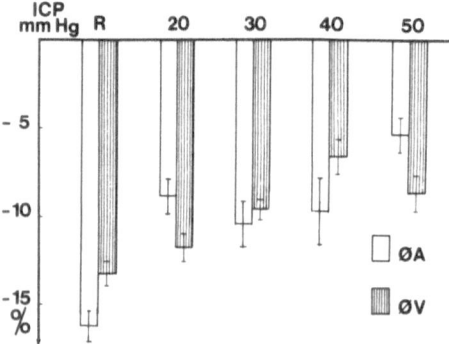

Fig. 3. Percent pial venous ($\emptyset V$) and arterial ($\emptyset A$) constriction during cervical sympathetic stimulation at a normal ICP (R) and in steps up to 50 mmHg

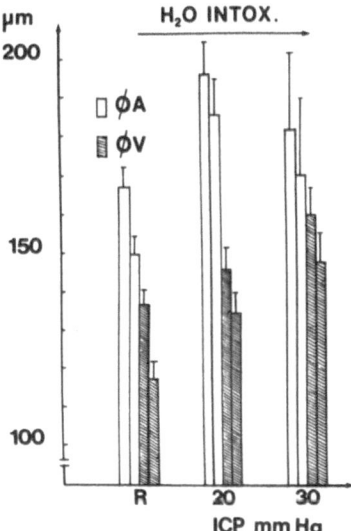

Fig. 4. Absolute mean calibres ± SEM of pial veins ($\emptyset V$) and pial arteries ($\emptyset A$) at various levels of ICP from resting conditions (R) in steps of 10 mm Hg up to 30 mm Hg. Elevation of ICP is achieved by water intoxication. The left bar of each pair represents the calibre before sympathetic stimulation, the right bar represents the calibre during stimulation

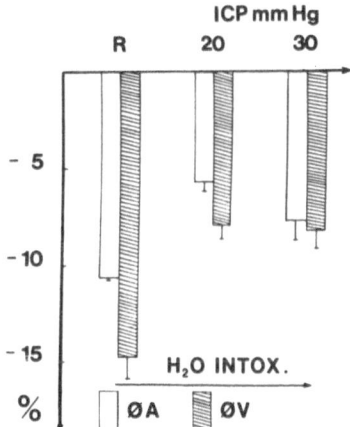

Fig. 5. Percent pial venous ($\emptyset V$) and arterial ($\emptyset A$) constriction during cervical sympathetic stimulation at normal ICP (R) and in steps up to 30 mm Hg

diameters decreased by $16.3 \pm 0.7\%$ under resting conditions, and by $5.4 \pm 2.0\%$ at an ICP 50 mm Hg (Table 1 and Fig. 3). Accordingly, pial arterial blood volume, calculated using the arterial radius, was decreased during sympathetic stimulation under resting conditions by 29.4%, by 25% at ICP 30 mm Hg, by 21.7% at 40 mm Hg and by 13.8% at ICP 50 mm Hg. However, the absolute decrease in pial arterial blood volume during

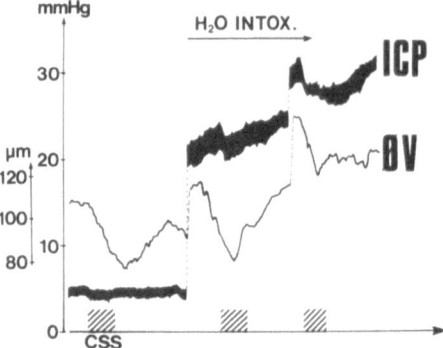

Fig. 6. Traces of an individual experiment of ICP and a 110 μm pial vein (∅ V). Hatched bars indicate the 90 second periods of cervical sympathetic stimulation (CSS). Under resting conditions with ICP 4 mm Hg, as well ICP of 20 and 30 mm Hg, both pial venous calibre and the ICP decrease during stimulation

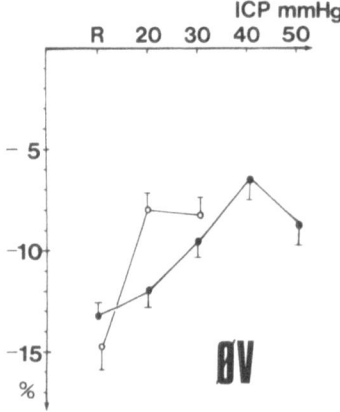

Fig. 7. Percent pial venous constriction during cervical sympathetic stimulation at ICP levels from resting conditions (R) up to 50 mmHg, induced either by brain oedema-water intoxication (open circles) or cisternal infusion of CSF

sympathetic stimulation did not change significantly at the high levels of ICP.

Pial veins. 90 pial venous portions with resting diameters from 36 to 204 μm were studied. As shown in Fig. 2, venous diameters decreased from a mean of 100.4 ± 4.0 μm under resting conditions to 84.5 ± 3.9 μm at an ICP of 50 mm Hg.

Expressed as percentages, the constriction due to cervical sympathetic stimulation was $13.2 \pm 0.7\%$ under resting conditions, $9.5 \pm 1.2\%$ at an ICP of 40 mm Hg, and $11.1 \pm 1.2\%$ at an ICP of 50 mm Hg (Table 1, Fig. 3).

Table 1 a

n = 351	Sympathetic stimulation CSF-infusion ICP mm Hg				
	R	S*	20	S*	30
S*	⌐————————— ** —————————⌐				
ØA	-16.3 ± 0.7 n = 47	**	-8.9 ± 1.1 n = 33	n.s.	-10.4 ± 1.4 n = 46
S*	**		n.s.		*
ØV	-13.2 ± 0.7 n = 90	n.s.	-11.9 ± 0.8 n = 65	n.s.	-9.6 ± 0.7 n = 70
S*	∟————————— ** —————————∟				

S* Statistical comparison. ** = $p < 0.01$.

Table 1 b

n = 281	Sympathetic stimulation CSF-infusion ICP mm Hg		
	R	40	50
S*	⌐————————————— ** —————————————⌐ ⌐————— ** —————⌐		
ØA	-16.3 ± 0.7 n = 47	-9.7 ± 2 n = 38	-5.4 ± 2 n = 13
ØV	-13.2 ± 0.7 n = 90	-6.6 ± 1 n = 71	-8.7 ± 1 n = 60
S*	∟————— ** —————∟ ∟————————————— ** —————————————∟		

S* Statistical comparison.** = $p < 0.01$.

Group B. $PaCO_2$ was 29.8 ± 1.6 mm Hg and PaO_2 112.6 ± 8.0 mm Hg. MAP increased from a resting level of 129.6 ± 10.8 mm Hg to 144.1 ± 11.3 mm Hg at an ICP of 20 mm Hg and to 153.4 ± 10.6 mm Hg at an ICP of 30 mm Hg. The mean resting ICP was 3.4 ± 0.5 mm Hg. At the beginning of the sympathetic stimulation following the induction of brain

Table 2

n = 270	Sympathetic stimulation H₂O-intoxication ICP mm Hg				
	R	S*	20	S*	30
S*	↓ ————————————— .n.s. ————————————— ↓				
∅A	—10 ± 0.5 n = 70	**	—5.8 ± 0.5 n = 51	n.s.	—7.5 ± 1.0 n = 22
S*	**		n.s.		n.s.
∅V	—14.7 ± 1.1 n = 60	**	—7.9 ± 0.7 n = 40	n.s.	—8.4 ± 0.9 n = 27
S*	↑ ————————————— ** ————————————— ↑				

S* Statistical comparison, ** = p < 0.01.

oedema the mean ICP was 20.6 ± 0.5 mm Hg and 31.1 ± 1.5 mm Hg, respectively.

Pial arteries. Absolute resting diameters of 70 pial arterial portions studied varied from 76 to 230 μm (mean 167.4 ± 5.3 μm) (Fig. 4); an ICP increase to 20 mm Hg resulted in a mean dilatation to 196.0 ± 8.5 μm; at an ICP level of 30 mm Hg, the mean arterial calibre was 181.2 ± 19.7 μm. The percentage reduction in diameter caused by sympathetic stimulation was $10.6 \pm 0.1\%$ at rest, $5.7 \pm 0.5\%$ at an ICP of 20 mm Hg and $7.7 \pm 1.0\%$ at an ICP of 30 mm Hg (Fig. 5).

Pial veins. 60 pial venous portions with the resting diameters ranging from 95 to 208 μm (mean 136.7 ± 3.9 μm) were monitored. During sympathetic stimulation under resting conditions the pial veins constricted significantly to 116.9 ± 3.9 μm (Fig. 4), *i.e.* by $14.7 \pm 1.1\%$ (Fig. 5). The percentage venoconstriction was significantly reduced at an ICP of 20 mm Hg, and at 30 mm Hg, it was still about 8% (Table 2, Fig. 5). An individual experiment is shown in Fig. 6. A comparison of the venous reactions in group A and group B shows a marked similarity under resting conditions and an ICP of 30 mm Hg (Fig. 7).

Comparison of the Reaction of ICP in Group A and Group B

Intracranial pressure was decreased by 15 to 20% during sympathetic stimulation under resting conditions (Fig. 8). At ICP levels of 20 and 30 mm Hg the resulting percent decrease in ICP remained unchanged with

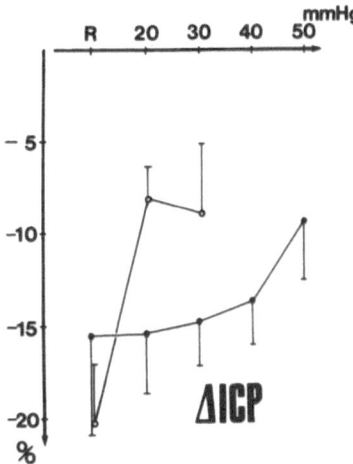

Fig. 8. Percent decrease of ICP during cervical sympathetic stimulation at various ICP levels from resting conditions (R) up to 50 mmHg, induced either by brain oedema-water intoxication (open circles) or cisternal infusion of mock CSF

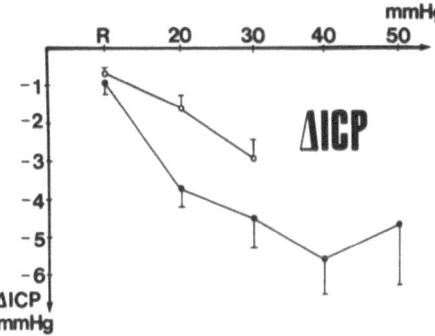

Fig. 9. Mean absolute decrease in ICP (mmHg) during cervical sympathetic stimulation at various levels of ICP from resting conditions (R) up to 50 mmHg, induced either by brain oedema-water intoxication (open circles) or cisternal infusion of mock CSF

CSF infusion and decreased insignificantly during brain oedema ("water intoxication"). At an ICP level of 30 mm Hg, the difference between the two models was statistically not significant. At an ICP of 50 mm Hg, induced by infusion of CSF, the percent ICP reduction achieved the same level as that observed at an ICP of 30 mm Hg induced by acute brain oedema (Fig. 8). This somewhat spindle-shaped difference in behaviour of the ICP in the two models can also be seen from the graphs showing absolute changes in ICP during sympathetic stimulation (Table 3, Fig. 9). The absolute response increased with each step of increased ICP, reaching 5 to 6 mm Hg at the maximum level of ICP.

Table 3

n = 42	Sympathetic stimulation/Δ % ICP ICP mm Hg				
	R	S*	20	S*	30
S*	⌐————————————— n.s. —————————————⌐				
H₂O-intox.	−20.2 ± 3.2 n = 7		−8.1 ± 1.8 n = 8		−8.9 ± 3.9 n = 4
S*	n.s.		*		n.s.
CSF-infus.	−15.5 ± 5.3 n = 8		−15.4 ± 3.2 n = 7		−14.7 ± 2.4 n = 8
S*	∟————————————— n.s. —————————————⌐				

S* Statistical comparison. * = $p < 0.05$.

Discussion

In agreement with previous reports[3,4,8] pial veins constricted significantly during sympathetic stimulation at physiological levels of ICP. Since larger vessels respond more than smaller ones[7], different mean resting diameters can probably explain the difference in the percentage constriction of pial arteries and veins in the present study as compared with a previous study. Although the percent constriction was less at the higher levels of ICP, the absolute ICP reduction during sympathetic stimulation rose, suggesting a potential though limited sympathoadrenergic regulation of cerebral blood volume not only at normal but also at elevated ICP, especially at levels up to 30 mm Hg.

The physiological importance of these data remains to be elucidated. An increased spontaneous sympathetic activity was recently reported to occur at higher ICP levels which causes a fall in cerebral perfusion pressure[13]. Responses of blood vessels and ICP to sympathetic stimulation in our series may therefore partly have been masked by the spontaneous rise of sympathetic tone arriving via the contralateral sympathetic nerve.

Compared with variations in CBV induced by different levels of carbon dioxide tension in the blood, the sympathoadrenergic activity is likely to have a much smaller impact on the intracranial pressure-volume relationship. The clinical relevance of the neurogenic regulation of the cerebral blood volume for a reduction in an elevated ICP may be limited.

In conclusion, the sympathoadrenergic effect on cerebral veins may play a role for the regulation of cerebral blood volume under circumstance of normal as well as increased ICP. The clinical impact of this regulatory mechanism on the normalization of an elevated ICP is, however, probably limited.

Acknowledgement

We are grateful to Mr. H. Leiner for technical assistance and to Miss S. Auer for statistical calculations.

This work was supported by the Austrian *Fonds zur Förderung der wissenschaftlichen Forschung* (Project No. 4368).

References

1. Auer, L. M.: The pathogenesis of hypertensive encephalopathy. Acta neurochir. (Wien), Suppl. *27* (1978), 1—111.
2. Auer, L. M., Haydn, F.: Multichannel videoangiometry for continuous measurement of pial microvessels. Acta Neurol. Scand. *60*, Suppl. 72 (1979), 208—209.
3. Auer, L. M., Johansson, B. B.: Pial venous constriction during cervical sympathetic stimulation in the cat. Acta Physiol. Scand. *110* (1980), 203—205.
4. Auer, L. M., Johansson, B. B.: Neurogenic regulation of pial vessels in the cat. In: Pathophysiology and Pharmacotherapy of Cerebrovascular Disorders (Betz, E., *et al.*, eds.), pp. 386—390. Baden-Baden: Gerhard Witzstrock Verlag. 1980.
5. Auer, L. M., Johansson, B. B., MacKenzie, E. T.: Cerebral venous pressure during actively induced hypertension and hypercapnia in cats. Stroke *11* (1980), 180—183.
6. Auer, L. M., Johansson, B. B.: Cervical sympathetic nerve stimulation decreases intracranial pressure in the cat. Acta Physiol. Scand. *113* (1981), 565—566.
7. Auer, L. M., Johansson, B. B., Lund, S.: Reaction of pial arteries and veins to sympathetic stimulation in the cat. Stroke *12* (1981), 528—531.
8. Auer, L. M., Johansson, B. B.: Extent and timecourse of pial venous and arterial constriction during cervical sympathetic stimulation in cats. In: The Cerebral Veins (Auer, L. M., Loew, F., eds.), pp. 131—136. Wien-New York: Springer. 1983.
9. Auer, L. M., Sayama, I., Johansson, B. B., Leber, K.: Sympathoadrenergic modulation of cerebral blood volume during increased ICP. Proc. of the Vth International Symposium on Intracranial Pressure, Tokyo (1982), in press.
10. Häggendal, E., Johansson, B. B.: Effect of increased intravascular pressure on the BBB to protein in dogs. Acta Neurol. Scand. *48* (1972), 271—275.
11. Symon, L., Crockard, H. A., Juhasz, J.: Some aspects of cerebrovascular resistance in raised intracranial pressure: an experimental study. In: Intracranial Pressure II (Lundberg, N., *et al.*, eds.), pp. 257—262. Berlin-Heidelberg-New York: Springer. 1975.
12. Ulrich, K., Auer, L. M., Kuschinsky, W.: Cat pial venoconstriction by topical microapplication of norepinephrine. J. Cerebr. Blood Flow Metab. *2* (1981), 109—111.
13. Yamamoto, S., Higashi, S., Fujii, H., Hayashi, M., Ito, H.: Vasomotor response in acute increased intracranial pressure. Proc. of the Vth International Symposium on Intracranial Pressure, Tokyo (1982), in press.

Author's address: Univ.-Prof. Dr. L. M. Auer, Department of Neurosurgery, Graz University, A-8036 Graz, Austria.

[1] Department of Physiology, University of Munich, Munich, Federal Republic of Germany, and [2] Department of Neurosurgery, Graz University, Graz, Austria

In vivo Study of Pial Venous Reactions to Norepinephrine

W. Kuschinsky[1], K. Ulrich[1], and L. M. Auer[2]

Summary

The demonstration of perivascular noradrenergic nerve fibres in pial veins has triggered investigations on the functional significance of this innervation. One approach to demonstrate such a functional significance of a noradrenergic innervation in pial veins is the demonstration of receptors for norepinephrine, which mediate vascular reactions. In the present study, therefore, the reactions of pial veins to the perivascular microapplication of norepinephrine (10^{-9} to 10^{-3} M) were investigated. A significant constriction was found with concentrations of norepinephrine ranging from 10^{-7} to 10^{-3} M. The maximal constriction of the pial veins was 22% at the highest concentration of norepinephrine. Additional studies using α_1 and α_2 receptor stimulating and blocking substances showed that the norepinephrine induced constrictions cannot be completely classified according to the concept of α_1 and α_2 receptors.

Keywords: Pial veins; norepinephrine; topical application.

Introduction

Constrictions have been demonstrated in pial arteries in vitro[6] and in vivo when norepinephrine was applied from the perivascular side[12,13,18]. The morphological basis for such a vascular reaction has been given by the demonstration of perivascular noradrenergic nerve fibres[7,11,15,16]. Pial veins were recently demonstrated to constrict significantly more than pial arteries during cervical sympathetic stimulation[1-4]. This pial venoconstriction during sympathetic stimulation could be due to 1. passive venous collapse as a consequence of a decreased intravascular pressure during constriction on the arterial side or 2. active contraction of pial venous smooth muscle cells induced by a release of norepinephrine from nerve fibres located around the pial veins. The question as to which of these mechanisms is effective can be solved by the perivenous microapplication of norepinephrine. This issue may be relevant for a better understanding of the possible influence of the sympatho-adrenergic system on the regulation of cerebral blood volume.

Material and Methods

Five cats were anaesthetized with 40 mg/kg α-chloralose, tracheotomized, immobilized (Gallamine, 15 mg/kg/h, i.v.) and artificially ventilated with room air. The end-tidal CO_2 concentration and arterial blood pressure were monitored. An intravenous infusion of 2.5 ml/kg/h of Tyrode's solution was given. The body temperature was kept between 37 and 38 °C. An open cranial window was made parietally, the dura opened and reflected, and the pial surface protected by a layer of 1–2 cm of paraffin oil. Using the microapplication technique[18], 10^{-9}—10^{-3} M norepinephrine hydrochloride (Arterenol®, Sigma) in artificial CSF was administered to 26 pial venous portions. The solution had the following composition: Na^+, 156 mM; K^+, 3 mM; Ca^{2+}, 1.5 mM; HCO_3^-, 15 mM; Cl^-, 147 mM; pH 7.28. To prevent autoxidation of norepinephrine, all solutions were bubbled with 95% N_2 and 5% CO_2 (equilibrated with water). In addition, each microinjection pipette was filled with the corresponding mock CSF immediately before microinjection. This mock CSF, 1–4 µl, was injected over 1 min. Photographic documentation was performed with a Wild-Heerbrugg stereo-zoom microscope equipped with a photographic camera. Photographs were taken before and 20, 40, and 60 sec after starting the injection of the test solution. Preinjection resting diameters of the 26 pial veins ranged between 80 and 498 µm (mean, 242 µm). Venous reactions to topical norepinephrine were calculated as a percentage of the resting diameters and they were evaluated statistically by the nonparametric Wilcoxon Matched Pairs Signed Ranks test. An overall value of $p < 0.05$ was chosen as the level of significance according to the procedure established by Bonferroni-Holm for multiple statistic testing (Holm, 1979). The mock CSF, which served as the solvent, had a mean dilating effect of 1.7 ± 2.9 (SEM) %. For the presentation of the data in the figures, this effect was subtracted from the respective norepinephrine effect. Six veins in which the solvent itself induced a reaction of more than ± 10% of control were not included in the results, although their reactions to norepinephrine were comparable to the reactions reported here. In order to obtain information about the general reactivity of the brain vessels, the reactions of pial arteries to acidic and alkaline mock CSF[14] were tested before and several times during the experiment. The pial arteries showed the well-known constriction to alkaline solutions and dilatation to acidic ones.

Results

Figs. 1 and 2 show the mean venous reactions to the different concentrations of perivascular norepinephrine. In Fig. 1 the concentration-response curve of the mean values is shown. Constriction started at 10^{-7} M norepinephrine and was maximal at 10^{-3} M, rising to $-21.8 \pm 2.1\%$. Venous constriction was statistically significant at norepinephrine concentrations of 10^{-7}—10^{-3} M when compared to the reaction to the solvent itself. Fig. 2 depicts the constriction response at 20, 40, and 60 s after starting the injection of norepinephrine: constriction was always strongest after 60 s; at 10^{-3} M norepinephrine, the 40- and 60-s reactions were identical.

The constrictory action of norepinephrine points to α receptors as main mediators of venoconstriction. For a more detailed influence on these receptors (*e.g.* under therapeutic aspects), it appears reasonable to investigate whether a subclassification into postsynaptic α_1 and α_2 receptors is possible, as suggested for various types of vascular smooth muscle cells[17]. The question, which of these receptors mediate the norepinephrine induced venoconstriction was tested using the microapplication technique and

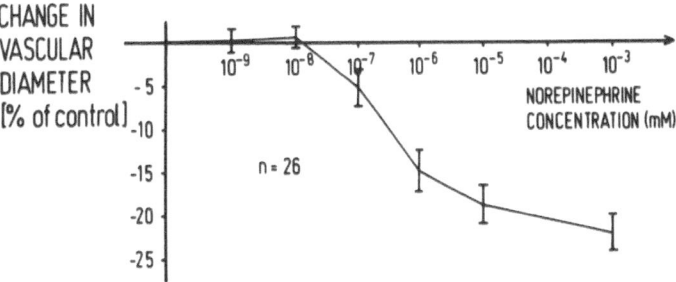

Fig. 1. Average concentration-response curve of pial venous constriction to perivascular norepinephrine: each value represents the mean of 3 measurements taken at 20, 40, and 60s after application. Ordinate: percentage pial venoconstriction, $\varnothing V$, pial venous diameter; *n* number of vessel portions investigated at each step of norepinephrine concentration

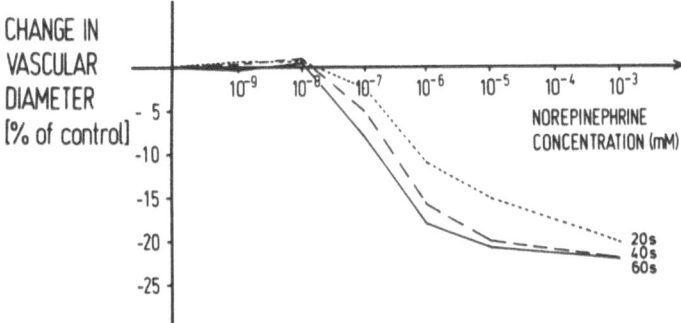

Fig. 2. Percentage venoconstriction at 20, 40, and 60s after starting the perivascular microapplication of norepinephrine. Ordinate: percentage pial venoconstriction. $\varnothing V$, pial venous diameter

adding either Prazosin (α_1-blocker) or Yohimbine (α_2-blocker) to the norepinephrine containing injection solution. Both drugs induced a concentration dependent decrease of the norepinephrine induced constrictions. Besides the action of α_1 and α_2 receptor blocking agents, the effect of the drugs which are known as α_1 or α_2 receptor stimulators was also investigated. α_1 receptor stimulation by phenylephrine induced weak constrictions, whereas the α_2 stimulator oxymetazoline had no consistent constrictory effect on pial veins.

Discussion

The present data demonstrate pial venoconstriction to perivascular norepinephrine and therefore support the hypothesis that pial veins do actively constrict to cervical sympathetic stimulation rather than collapse due to pial arterial constriction. Stronger constriction of large compared with small vessels fits with the previous experiments showing pial venoconstriction during cervical sympathetic stimulation [1-4]. The degree of pial venous constriction to perivascular norepinephrine can be compared to the degree of pial arterial constriction obtained in a former series under comparable experimental conditions [13]. Although a direct statistical comparison cannot be performed, since different concentrations of norepinephrine were used, it can be stated that the percentage reactions of pial arteries and veins are in a comparable range.

Our studies concerning the effect of norepinephrine agree well with parallel studies performed using the same technique [8]. These authors even claimed a slightly higher sensitivity of veins to norepinephrine. These data hence support the present view that the sympatho-adrenergic system may play a role in the regulation of cerebral blood volume via the control of the cerebral venous tone.

The results concerning α_1 and α_2 receptor agonists and antagonists are in accordance with the concept of an unusual type of α receptor in the pial veins, as has been postulated already for pial arteries [5,9].

References

1. Auer, L. M., Johansson, B. B.: Neurogenic regulation of pial vessels in the cat. In: Pathology and Pharmacotherapy of Cerebrovascular Disorders (Betz, E., Grote, I., Heuser, D., Wüllenweber, R., eds.), pp. 386—390. Baden-Baden: Gerhard Witzstrock Verlag. 1980.
2. Auer, L. M., Kuschinsky, W., Johansson, B. B., Edvinsson, L.: Sympatho-adrenergic activity of cat pial veins. J. Cerebr. Blood Flow Metabol. (Suppl. 1) *1* (1981), 311—312.
3. Auer, L. M., Kuschinsky, W., Johansson, B. B., Edvinsson, L.: Sympatho-adrenergic influence on pial veins and arteries in the cat. In: Cerebral Blood Flow: Effects of Nerves and Neurotransmitters (Heistad, D. D., Marcus, M. L., eds.), pp. 169—176. Elsevier: North-Holland, Inc. 1982.
4. Auer, L. M., Johansson, B. B., Lund, S. T.: Reaction of pial arteries and veins to sympathetic stimulation in the cat. Stroke *12* (1981), 528—531.
5. Bevan, J. A., Bevan, R. D.: Sympathetic control of the rabbit basilar artery. In: Neurogenic Control of the Brain Circulation (Owman, Ch., Edvinsson, L., eds.), pp. 285—293. Vol. *30*. Oxford-New York: Pergamon Press. 1977.
6. Edvinsson, L.: Neurogenic mechanisms in the cerebrovascular bed. Autonomic nerves, amine receptors and their effect on cerebral blood flow. Acta Physiol. Scand. Suppl. *427* (1975), 1—35.
7. Edvinsson, L., Nielsen, K. C., Owman, Ch., Sporrong, B.: Cholinergic mechanisms in pial vessels. Histochemistry, electron microscopy and pharmacology. Z. Zellforsch. *134* (1972), 311—325.
8. Edvinsson, L., McCulloch, J., Uddman, R.: Sympathetic and peptidergic mechanisms in cerebral veins. J. Cerebr. Blood Flow Metabol. (Suppl. 1) *1* (1981), 327—328.
9. Edvinsson, L., Owman, Ch.: Pharmacological characterization of postsynaptic vasomotor receptors in brain vessels. In: Neurogenic Control of the Brain Circulation (Owman, Ch., Edvinsson, L., eds.), Vol. *30*, pp. 167—183. Oxford-New York: Pergamon Press. 1977.
10. Holm, St.: A simple sequentially rejective multiple test procedure. Scand. Statist. *6* (1979), 65—70.

11. Iwayama, T., Furness, J. B., Bumstock, G.: Dual adrenergic and cholinergic innervation of the cerebral arteries of the rat: An ultrastructural study. Circ. Res. (1970), 635—646.
12. Kuschinsky, W., Wahl, M.: Alpha receptor stimulation by endogenous and exogenous norepinephrine and blockade by phentolamine in pial arteries of cats. Circ. Res. *37* (1975), 168—174.
13. Kuschinsky, W., Wahl, M.: Interactions between perivascular norepinephrine and potassium or osmolarity on pial arteries of cats. Microvasc. Res. *14* (1977), 173—180.
14. Kuschinsky, W., Wahl, M., Bosse, O., Thurau, K.: Perivascular potassium and pH as determinants of local pial arterial diameter in cats. Circ. Res. *31* (1972), 240—247.
15. Nelson, E., Rennels, M.: Innervation of intracranial arteries. Brain *93* (1970), 475—490.
16. Nielsen, K. C., Owman, Ch.: Adrenergic innervation of pial arteries related to the circle of Willis in the cat. Brain Res. *6* (1967), 773—776.
17. Timmermans, P. B. M. W. M., van Zwieten, P. A.: The postsynaptic α-adrenoceptor. J. Auton. Pharmac. *1* (1981), 171—183.
18. Wahl, M., Kuschinsky, W., Bosse, O., Olesen, J., Lassen, N. A., Ingvar, D. H., Michaelis, J., Thurau, K.: Effect of 1-norepinephrine on the diameter of pial arterioles and arteries in the cat. Circ. Res. *31* (1972), 248—256.

Authors' addresses: Prof. Dr. W. Kuschinsky, cand. med. K. Ulrich, Department of Physiology, University of Munich, Pettenkoferstrasse 12, D-8000 München 2, Federal Republic of Germany, Univ.-Prof. Dr. L. M. Auer, Department od Neurosurgery, Graz University, A-8036 Graz, Austria.

[1] Department of Oto-Rhino-Laryngology, General Hospital, Malmö, Sweden, [2] Department of Clinical Pharmacology, University Hospital, Lund, Sweden, [3] The Wellcome Surgical Institute, University of Glasgow, Glasgow, Scotland

Differences in Reactivity of Pial Arteries and Veins to Local Chemical Factors and Putative Neurotransmitters

R. Uddman[1], L. Edvinsson[2], J. McCulloch[3], and P. Watt[3]

Summary

The vasomotor responses of individual pial arteries and veins were examined in anaesthetised cats. The reactions of the vessels lying on the convexity of the cerebral cortex to perivascular microinjections of mock cerebrospinal fluid (CSF) containing varying concentrations of potassium (K^+), bicarbonate (HCO^-_3), noradrenaline, vasoactive intestinal peptide or substance P were studied. The microapplication of mock CSF at different pH values evoked significant changes in the calibre of both pial arteries and veins. The microapplication of mock CSF containing different concentrations of potassium evoked only small alterations in the calibre of the pial veins, whereas the pial arteries showed marked alterations in calibre. With a concentration of 40 mM of potassium, both pial arteries and veins constricted significantly. The perivascular microapplication of noradrenaline resulted in pronounced concentration-dependent reductions in the calibre of both pial arteries and veins. The veins were more sensitive to NA than the arteries. These reductions in the diameter of the cerebral vessels were attenuated by the concomitant microapplication of phentolamine (10^{-6} M). The perivascular microinjection of substance P and vasoactive intestinal peptide increased the calibre of the pial arteries and, to a lesser extent, that of pial veins.

Keywords: Cerebral veins; neutrotransmitters; potassium; bicarbonate.

Introduction

The main resistance function in the brain circulation is confined to small arteries and arterioles, and the tone of these vessels thus contributes to the regulation of the cerebral blood flow. On the other hand, cerebral blood volume is largely determined by capacitance vessels, mainly consisting of small veins, venules and sinusoids. Hence the tone of these vessels is an important modulator of cerebral blood volume.

In a recent study we have demonstrated the occurrence and distribution of noradrenaline (NA), vasoactive intestinal peptide (VIP) and substance P (SP) nerve fibres around pial arteries and veins [2]. This rich supply of nerve fibres indicates a nervous control of the cerebrovascular bed. So far cerebral arteries have been studied extensively whilst cerebral veins have received only little attention. In contrast, the peripheral venous circulation has been the subject of much attention and both putative neurotransmitters and local chemical factors have been found to elicit potent effects [5].

In the present series of experiments we have investigated the effects of changes in the bicarbonate and potassium concentrations, and the putative neurotransmitters (NA, SP, VIP) on the reactivity of the pial arteries and veins in situ.

Material and Methods

The experiments were performed on 35 cats of either sex, weighing between 2 and 4 kg using a method similar to that described by Wahl *et al.*[7]. The cats were anaesthetized initially with a mixture of alfaxolone (6.75 mg/kg) and alfadolone acetate (6.75 mg/kg, i.v.). The animals were intubated and connected to an intermittent positive-pressure ventilation pump, delivering approximately 25% oxygen/75% nitrogen in an open circuit. The right femoral artery and vein were cannulated to permit the continuous measurement of arterial blood pressure and the administration of fluid or anaesthetic agents. Anaesthesia was maintained during the subsequent course of the experiment with α-chloralose (60 mg/kg, i.v.).

The animals were maintained normocapnic throughout the course of the experiments. The end-tidal concentration of carbon dioxide was monitored continuously by means of an infrared analyzer, and samples of arterial blood were taken frequently during the experiment for the estimation of arterial pH, oxygen and carbon dioxide tensions. All the parameters were close to resting levels throughout the experiments.

The animals were placed in a stereotaxic frame. A longitudinal incision was made in the scalp, which was then retracted and ligated on to a metal ring in such a way that it formed an intact pool over the calvarium. A craniotomy was made over the left parietal cortex and the exposed dura was bathed with warmed mineral oil maintained at 38 °C. Surgical manipulations were performed with the aid of a stereomicroscope with a zoom lens, the field being illuminated by fibre optics. The dura was removed carefully, and any bleeding from the cut dural edges was sealed using bipolar diathermy.

Vascular calibre was measured using an image-splitting technique. Individual pial vessels on the convexity of the brain were viewed in focus through the microscope at a magnification of either × 40 or × 70. The image was passed through a Vickers image-splitting eyepiece to a closed circuit television camera and displayed on a television monitor. Vascular diameter was measured from the degree of shear applied to the image splitter which had been calibrated against wire and thread of known diameters.

The solutions, which were microinjected, were based on a mock cerebrospinal fluid of the following composition Na^+ 154 mM, K^+ 3 mM, Ca^{++} 2.5 mM, Cl^- 149.9 mM, HCO_3 12.1 mM, the pH of which was 7.2 when aerated by a gas mixture of 5% CO_2/95% O_2. Solutions containing the various concentrations of potassium were prepared by substitution of this ion for sodium. Solutions containing the various concentrations of bicarbonate were prepared by substitution of this ion for chloride in the mock CSF. All solutions were aerated with 5% CO_2/95% O_2 and were freshly prepared on the days of the experiments. Noradrenaline, SP or VIP were dissolved in mock cerebrospinal fluid immediately prior to use.

Glass micropipettes with a tip diameter of 8–10 μm were filled with cerebrospinal fluid under mineral oil. The micropipettes were inserted through the arachnoid into the perivascular space close to a cerebral artery or vein using a micromanipulator. A minute amount of the mock cerebrospinal fluid (about 5 μl) containing the agents tested were injected with extreme care into the perivascular space over 15–30 sec, and the resulting alterations in vascular calibre were monitored for periods of up to 3 min following the injection. The observed changes in calibre in response to the microinjection of the various test solutions (expressed as percent changes of the diameter of the vessel prior to the administration) were compared to those following the administration of mock cerebrospinal fluid alone by means of an analysis of variance[4].

Results

Alterations of Perivascular pH

Various concentrations of bicarbonate in the mock CSF resulted in changes in pH. Perivascular microinjection of mock CSF at different pH levels evoked significant changes in pial arterial and venous calibre (Fig. 1).

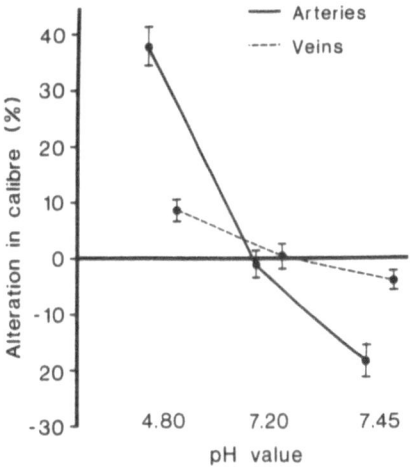

Fig. 1. Alterations in the calibre of pial arteries and veins to perivascular microapplication of mock CSF at varying pH values. Data presented as means ± SE

Bicarbonate-free CSF, pH 4.80, resulted in increases in the calibres of the vessels. This response and that to CSF containing 22 mM bicarbonate (pH 7.80) (reduction in calibre) were proportionally smaller in pial veins than in pial arteries.

Alteration of Perivascular Potassium Concentration

The perivascular microinjection of CSF containing potassium (0 mM, 7 mM, 10 mM) failed to alter the calibre of the pial veins, whereas significant alterations in arterial calibre were observed with each of these concentrations of potassium (Fig. 2). Cerebrospinal fluid containing high concentrations of potassium (40 mM) induced significant constrictions of both pial arteries and veins $47 \pm 6\%$ and $14 \pm 1\%$, respectively.

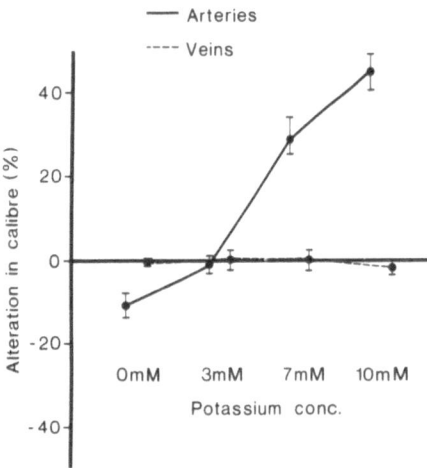

Fig. 2. Alterations in the calibre of pial arteries and veins to perivascular microapplication of mock CSF containing varying concentrations of potassium. Data are presented as means ± SE

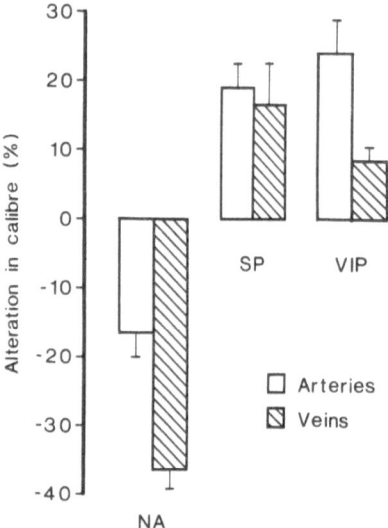

Fig. 3. Alterations in the vascular diameter of pial arteries and veins on the microapplication of mock CSF containing 10^{-5} M noradrenaline, 10^{-7} M substance P and 10^{-6} M vasoactive intestinal peptide. Data are given as means ± SE

Vasomotor Responses to the Perivascular Administration of NA, SP and VIP

The perivascular micro-injection of NA resulted in pronounced dose-dependent reductions in the calibre of both arteries and veins. Pial veins displayed proportionately greater responses to NA than pial arteries (Fig. 3). The micro-application of SP and VIP resulted in significant increases in arterial and venous calibre (Fig. 3). The magnitude of the responses in the veins to VIP was rather small.

Discussion

The present series of experiments provides a comprehensive and systematic comparison between the reactivity of cerebral arteries and veins to local chemical factors and some putative neurotransmitters[2,3]. The responses of pial arteries and veins to local alterations in the concentration of potassium and bicarbonate in the perivascular fluid revealed that the pial veins were less sensitive (at least with respect to the magnitude of the calibre alterations observed), than were pial arteries. However, the pial veins were not insensitive to changes in the perivascular concentrations of these ions, as consistent alterations in their calibre were observed with acidic solutions and with solutions containing high concentrations of potassium.

Similarly, the micro-injection of SP and VIP resulted in relatively small increases in calibre of arteries and veins. In contrast, the perivascular micro-application of NA resulted in more pronounced dose-dependent reductions in the diameter of pial veins $(32 \pm 3\%)$ than arteries $(22 \pm 3\%)$. Furthermore, the veins were more sensitive to NA than the pial arteries under similar conditions. These findings are in agreement with those of other authors, showing that pial veins respond more markedly to sympathetic nerve stimulation or micro-application of NA than arteries[1,6].

Taken together, these results are consistent with the hypothesis that local alterations in the chemical composition of the perivascular fluid are of greater importance in the control of cerebrovascular resistance, whereas neurogenic factors may be of greater importance in the regulation of cerebrovascular capacitance.

Acknowledgements

This study was supported by grants from the Swedish Medical Research Council (04X-5958) and the Swedish Heart Association.

References

1. Auer, L. M., Johansson, B. B.: Neurogenic regulation of pial vessels in the cat. In: Pathophysiology and Pharmacotherapy of Cerebrovascular Disorders (Betz, E., *et al.*, eds.), pp. 386—390. Baden-Baden: Gerhard Witzstrock Verlag. 1980.
2. Edvinsson, L., McCulloch, J., Uddman, R.: Feline cerebral veins and arteries: Comparison of autonomic innervation and vasomotor responses. J. Physiol. *325* (1982), 161—173.

3. McCulloch, J., Edvinsson, L., Watt, P.: Comparison of the effects of potassium and pH on the calibre of cerebral veins and arteries. Pflügers Arch. *393* (1982), 95—98.
4. Scheffé, H.: The Analysis of Variance. New York: Wiley. 1959.
5. Shepherd, J. T., Vanhoutte, P. M.: Veins and their Control. Philadelphia: Saunders. 1975.
6. Ulrich, K., Auer, L. M., Kuschinsky, W.: Cat pial veno-constriction by topical microapplication of norepinephrine. J. Cerebr. Blood Flow Metabol. *2* (1982), 109—111.
7. Wahl, M., Kuschinsky, W., Bosse, O., Neiss, A.: Micropuncture evaluation of β-receptors in pial arteries of cats. Pflügers Arch. *348* (1974), 293—303.

Author's address: R. Uddman, M.D. Ph.D., Department of Oto-rhino-laryngology, University of Lund, Malmö General Hospital, S-214 01 Malmö, Sweden.

Departamento de Fisiología, Facultad de Medicina, Universidad Autónoma, Madrid, Spain

Cerebral Venoconstriction Induced by Norepinephrine and Vasopressin

G. Dieguez, M. V. Conde, J. L. Moya, B. Gómez, and S. Lluch

Summary

The cumulative application of norepinephrine (10^{-10} to $3 \cdot 10^{-4}$ M) and vasopressin (10^{-13} to $3 \cdot 10^{-7}$ M) produced dose-dependent contractile responses in isolated goat pial veins. The control dose-response curve for norepinephrine was shifted 61-fold to the right by phentolamine (10^{-7} M), and (1-Deaminopenicillamine, 4-valine, 8-D-arginine)-vasopressin (10^{-6} M) produced a 10-fold shift to the right of the control dose-response curve for vasopressin. Under our experimental conditions, the resting tension (75 mg) represents an effective transmural pressure of 2.61 ± 0.10 mmHg, and the maximal effects induced by norepinephrine and vasopressin represented an increase in transmural pressure of 190 and 185%, respectively, above the resting values. The results indicate that norepinephrine and vasopressin produce a considerable constriction in goat cerebral veins by the activation of specific receptor sites. This venoconstriction may have significant effects on cerebral blood volume and intracranial pressure in certain physiological and pathophysiological states in which plasma levels of these vasoactive agents are elevated.

Keywords: Cerebral veins; norepinephrine; vasopressin.

Introduction

In recent years it has become evident that cerebral veins contract in response to adrenergic stimuli and may be involved in regulating cerebral blood volume[2,4]. In the present experiments we have studied the pattern of the response of isolated goat pial veins to norepinephrine and vasopressin in the absence and in the presence of specific antagonists. Previous observations from our laboratory have indicated the presence of specific receptors for norepinephrine and vasopressin in goat cerebral arteries[3,11].

Material and Methods

Nine female goats, weighing 32–46 kg were used. The animals were sacrificed by injecting intravenously a saturated solution of potassium chloride and the brain was immediately removed. Cylindrical segments, 4 mm in length and 0.65 ± 0.03 mm in external diameter, were cut from the pial veins obtained from the lateral surface of the brain hemispheres. Two stainless steel pins, 150 μm in diameter, were introduced through the venous lumen. Each segment was prepared for isometric recording as previously described[11]. A micrometer head measured the displacement of the transducer and, at the same time, the distance between the pins at any given time. A resting tension of 75 mg was applied to the tissue and the segments were allowed to equilibrate for 40–60 minutes before any drug was added. Dose-response curves for norepinephrine (10^{-10} to $3 \cdot 10^{-4}$ M) and vasopressin (10^{-13} to $3 \cdot 10^{-7}$ M) were determined in a cumulative manner in the absence and in the presence of phentolamine (10^{-7} M) or the vasopressin-antagonist (1-Deaminopenicillamine, 4-valine, 8-D-arginine)-vasopressin (dPVDAVP, 10^{-6} M), respectively. The antagonists were added to the organ bath 10 min before the agonists. The geometric mean ED_{50} and pA_2 values were obtained using classical pharmacological methodology[1, 5, 6]. The experimental arrangement enabled the wall tension and internal circumference length of the vessel to be controlled and, assuming that the venous wall was sufficiently thin for Laplace's equation to apply, we calculated the effective transmural pressure under resting conditions and during the maximal effects induced by norepinephrine and vasopressin. The formula of Laplace's equation is,

$$P = 2 \pi T/L$$

where T, venous wall tension is the circumferential wall force per unit length given by $F/2$ g. F is the force applied to the tissue under resting conditions and during the maximal contraction produced by norepinephrine and vasopressin, and g is the vascular segment length. L refers to the internal circumference length corresponding to a wall tension T,

$$L = d(2 + \pi) + 2l$$

where d is diameter of pins and l is distance between the inner edges of pins measured by displacement of the micrometer head. The wall tension, T, is expressed in mg/mm and transmural pressure, P, in mg/mm². To express P in mmHg the values in mg/mm² were divided by 13.6.

Drugs used were: norepinephrine bitartrate (Sigma); arginine-vasopressin (Sigma); phentolamine methanesulfonate (Regitin, Ciba), and (1-Deaminopenicillamine, 4-valine, 8-D-arginine)-vasopressin (Bachem).

Statistical analysis was done by means of the Student's t test and a probability value of less than t 5% was considered significant.

Results

Figs. 1 and 2 illustrate actual recordings and dose-response curves, respectively, showing the contractile effects induced by norepinephrine and vasopressin on isolated goat cerebral veins in the absence and in the presence of their specific antagonists. The maximal effects produced, as well as the ED_{50} values are given in Table 1.

Phentolamine (10^{-7} M) displaced the control curve for norepinephrine 61-fold to the right. The maximal contractions obtained in the absence and in the presence of phentolamine were not significantly different ($P > 0.20$). From the dose ratio of 61 observed in the presence of phentolamine, an apparent Kb value of $1.7 \cdot 10^{-9}$ M was obtained. The negative logarithm of this Kb value is 8.77.

Vasopressin produced a contractile response in a dose-dependent manner $3 \cdot 10^{-13}$ M and 10^{-8} M being the threshold and the maximally

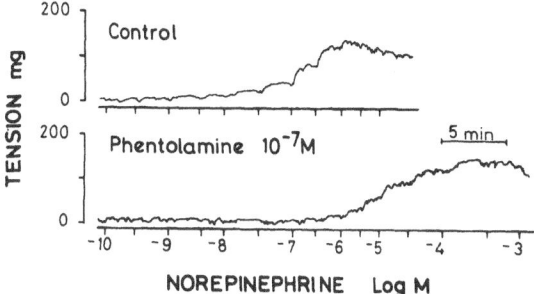

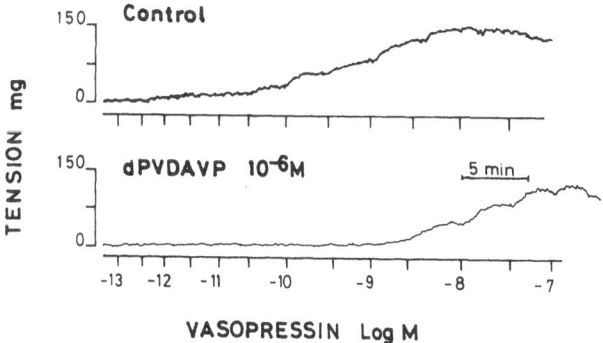

Fig. 1. Experimental tracings showing contractile response of goat pial veins to norepinephrine and vasopressin

effective doses respectively. The dose-response curve for vasopressin, determined in the presence of dPVDAVP (10^{-6} M), was shifted 10-fold to the right of the control curve. dPVDAVP did not modify the maximal contraction induced by vasopressin significantly (P > 0.20). The Kb value was $1.1 \cdot 10^{-7}$ M and their negative logarithm is 6.9.

Table 2 summarizes the values of wall tension and effective transmural pressure, calculated under resting conditions and during the maximal contraction induced by norepinephrine and vasopressin. Under our experimental conditions, the resting tension represents an effective transmural pressure of 2.61 ± 0.10 mm Hg, and the maximal effects induced by norepinephrine and vasopressin represented an increase in transmural pressure of 190 and 185%, respectively, above the resting values.

Discussion

The present results indicate that goat cerebral veins have a high ability to constrict in response to norepinephrine and vasopressin. A contractile response to norepinephrine has been also described in cat pial veins[2] as well as in other venous beds[9,10].

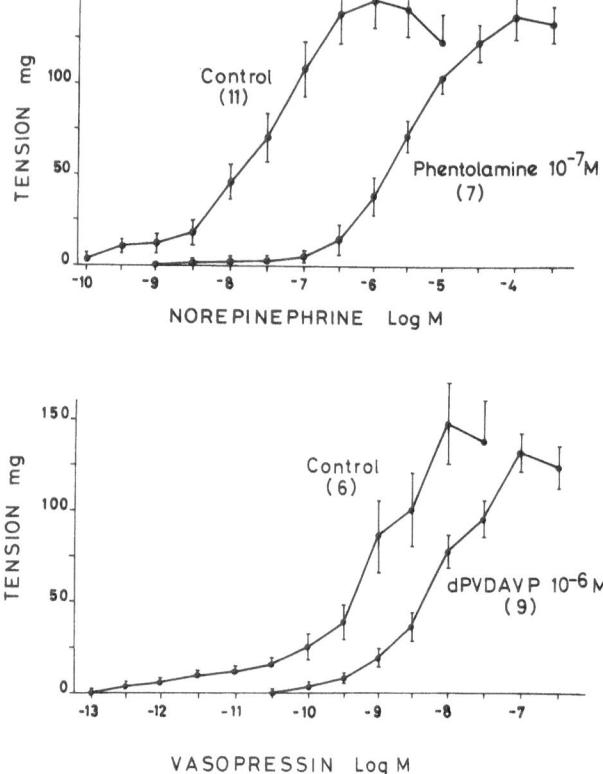

Fig. 2. Dose-response curves for norepinephrine and vasopressin in goat pial veins. Values are means ± SEM. Numbers of experiments are indicated in parentheses

Phentolamine (10^{-7} M) produced a 61-fold shift to the right of the control dose-response curve for norepinephrine, presumably due to an equilibrium competitive agonist-antagonist interaction. The negative logarithm of the apparent Kb values (8.77) found in the present experiments is higher than that measured in goat cerebral arteries (7.23)[11]. From a comparative analysis, it appears that the affinity of the adrenergic receptors of goat cerebral veins for norepinephrine is higher than that observed in cerebral arteries of the same animal species.

The contraction elicited by vasopressin was antagonized by dPVDAVP (10^{-6} M) in an equilibrium competitive manner. This compound does not interfere with the venoconstriction produced by norepinephrine, angiotensin II, and potassium chloride (unpublished results). This suggests that cerebral venoconstriction is due to activation of specific receptors to this neuropeptide. It is interesting to note that contractile responses are elicited with vasopressin concentrations which closely resemble the circulating concentrations observed under physiological conditions (*i.e.*

Table 1. *Maximal Contractions and ED_{50} Values of Goat Cerebral Veins to Norepinephrine (NE) and Vasopressin (VP)*

	Maximal contraction (mg)	ED_{50} (M)
NE	145 ± 15.12 (11)	$5.4 \pm 1.57 \cdot 10^{-8}$ (11)
VP	147 ± 20.1 (6)	$1.7 \pm 1.06 \cdot 10^{-9}$ (6)

Values are means ± SEM. Number of experiments are indicated in parentheses.

Table 2. *Wall Tension (T) and Effective Transmural Pressure (P) in Goat Cerebral Veins Under Resting Conditions (Control) and During Maximal Contraction Induced by NE and VP*

	Control	NE	VP
T mg/mm	9.4 (17)	27.5 ± 1.89 (11)	27.8 ± 5.13 (6)
P mm Hg	2.6 ± 0.10 (17)	7.6 ± 0.52 (11)	7.5 ± 1.21 (6)

Values are means ± SEM. Number of experiments are indicated in parentheses.

10^{-12} to 10^{-11} M)[7,8]. The maximum developed tension observed in goat cerebral veins in response to vasopressin is lower than that developed in human and goat cerebral arteries[3], but ED_{50} values are similar in the three types of vessels.

Our results also indicate that the maximal contractions induced by norepinephrine and vasopressin represent large increases in transmural pressure. These increments may reflect marked constrictor effects in the in vivo situation.

In conclusion, the present results indicate that goat cerebral veins possess a considerable ability to constrict in response to low concentrations of norepinephrine and vasopressin through activation of specific receptors. This venoconstriction could have significant effects on cerebral blood volume and intracranial pressure in certain physiological and pathophysiological conditions in which the plasma levels of these substances are elevated.

Acknowledgement

This investigation was supported in part by the Comisión Asesora de Investigación Científica y Técnica and Ministerio de Sanidad. We are indebted to Miss M. E. Martínez for technical assistance.

References

1. Arunlakshana, O., Schild, H. O.: Some quantitative uses of drug antagonists. Brit. J. Pharmacol. *14* (1959), 48—58.
2. Auer, L. M., Johansson, B. B., Kuschinsky, W., Edvinsson, L.: Sympathoadrenergic activity of cat pial veins. J. Cerebr. Blood Flow Metabol. 1, Suppl. *1* (1981), 311—312.
3. Conde, M. V., Dieguez, G., Gómez, B., Lluch, S.: Vasoconstriction of isolated human and goat cerebral arteries in response to vasopressin. Neurosciences Letters Suppl. *10* (1982), 121.
4. Edvinsson, L., McCulloch, J., Uddman, R.: Sympathetic and peptidergic meachisms in cerebral veins. J. Cerebr. Blood Flow Metab. *1*, Suppl. 1 (1981), 327—328.
5. Fleming, W. W., Westfall, D. P., Lande, I. S., de la, Jellet, L. B.: Log-normal distribution of equieffective doses of norepinephrine and acetylcholine in several tissues. J. Pharmacol. exp. Ther. *181* (1972), 339—345.
6. Furchgott, R. F.: The pharmacological differentiation of adrenergic receptors. Ann. N. Y. Acad. Sci. *139* (1967), 553—570.
7. Hammer, M., Ladefoged, J., Ølgaard, K.: Relationship between plasma osmolality and plasma vasopressin in human subjects. Amer. J. Physiol. *238* (1979), E313—E317.
8. Larsson, B., Olsson, K., Fyhrquist, F.: Vasopressin release induced by hemorrhage in the goat. Acta physiol. scand. *104* (1978), 309—317.
9. Pearce, W. J., Bevan, J. A.: The possible influence of extracerebral venoconstriction of cerebral hemodynamics. J. Cerebr. Blood Flow Metabol. *1*, Suppl. 1 (1981), 325—326.
10. Sutter, M. C.: The pharmacology of isolated veins. Brit. J. Pharmacol. *24* (1965), 742—751.
11. Urquilla, P. R., Marco, E. J., Lluch, S.: Pharmacological receptors of the cerebral arteries of the goat. Blood Vessels *12* (1975), 53—67.

Authors' address: G. Dieguez, M.D., M. V. Conde, Ph.D., J. L. Moya, M.D., B. Gómez, M.D., and S. Lluch, M.D., Departamento de Fisiología, Facultad de Medicina, Universidad Autónoma, Arzobispo Morcillo 1, Madrid 34, Spain.

Department of Neurology, School of Medicine, Keio University, Shinjuku-ku, Tokyo, Japan

Autonomic Nervous Action Potential of Cerebral Veins

S. Komatsumoto, F. Gotoh, K. Shimazu, M. Ichijo, K. Yamashita, N. Araki, and J. Hamada

Summary

1. Averaged action potential waves were recorded from cerebral arteries and from cerebral veins during cervical sympathetic stimulation. The averaged potential waves were suppressed after the administration of hexamethonium or tetrodotoxin.

2. Discharges of the action potentials both from the cerebral artery and from the cerebral vein increased during hypotension induced by exsanguination, and decreased during hypertension induced by reinfusion of the blood.

The above data suggest that not only the cerebral artery but also the cerebral vein is, in part, functionally under the control of the cervical sympathetic nervous system.

Keywords: Action potential; cerebral vein; superior cervical ganglion; tetrodotoxin.

Introduction

Innervation of the autonomic nervous system in the cerebral veins has been confirmed histochemically[7]. Its functional significance, however, still remains controversial. In the cat we have previously demonstrated autonomic nervous action potentials directly from the cerebral arterial walls and their response to changes in cerebral perfusion pressure[5]. The present study focussed on an attempt to record the autonomic nervous action potentials from the cerebral veins in the cat.

Materials and Methods

Twelve adult cats were anaesthetized with 1.0% α-chloralose and 10% urethane followed by artificial respiration. The rectal temperature was maintained at between 37° and 38 °C by the use of a heating pad. A burr hole was cut in the skull. The pial arteries and veins were exposed and covered with mineral oil, which was used for the maintenance of local temperature and for insulation of the vessels. Blood pressure was monitored continuously by means of a Statham strain gauge connected to a catheter inserted into the descending aorta through a femoral artery. A femoral vein was also cannulated to administer agents.

The cervical sympathetic chain was isolated and severed just below the superior cervical ganglion. A bipolar platinum electrode for stimulation was placed on the cranial end of preganglionic nerve fibres.

1. Averaged potential waves from the arteries and veins were recorded during preganglionic stimulation of cervical sympathetic fibres on the ipsilateral side. The action potentials were recorded using fine bipolar platinum electrodes, highly sensitive pre-amplifiers, band-pass filters and a data analyzing computer. The sympathetic nerve was stimulated with square wave pulses of 5 volts and 300 μsec duration at 3 Hz. The averaged potential waves were analyzed by the program of signal averaging.

2. Discharges of the action potentials in response to changes in blood pressure were recorded using the same system. Mass discharges were analyzed with the program of pulse density variation. Changes in cerebral perfusion pressure were induced by exsanguination and reinfusion of the blood.

Results

1. Averaged action potential waves during cervical sympathetic stimulation.

We attempted to record the averaged action potential waves from cerebral vessels and other parts of the brain.

The averaged potential waves were obtained from the ipsilateral cerebral arteries and veins by repeated sympathetic stimulation (Fig. 1), but the waves were not evoked from any of the following; the pial arteries on the contralateral side, the avascular area of the brain cortex and the temporal muscles. The arterial averaged potential waves had a large peak with a time delay of 5 to 15 msec followed by several smaller peaks. The venous averaged potential waves, however, showed a similar but less well defined peak with a greater variation.

Fig. 2 represents the effect of the tetrodotoxin (25 μg i.v.) on the arterial and venous averaged potential waves during cervical sympathetic stimulation. After the administration of tetrodotoxin, both arterial and venous averaged potential waves were markedly suppressed. Similar results were observed in two out of three cats, and in the other cat the degree of reduction in the arterial and venous peaks was smaller.

The effect of the ganglion blockade by hexamethonium (25 mg i.v.) on the averaged potential waves during cervical stimulation was investigated in two cats. The height of the arterial and venous peaks was reduced by hexamethonium (Fig. 3). The same result was obtained in the other cat.

2. Effect of changes in blood pressure on the action potential.

The discharges of the arterial action potentials increased during hypotension induced by exsanguination, and decreased during hypertension induced by reinfusion of the blood. These responses of the arterial potentials were maintained in all 7 cats used in the present study.

Fig. 4 shows the response of the venous discharges to changes in blood pressure. The responses were similar to those of the arterial discharges. The same venous responses were obtained in 5 out of the 7 cats.

PIAL ARTERY

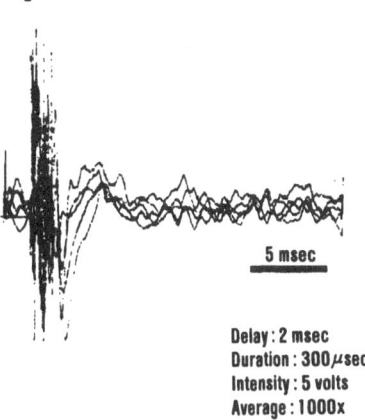

5 msec

Delay : 2 msec
Duration : 300 μsec
Intensity : 5 volts
Average : 1000x

PIAL VEIN

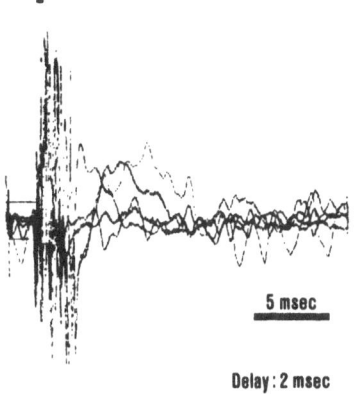

5 msec

Delay : 2 msec
Duration : 300 μsec
Intensity : 5 volts
Average : 1000x

Fig. 1. Averaged action potential waves from the ipsilateral pial arteries and veins during cervical sympathetic stimulation

Discussion

The first part of the present study demonstrated that repeated sympathetic stimulation evoked the averaged potential wave from the cerebral vessels on the ipsilateral side, but not from the contralateral side. Furthermore, the averaged potential waves from both the ipsilateral arteries and veins were suppressed either by blockade of autonomic ganglia or nerve fibres. These data suggest that the observed potential does not originate from the artifactual electrical leakage but is derived from the nerve conducted potential through the superior cervical ganglia.

PIAL ARTERY   PIAL VEIN

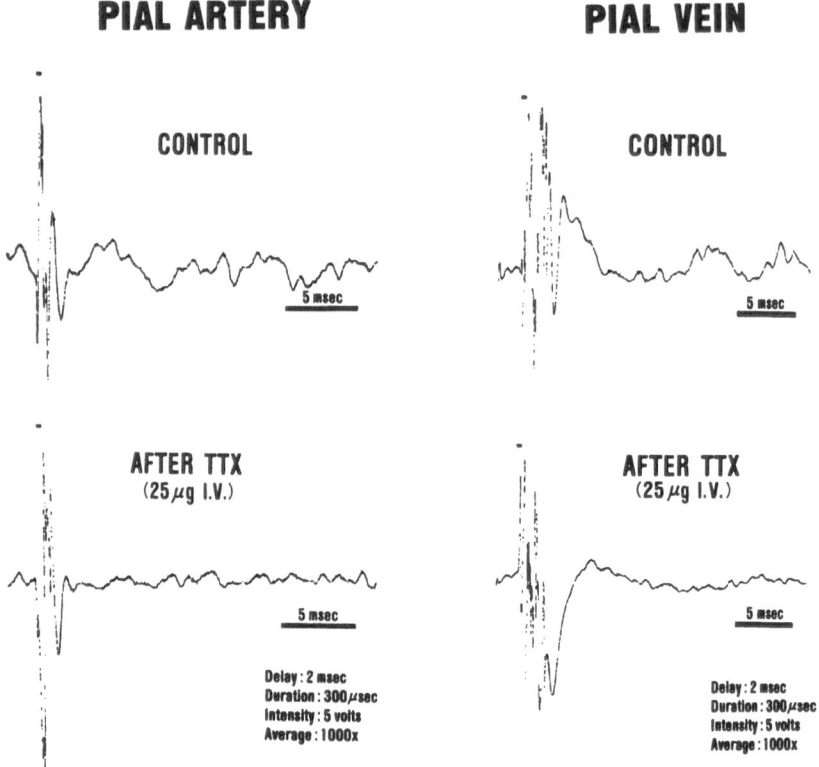

Fig. 2. Effect of tetrodotoxin on the averaged action potential waves of the pial artery and vein

There is abundant anatomical evidence that the cerebral arteries are richly innervated from the superior cervical ganglion[2]. These anatomical findings are further supported by our functional evidence of the innervation. In the cerebral veins, however, adrenergic terminals have been demonstrated histochemically[7], but the origin of their innervation has remained unknown. The present study gives confirmatory evidence of the innervation of the cerebral veins. Recently Auer *et al.*[1] reported constriction of the cerebral veins following preganglionic cervical stimulation. Their results are in accordance with our present data. Thus, at least a part of the venous innervation seems to originate in the superior cervical ganglion.

In an effort to clarify the functional role of the action potential derived from the artery and vein, the effect of changes in cerebral perfusion pressure on the discharges of the action potentials were investigated in the second part of the present study. In our previous studies[4,6], the responses of the action potentials from the pial artery to alteration of blood pressure disappeared after superior cervical ganglionectomy, and also after the administration of fusaric acid (a potent inhibitor of dopamine-β-

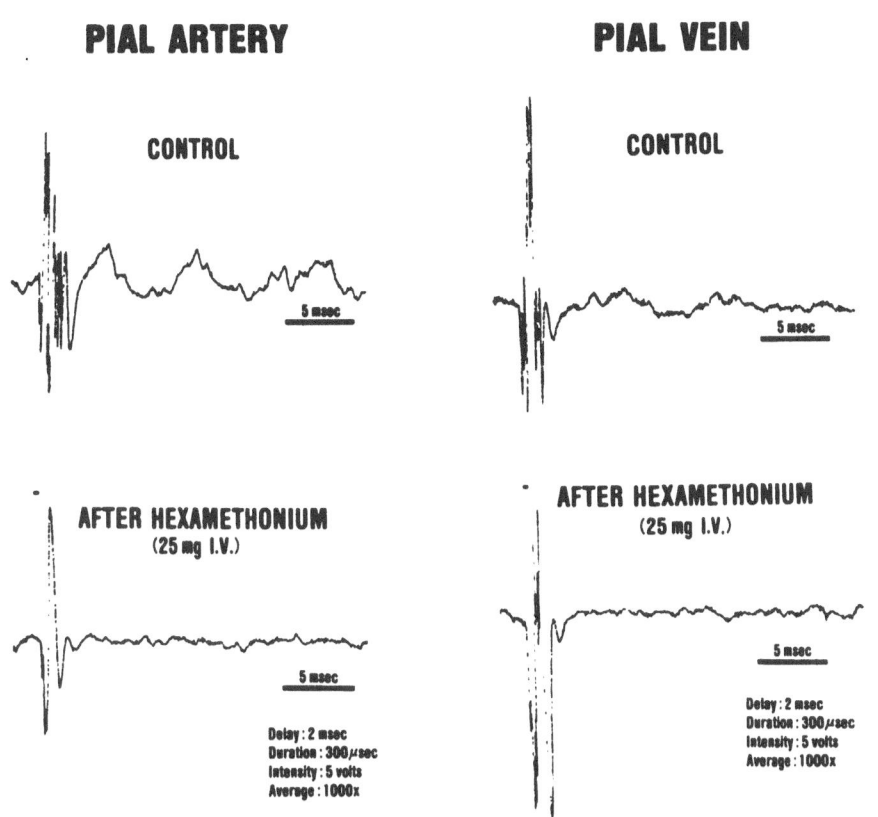

Fig. 3. Effect of hexamethonium on the averaged action potential waves of the pial artery and vein

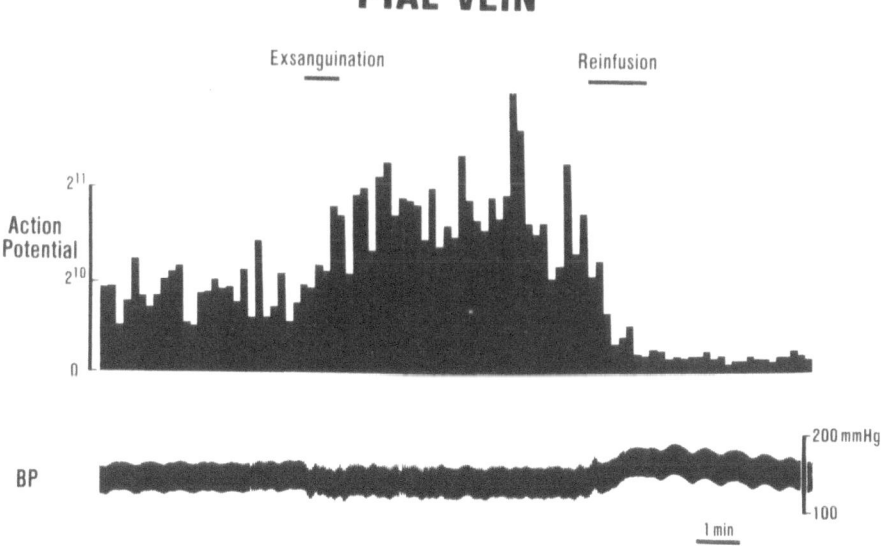

Fig. 4. Responses of discharges from the pial vein to changes in blood pressure

hydroxylase). These previous data suggested that the noradrenergic nervous system of the pial arteries was closely related to the autoregulatory mechanism of cerebral blood flow.

In the present study, changes in the pial venous discharges by alterations in the blood pressure were also observed. Gotoh *et al.*[3] reported that the vasomotor activity of the pial vein during alteration of blood pressure was influenced neurogenically but the degree of vasomotor activity was less marked in the veins than in the pial arteries. These data suggest that at least a part of the action potentials from the cerebral veins play a role in the vasomotor activity of the veins in response to changes in perfusion pressure.

It can be concluded that the cerebral venous vessels are, in part, functionally under the control of the cervical sympathetic nervous system.

References

1. Auer, L. M., Johansson, B. B.: Pial venous constriction during cervical sympathetic stimulation in the cat. Acta Physiol. Scand. *110* (1980), 203—205.
2. Edvinsson, L., MacKenzie, E. T.: Amine mechanisms in the cerebral circulation. Pharmacol. Rev. *28* (1976), 275—348.
3. Gotoh, F., Muramatsu, F., Fukuuchi, Y., Amano, T., Tanaka, K.: Neurogenic and chemical influences on the diameter of pial veins. In: The Cerebral Veins (Auer, L. M., Loew, F., eds.), pp. 179—186. Wien-New York: Springer. 1983.
4. Komatsumoto, S., Gotoh, F., Shimazu, K., Ichijo, M., Araki, N.: Effect of sympathetectomy on pial arterial autonomic action potentials. J. Cerebr. Blood Flow Metabol. Suppl. *1*, 1 (1981), 315—316.
5. Shimazu, K., Gotoh, F., Nakajima, S., Komatsumoto, S., Ichijo, M.: Demonstration of autonomic action potential from cerebral artery. Acta Neurol. Scand. Suppl. 72, *60* (1979), 98—99.
6. Shimazu, K., Gotoh, F., Komatsumoto, S., Ichijo, M., Araki, N.: Effect of inhibition of dopamine-β-hydroxylase on the action potential of a cerebral artery. In: Cerebral Microcirculation and Metabolism (Cervós-Navarro, J., Fritschka, E., eds.), pp. 279—283. New York: Raven Press. 1981.
7. Owman, C., Falck, B., Mchedlishvili, G. I.: Adrenergic structures of the pial arteries and their connections with the cerebral cortex. Fed. Proc. 25 (translation suppl. 1966), T612—T614.

Authors' address: Satoru Komatsumoto, M.D., Fumio Gotoh, M.D., Kunio Shimazu, M.D., Makoto Ichijo, M.D., Kazuhiko Yamashita, M.D., Nobuo Araki M.D., and Junichi Hamada, M.D., Department of Neurology, School of Medicine, Keio University, 35 Shinanomachi, Shinjuku-ku, Tokyo 160, Japan.

Department of Neurology, School of Medicine, Keio University, Shinjuku-ku, Tokyo, Japan

Neurogenic and Chemical Influences on the Diameter of Pial Veins

F. Gotoh, F. Muramatsu, Y. Fukuuchi, T. Amano, and K. Tanaka

Summary

Neurogenic and chemical influences on the diameter of the pial vessels were assessed in 16 cats using a microphotographic technique. Hypotension, induced by exsanguination, resulted in a marked autoregulatory dilatation of the larger arteries, but the changes in the veins and the smaller arteries were less marked. Hypertension, induced by reinfusion of the blood, resulted in a marked autoregulatory constriction of the larger arteries, but the changes in the veins and the smaller arteries were less. Hexamethonium (5 mg/kg i.v.) caused minimal changes in the vessel diameters. 5-Hydroxydopamine (5 mg/kg i.v.) resulted in a significant dilatation of the arteries and the smaller veins, but did not dilate the larger veins. Papaverine (5 mg/kg i.v.) had effects on the pial vessels similar to those of 5-hydroxydopamine. Inhalation of 5% CO_2 caused a marked dilatation of the smaller veins and arteries, but did not show significant effects on the larger vessels. Hyperventilation resulted in a significant constriction of the small veins and arteries, but did not cause a significant change in the diameters of the larger vessels.

These data suggest that the diameter of the pial veins was influenced both neurogenically and chemically, but also that the effects of these two agents were less marked in the veins than in the arteries.

Keywords: Pial veins; neurogenic, chemical influences.

Introduction

Neurogenic and chemical influences on the pial arteries have been studied by many investigators and considerable information concerning the physiological and pathological responses of the pial arteries has been accumulated[5,7]. The present knowledge concerning the pial veins, however, is comparatively scarce.

The purpose of the present study was to assess the neurogenic and chemical influences on the diameter of the pial veins, and to compare them with those on the pial arteries.

12*

Materials and Methods

Sixteen adult cats of both sexes ranging in weight from 2.5 to 3.9 kg were used in the experiments. The animals were anaesthetized with an intraperitoneal injection of 30–50 mg/kg of pentobarbital and supplemented with the local application of 1% procaine. Each animal was immobilized and mechanically ventilated via a tracheotomy and a tracheal cannula. Blood pressure was recorded via a polyethylene tube placed in a femoral artery by means of a strain gauge transducer and a polygraph. The skull and dura were opened and the exposed pial surface was completely covered with mineral oil. Changes in the diameter of the pial veins and arteries of 15 to 250 μ in diameter were measured by means of the microphotographic technique before and after various procedures. The procedures used in this study were exsanguination and reinfusion of the blood, inhalation of 5% CO_2, hyperventilation and intravenous injections of 5 mg/kg of papaverine, 5 mg/kg of hexamethonium and 5 mg/kg of 5-hydroxydopamine.

Microphotographs of the pial vessels were taken by means of an inverted 24 mm-wide Nikkor lens attached to a Nikon F camera via a bellows, or a Nikon microscope with a photographic attachement. The photographs were enlarged so that 300 × magnification was achieved in the prints.

Results

Alteration of blood pressure: As a quantitative expression of the changes in diameter brought about by blood pressure alteration, the vasomotor index (VMI) was calculated using the following formula: VMI $(\mu/\text{mm Hg}) = - \Delta D / \Delta MABP$ (ΔD: change in diameter, $\Delta MABP$: change in mean arterial blood pressure).

Fig. 1 a shows the effects of hypotension, induced by exsanguination, on the VMI of the pial veins and arteries. Induced hypotension resulted in a marked dilatation of the large arteries, but the changes in the veins and the smaller arteries were less marked. In general, the VMI of the veins was smaller than that of the arteries and showed no definite correlation with the diameter of the vessels. On the contrary, the VMI of the arteries was dependent on the diameter of the vessels, that is, the larger the diameter, the larger the VMI.

Fig. 1 b shows the effects of hypertension induced by reinfusion of the blood. Induced hypertension resulted in a marked constriction of the larger arteries but the changes in the veins and the smaller arteries were less marked.

Table 1 summarizes the VMI of the arteries and veins according to their sizes.

Ganglion blockade: Administration of hexamethonium caused no definite tendency to change the vessel diameter regardless of the size or kind of vessel. The data concerning hexamethonium in Table 2 shows percentage change of the diameter of the pial vessels according to their sizes five minutes after the injection.

False sympathetic transmitter: Fig. 2 shows changes in the pial vessel diameter 30 minutes after intravenous injection of 5-hydroxydopamine, a false sympathetic transmitter, which replaces noradrenergic vesicles in the nerve terminals. The increase in the diameter of the smaller veins was

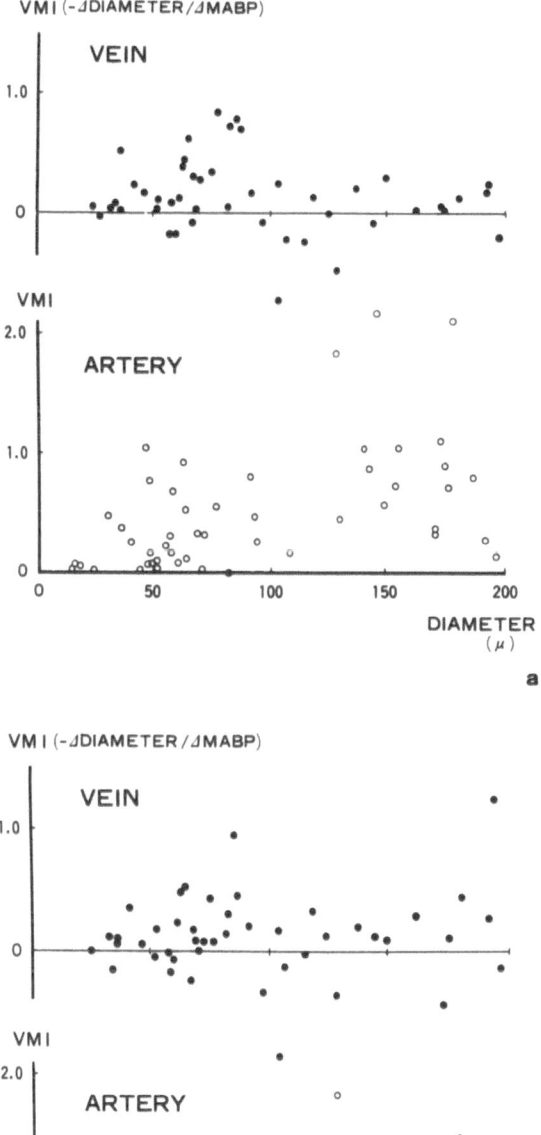

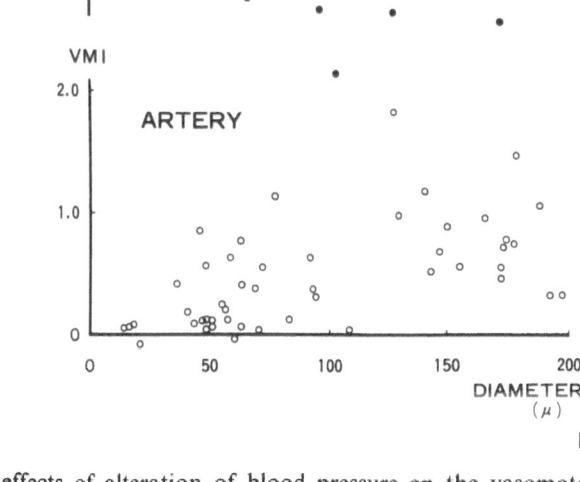

Fig. 1. The effects of alteration of blood pressure on the vasomotor index (VMI). a) Hypotension induced by exsanguination. b) Hypertension induced by reinfusion of the blood

Table 1. *VMI of the Pial Veins and Arteries*

Vessel diameter (D) in micrometers	D < 50	50 ≦ D < 100	100 ≦ D < 150	150 ≦ D < 200
Exsanguination				
VMI of Vein	0.14 ± 0.06 n = 8	0.26 ± 0.07 n = 21	0.14 ± 0.11* n = 9	0.09 ± 0.06* n = 8
VMI of Artery	0.27 ± 0.09 n = 13	0.33 ± 0.06 n = 18	1.13 ± 0.31 n = 6	0.76 ± 0.15 n = 12
Reinfusion				
VMI of Vein	0.07 ± 0.06 n = 7	0.16 ± 0.06 n = 21	0.04 ± 0.12* n = 9	0.24 ± 0.17* n = 8
VMI of Artery	0.20 ± 0.07 n = 12	0.35 ± 0.07 n = 18	0.87 ± 0.25 n = 6	0.73 ± 0.10 n = 12

VMI: Vasomotor Index (mean ± S.E.).
n: Number of vessels.
* Statistically significant difference compared to arteries of the same size.

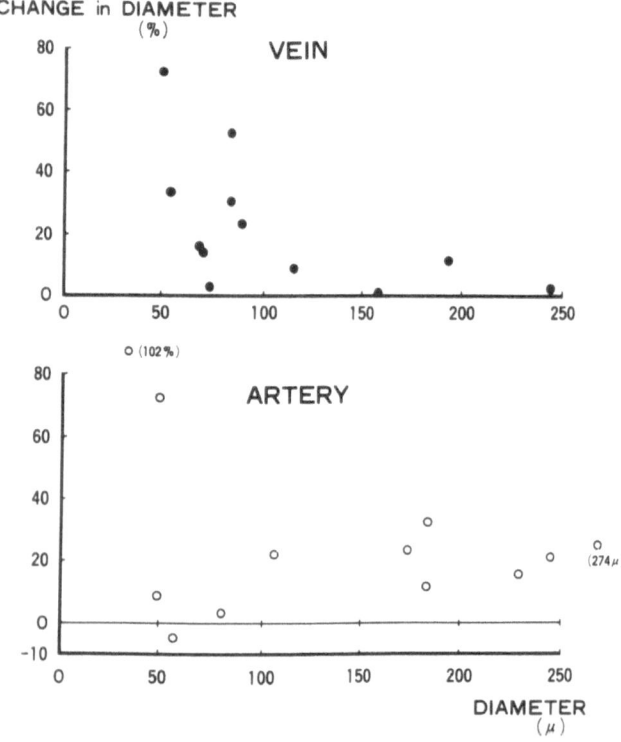

Fig. 2. Changes in the pial vessel diameter 30 minutes after intravenous injection of 5-hydroxydopamine

Table 2. *Percent Changes in the Diameters of Pial Veins and Arteries*

Vessel diameter (D)	D < 100 μ	D ≧ 100 μ
Hexamethonium		
Vein	1.6 ± 3.6%	3.8 ± 2.4%
	n = 10	n = 10
Artery	0.7 ± 2.1%	—2.0 ± 1.5%
	n = 9	n = 14
5-Hydroxydopamine		
Vein	30.6 ± 8.1%	5.9 ± 2.7%*
	n = 8	n = 4
Artery	36.3 ± 21.6%	21.5 ± 2.5%
	n = 5	n = 7
Papaverine		
Vein	34.4 ± 18.5%	9.7 ± 4.5%*
	n = 10	n = 8
Artery	32.8 ± 3.9%	28.7 ± 6.1%
	n = 17	n = 7
5%CO_2 Inhalation		
Vein	12.0 ± 3.5%	5.9 ± 10.1%
	n = 7	n = 3
Artery	11.9 ± 2.5%	5.6 ± 1.9%
	n = 9	n = 2
Hyperventilation		
Vein	—12.6 ± 4.8%	—4.8 ± 2.3%
	n = 6	n = 4
Artery	—6.9 ± 3.4%	0.6 ± 1.3%
	n = 9	n = 5

* Statistically significant difference compared to arteries.

significant and was not different from that of the smaller arteries, but the change in the larger veins was far smaller than that of the larger arteries, as shown in Table 2.

Papaverine: Fig. 3 shows the effects of intravenous injection of papaverine on the diameter of the pial veins and arteries. The smaller veins markedly dilated five minutes after the administration of the drug, but the changes in the larger veins were less marked. On the other hand, the pial arteries were considerably dilated by the drug regardless of their sizes (Table 2).

Changes in arterial carbon dioxide tension: Fig. 4 indicates the effects of changes in arterial carbon dioxide tension by 5% CO_2 inhalation or by hyperventilation on the diameter of the pial veins and arteries.

The smaller veins were markedly dilated during CO_2 inhalation and significantly constricted during hyperventilation, but the larger veins did not show a significant change in the diameter during the same procedure. These changes were similar to those of the arteries that have been reported previously[3].

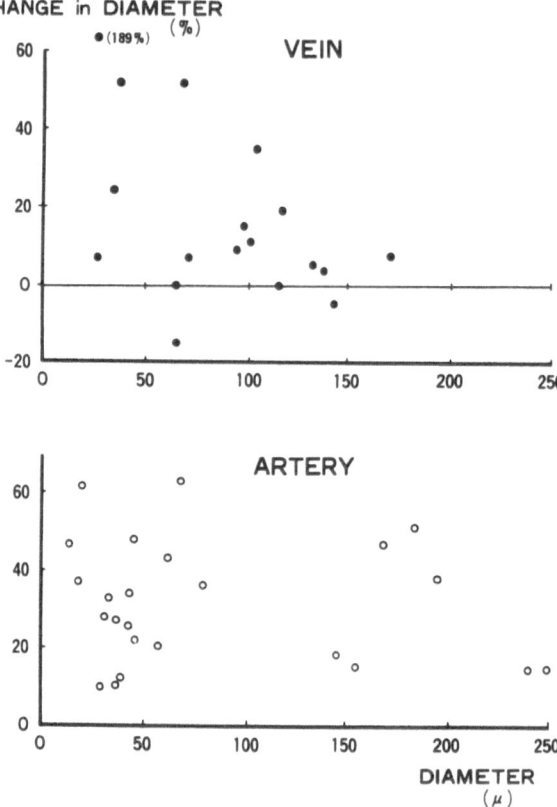

Fig. 3. Changes in the pial vessel diameter five minutes after intravenous injection of
papaverine

Discussion

In 1938, Forbes *et al.*[2] observed constriction of the pial vessel as a result
of the local application of adrenaline, and stimulation of the cervical
sympathetic nerve. After their observation, the physiological and pathological significance of the pial veins was neglected by investigators in the field of
the cerebral circulation for a long time. However, advances in the
knowledge concerning the innervation of the cerebral arteries and veins of
late have stimulated the interest of investigators in the neurogenic influences
on the cerebral veins. Recently, Auer *et al.*[1] demonstrated significant
constriction of the pial veins during cervical sympathetic stimulation.
Dilatation of the smaller pial veins after the administration of 5-hydroxy-
dopamine in the present study seems to be consistent with their data, in
which the cerebral veins were considered to be under direct neurogenic
control. But it is difficult to explain the lack of dilatation of the larger veins
in spite of the marked dilatation of the larger arteries by the false
neurotransmitter, because the larger veins are known to have adrenergic

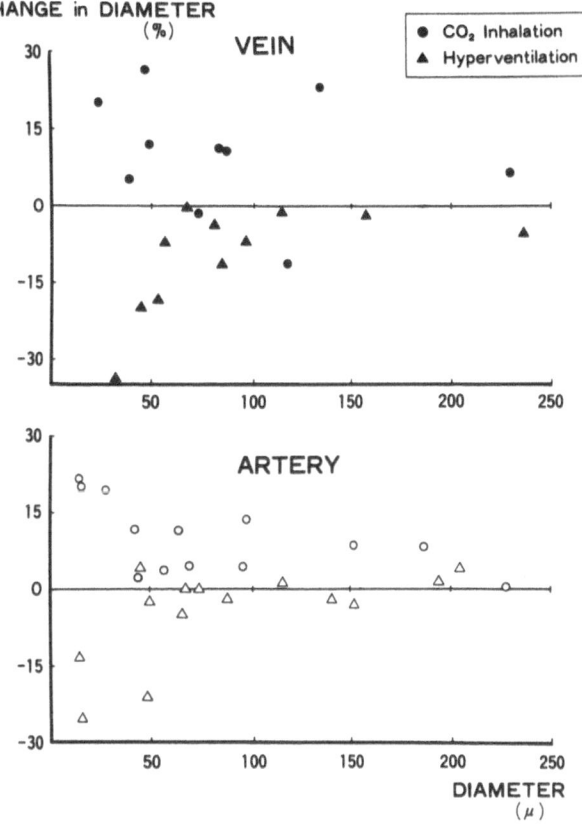

Fig. 4. Effects of changes in arterial carbon dioxide tensions, by 5% CO_2 inhalation or hyperventilation, on the pial vessel diameter

innervation[4,6]. Dilatation of the smaller veins by CO_2 inhalation or by the administration of papaverine, and their constriction by hyperventilation in the present study seem to imply that the changes in the diameter of the smaller veins are in parallel with the changes in the cerebral blood flow. The larger veins were relatively insentitive to various stresses. These data seem to favour the hypothesis that the changes in the diameter of the cerebral veins are secondary to those of the cerebral blood flow, rather than primary active changes, although the possibility of the latter cannot be completely ruled out.

References

1. Auer, L. M., Johansson, B. B.: Pial venous constriction during cervical sympathetic stimulation in the cat. Acta Physiol. Scand. *110* (1980), 203—205.
2. Forbes, H. S., Cobb, S. S.: Vasomotor control of cerebral vessels. Brain *61* (1938), 221—233.

3. Gotoh, F., Muramatsu, F., Fukuuchi, Y., Amano, T.: Dual control of cerebral circulation: Separate sites of action in the vascular tree in autoregulation and chemical control. Cerebral Circulation and Metabolism (Langfitt, T. W., *et al.*, eds.), pp. 43—45. Berlin-Heidelberg-New York: Springer. 1975.

4. Kajikawa, H.: Fluorescence histochemical studies on the distribution of adrenergic nerve fibres to intracranial blood vessels. Arch. Jap. Chir. *37* (1968), 473—484.

5. Kuschinsky, W., Wahl, M.: Local chemical and neurogenic regulation of cerebral vascular resistance. Physiol. Rev. *58* (1978), 656—689.

6. Owman, C., Falck, B., Mchedlishvili, G. I.: Adrenergic structures of the pial arteries and their connections with the cerebral cortex. Fed. Proc. 25 (translation supple. 1966), T 612—614.

7. Purves, M. J.: The Physiology of the Cerebral Circulation. London: Cambridge University Press. 1972.

Authors' address: F. Gotoh, M.D., Department of Neurology, School of Medicine, Keio University, 35 Shinanomachi, Shinjuku-ku, Tokyo 160, Japan.

Merck Institute for Therapeutic Research, West Point, Penn., Department of Pharmacology, School of Medicine, Center for the Health Sciences, University of California, Los Angeles, Calif., U.S.A.

The Facial Vein as a Temperature-Sensitive Sphincter Involved in Cerebral Thermoregulation

R. J. WINQUIST, R. A. ROWAN, and J. A. BEVAN

Summary

The buccal segment of the rabbit facial vein develops a maintained myogenic contraction which is dependent on the amount of wall stress and presence of extracellular calcium. The calcium influx sustaining tone is blocked by the inorganic but not the organic calcium antagonists. The tonic contraction is exquisitely sensitive to small changes in temperature in the range 33 to 44 °C. This venous segment also contains a preponderance of β-adrenergic receptors and receives a dense adrenergic innervation throughout the smooth muscle layers. Both neural and local factors influence tone in the facial vein which may act as a temperature sensitive sphincter aiding in the distribution of cooled nasal venous blood between central and superficial venous drainage systems. The calcium channels involved in the tonic contraction appear distinct from those previously characterized in vascular smooth muscle.

Keywords: Facial vein; cerebral thermoregulation.

The temperature of the brain is regulated by extracerebral vascular mechanisms[6]. Extracranial venous structures play an integral role in this process by cooling arterial blood flowing to the brain[6] as well as directly cooling ventral areas of the brain[3]. In this chapter we will discuss some of the unique properties of the facial vein which acts as a temperature-sensitive sphincter in the rabbit and possibly in man.

Temperature-Sensitive Tone in the Rabbit Facial Vein

Isolated rings from the buccal segment of the rabbit facial vein develop a sustained, intrinsic contraction in response to stretch of its wall[2]. The contraction is myogenic in origin developing independently of neural or hormonal influences. It is directly related to both the tangential wall stress and the concentration of extracellular calcium (Fig. 1). Possibly the most intriguing property of this venous segment is the extraordinary sensitivity of the developed tone to small changes in ambient temperature over the range 33–43 °C[13]. A rise or fall of only 1 °C from body core temperature causes a large increase or decrease in myogenic tone. These marked changes in the level of tone closely follow the rate of the change in temperature. The maximum temperature-induced tone was found to be approximately 80 percent of the maximum force that developed to exogenously added drugs.

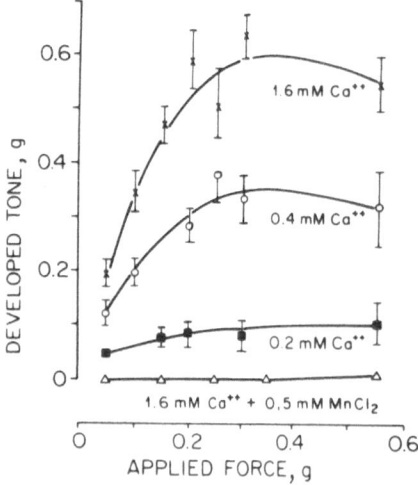

Fig. 1. The relationship between developed myogenic tone and the force applied to the vessel wall of isolated rabbit facial vein rings. Developed tone increases as the rings are increasingly stretched. Decreasing the concentration of extracellular calcium from 1.6 mM to 0.4 mM and 0.2 mM reduces the force developed by a particular level of applied force. Tone fails to develop when the rings are stretched in the presence of manganous chloride (from ref.[14])

Obligatory Role of Extracellular Calcium

The temperature-sensitive tone in the rabbit facial vein is rapidly and reversibly inhibited when specimens are washed in Ca^{++}-free buffer or exposed to the inorganic Ca^{++}-antagonists manganese or lanthanum[14]. The level of tone can be positively correlated with the concentration of Ca^{++} in the physiological buffer. An ultrastructural analysis has demonstrated a relative lack of cellular reticular structures[10]. We conclude that the temperature-sensitive tone in the facial vein requires a continuous presence and influx of extracellular Ca^{++} across the plasma membrane. Presumably, the effect of small changes of temperature on calcium influx would explain this vessel's dramatic response to this environmental stimulus.

An unexpected finding was the insensitivity of tone in this vein to the organic calcium entry blockers verapamil, diltiazem and nifedipine[1,12].

These compounds are believed effective in antgonizing voltage-operated calcium channels in vascular smooth muscle[5]. Thus, myogenic tone in the rabbit facial vein appears to be sustained by calcium influx through stretch-operated channels (SOCs) which may be distinct from receptor-operated and voltage-sensitive channels in smooth muscle cells. It is of interest to note that the intrinsic vascular tone of resistance vessels (as indicated by total peripheral resistance) in normotensive individuals[8] and in the rabbit cerebral artery[1] is also refractory to these compounds. It is not known if exquisite temperature sensitivity is a general characteristic of SOCs in vascular smooth muscle.

Beta-Adrenergic Relaxation in the Facial Vein

This buccal segment of the facial vein displays a beta-adrenoceptor-mediated relaxation in response to stimulation of the adrenergic nerves or exogenously added norepinephrine, that overshadows any concomitant effect of α-adrenoceptor activation[9]. Both the onset and initial phase of the response are rapid. These characteristics correlate well with structural and functional analyses: the adrenergic innervation ramifies throughout the smooth muscle layers and there is a relatively narrow adrenergic neuromuscular cleft[9,10]. There is convincing evidence that the beta-adrenergic receptors are aggregated at the postsynaptic membrane, *i.e.* closely apposed to sites of transmitter release[15].

Physiological Role of the Facial Vein

The buccal segment of the facial vein has the features of a temperature-sensitive sphincter. In situ, its tone would be expected to change markedly in response to the temperature of the blood draining the nasal turbinates. When the blood temperature rises due to an increase in ambient temperature the level of tone is elevated, shunting relatively cool nasal venous blood towards the large central sinuses at the base of the brain (Fig. 2). These sinuses would draw heat away from the brain[3] or cool carotid arterial blood by a countercurrent mechanism in animals with a carotid rete (*i.e.* cat, sheep)[6]. The finding that the cat facial vein also develops temperature-sensitive tone[12] argues for a thermoregulatory role for this vein in species with an internal carotid or carotid rete. The direction of blood flow along the human nasal ridge is altered by small changes in the ambient temperature[4]—an observation which is consistent with our thermoregulatory model. However, recent attempts to confirm the temperature sensitivity of human facial vein tone in vitro, which has many other attributes in common with the rabbit analogue, have not been successful[7].

The rapid neurogenic relaxation characteristic of the facial vein complements the temperature sensitivity of the SOCs. An increased sympathetic output elicited by hypothermia constricts most veins[11] but

A

B

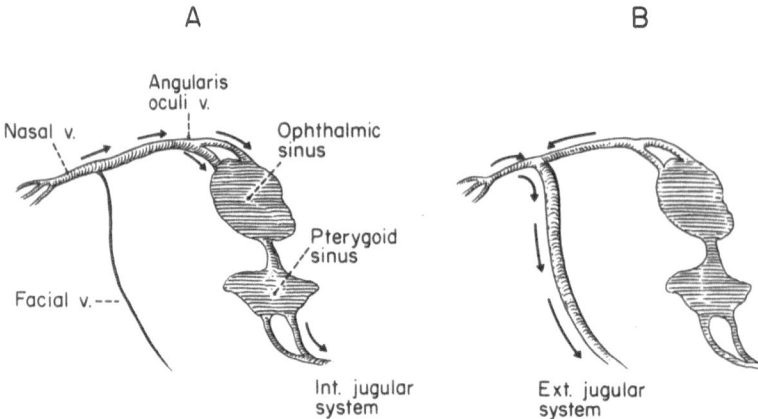

Fig. 2. The influence of the facial vein tone on the direction of venous blood flowing from the nose. A) An increase in the temperature-sensitive tone in the facial vein with a rise in ambient temperature directs more blood cooled in the nose into the central sinuses. This aids in cerebral thermoregulation. B) A fall in ambient temperature inhibits tone in the facial vein both via neurogenic and myogenic mechanisms. This results in drainage of the cooled venous blood into the external jugular system (from ref.[13]). Copyright 1980 by the American Association for the Advancement of Science

quickly relaxes the facial vein. A decrease in ambient temperature causes a concomitant fall in the myogenic response allowing drainage of cooled venous blood into the external jugular system. Thus both local and centrally mediated effects responding to a fall in ambient are complementary in this vessel.

References

1. Bevan, J. A.: Selective action of diltiazem on cerebral vascular smooth muscle in the rabbit: Antagonism of extrinsic but not intrinsic maintained tone. Amer. J. Cardiol. *49* (1982), 519—524.
2. Bevan, J. A., Pegram, B. L., Prehn, J. L., Winquist, R. J.: β-adrenergic receptor-mediated vasodilatation. In: Mechanisms of Vasodilatation (Vanhoutte, P., Leusen, I., eds.), pp. 258—265. Basel: Karger. 1978.
3. Caputa, M., Kadziela, W., Narebski, J., Tycynski, M.: Brain-cranial venous blood heat exchanger and hyperthermia in rabbits. Bull. Acad. Pol. Sci. Ser. Sci. Biol. *25* (1977), 695—698.
4. Caputa, M., Perrin, G., Cabanac, M.: Ecoulement sanguin reversible dans la veine ophtalmique: Mécanisme de refroidissement sélectif du cerveau humain. C.R. Acad. Sci. Paris *287* (1978), 1011—1014.
5. Fleckenstein, A.: Specific pharmacology of calcium in myocardium, cardiac pacemakers and vascular smooth muscle. Ann. Rev. Pharmacol. Toxicol. *17* (1977), 149—166.
6. Haywood, J. N., Baker, M. A.: A comparative study of the role of the cerebral arterial blood in the regulation of brain temperature in five mammals. Brit. Res. *16* (1969), 417—440.

7. Mellander, S., Andersson, P.-O., Afzelius, L.-E., Hellstrand, P.: Neural beta-adrenergic dilatation of the facial vein in man. Acta Physiol. Scand. *114* (1982), 393—399.
8. Pedersen, O. L.: Calcium blockade as a therapeutic principle in arterial hypertension. Acta Pharmacol. Toxicol. *49* (Suppl. 11) (1981), 1—32.
9 Pegram, B. L., Bevan, R. D., Bevan, J. A.: Facial vein of the rabbit. Neurogenic vasodilation mediated by β-adrenergic receptors. Circ. Res. *39* (1976), 854—860.
10. Rowan, R. A., Bevan, R. D., Bevan, J. A.: Ultrastructural features of the innervation and smooth muscle of the rabbit facial vein, and their relationship to function. Circ. Res. *49* (1981), 1140—1151.
11. Vanhoutte, P. M., Janssens, W. J.: Local control of venous function. Microvasc. Res. *16* (1978), 196—214.
12. Winquist, R. J., Baskin, E. B.: Unpublished results.
13. Winquist, R. J., Bevan, J. A.: Temperature sensitivity of tone in the rabbit facial vein: myogenic mechanism for cranial thermoregulation? Science *207* (1980), 1001—1003.
14. Winquist, R. J., Bevan, J. A.: In vitro model of maintained myogenic vascular tone. Blood Vessels *18* (1981 a), 134—138.
15. Winquist, R. J., Bevan, J. A.: Relative location of α- and β-adrenoceptors to sites of release of sympathetic transmitter in the rabbit facial vein. Circ. Res. *49* (1981 b), 486—492.

Author's address: R. J. Winquist, Merck Institute for Therapeutic Research, West Point, PA 19486, U.S.A.

Experimental Research Department and Second Institute of Physiology, Semmelweis University Medical School, Budapest, Hungary

Effect of Topically Administered Epinephrine, Norepinephrine and Acetylcholine on Cerebrocortical Vascular Volume and NAD/NADH Redox State*

E. Dóra

Summary

The brain cortex of chloralose anaesthetized, flaxedil-immobilized, and artificially ventilated cats was superfused with various concentrations of epinephrine (E), norepineph-rine (NE) and acetylcholine (Ach), dissolved in artificial cerebrospinal fluid (mock CSF). The cerebrocortical NADH fluorescence and vascular volume were measured by a microscope fluororeflectometer, while in some of the experiments the brain cortex was photographed simultaneously through a cranial window. Cerebrocortical vascular volume (CVV) was decreased and the NAD/NADH redox state shifted toward NADH oxidation by concentrations of E and NE as low as $3 \cdot 10^{-8}$ M, and the saturation of these responses occurred at $7.5 \cdot 10^{-7}$ M. The pial veins constricted much more markedly with E and NE than arterial vessels. CVV was increased by Ach as low in concentration as $7.3 \cdot 10^{-8}$ M, but the NAD/NADH redox state was not influenced by any concentration of Ach tested. Since the pial vessels were constricted and CVV was diminished by E and NE, and these vascular effects appeared at lower concentrations than is presumed to occur in the synaptic cleft, our results strongly support the regulating role of catecholamines in cerebral vascular resistance (CVR). The NADH oxidation, obtained with catecholamines, is attributed to increased tissue oxygen consumption. The finding that acetylcholine augmented CVV and cortical oxygen supply, but did not shift the redox state toward a more oxidized state, might indicate that the cerebral cortex of normoxic animals is biochemically not hypoxic.

Keywords: Catecholamines; acetylcholine; pial vessels; cerebrocortical vascular volume; NAD/NADH redox state; critical oxygen tension; hypoxia.

Introduction and Methods

Available data are inconsistent as far as to role of the autonomic nervous system and its humoral mediators are concerned in the regulation of

* This study was supported by the Research Council of Hungarian Academy of Sciences (No. 1-07-0304-02-0/K) and by the NIH Program Project Grant NS 10939.

Cerebral Veins   13

cerebral blood flow (CBF) and metabolism. On the other hand, it has not yet been explored whether those cortical microregions whose oxygen tension is close to zero[5] even in normoxic animals are biochemically hypoxic or not. To answer these questions we superfused the brain cortex with various concentrations of E, and NE, and Ach.

All drugs were dissolved in the mock CSF of Wahl and Kuschinsky[10]. The experiments were carried out on chloralose-anaesthetized, immobilized, and artificially ventilated cats. The cerebrocortical vascular volume and NADH fluorescence were measured by a microscope fluororeflectometer as described previously[2]. The exciting light (366 nm) of the fluororeflectometer was focussed on such a cortical field where the diameter of visible pial vessels was less than 50 μm. To assess the changes in the diameter of large pial vessels also (diameters more than 100 μm), in some of the experiments the brain cortex was photographed. The specificity of the catecholamines- and acetylcholine-induced cortical vascular reactions was tested with phenoxybenzamine and atropine, applied intravenously and topically to the surface of the brain cortex, respectively.

Results

Norepinephrine and epinephrine had similar effects on cerebrocortical vascular volume, NAD/NADH redox state, and the diameter of large pial vessels. In the concentration range of $3 \cdot 10^{-8} - 7.5 \cdot 10^{-7}$ M they dose-dependently decreased CVV and shifted the NAD/NADH redox state towards oxidation (Figs. 1 and 2). $7.5 \cdot 10^{-7}$ M NE decreased CVV and NAD reduction equally by about 8%. These are considerable changes if one bears in mind that, in the case of a properly functioning CBF autoregulation, the elevation of arterial blood pressure by approximately 40 mm Hg resulted in approximately 5% and 8% decreases in CVV and NAD reduction, respectively[2].

These effects of norepinephrine are also demonstrated in a typical experimental recording (Fig. 3). As can be seen in the kinetics of NE-induced vascular responses, an initial steep and a subsequently less steep vasoconstriction can be differentiated (increase in R). Corrected NADH fluorescence (CF) showed monophasic NADH oxidation (CF-decreased) during catecholamine superfusion. None of the concentrations of NE-influenced arterial blood pressure (BP) and the ECoG of the superfused brain cortex.

To see how larger pial vessels (diameter more than 100 μm) react to NE, the brain cortex was photographed in 4 experiments. It was found that the large pial veins constricted about twice as much as the arteries in response to the same concentration of NE. Such an example is shown in Fig. 4. The pictures were taken from the same experiment demonstrated in Fig. 3. The circle drawn in the control picture (mock CSF) represents the location and the size of the fluororeflectometrically monitored cortical area where the diameter of visible pial vessels was less than 50 μm. In the control picture a large vein and an artery can be seen whose diameters are approximately 350 μm and 170 μm, respectively. When $7.5 \cdot 10^{-7}$ M NE was applied, the diameters of these vessels decreased to about 170 μm and 130 μm, respectively.

The cholinergic agonist, acetylcholine, increased cerebrocortical vascular volume (reflectance decreased) dose-dependently in the concentration

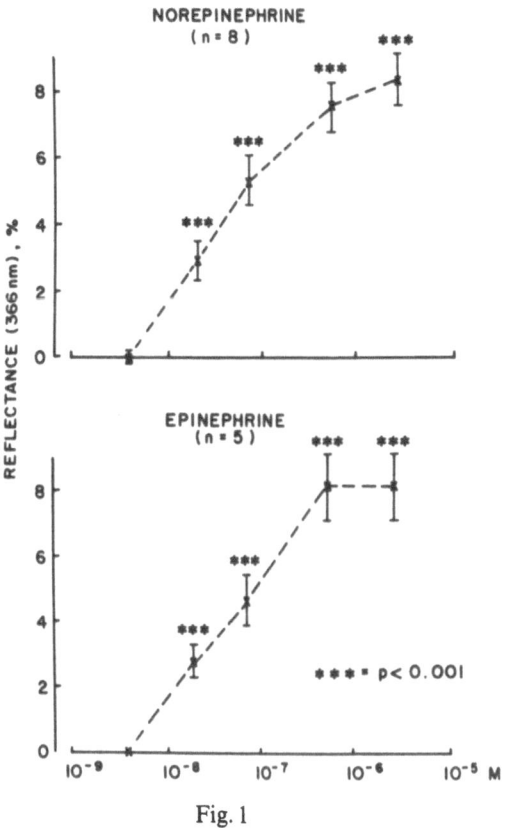

Fig. 1

Figs. 1 and 2. Effect of superfusion of the brain cortex with various concentrations of norepinephrine and epinephrine on cerebrocortical vascular volume (Fig. 1, reflectance) and corrected NADH fluorescence. The results are expressed as mean value ± SE. Asterisks show the significant changes, n the number of experiments evaluated. Note that reflectance is inversely related to changes in vascular volume (CVV). An increase in reflectance and a decrease in the corrected NADH fluorescence imply vasoconstriction and NADH oxidation, respectively

range of between $7.3.3 \cdot 10^{-8}$ M and $7.3 \cdot 10^{-4}$ M (Fig. 5). The NAD/NADH redox state (corrected NADH fluorescence) was not affected by any concentration Ach. $7.3 \cdot 10^{-6}$ M Ach resulted in an approximately 8% increase in vascular volume, which is about $1/3$ of the vascular reaction obtained in another study during epileptic seizures[4].

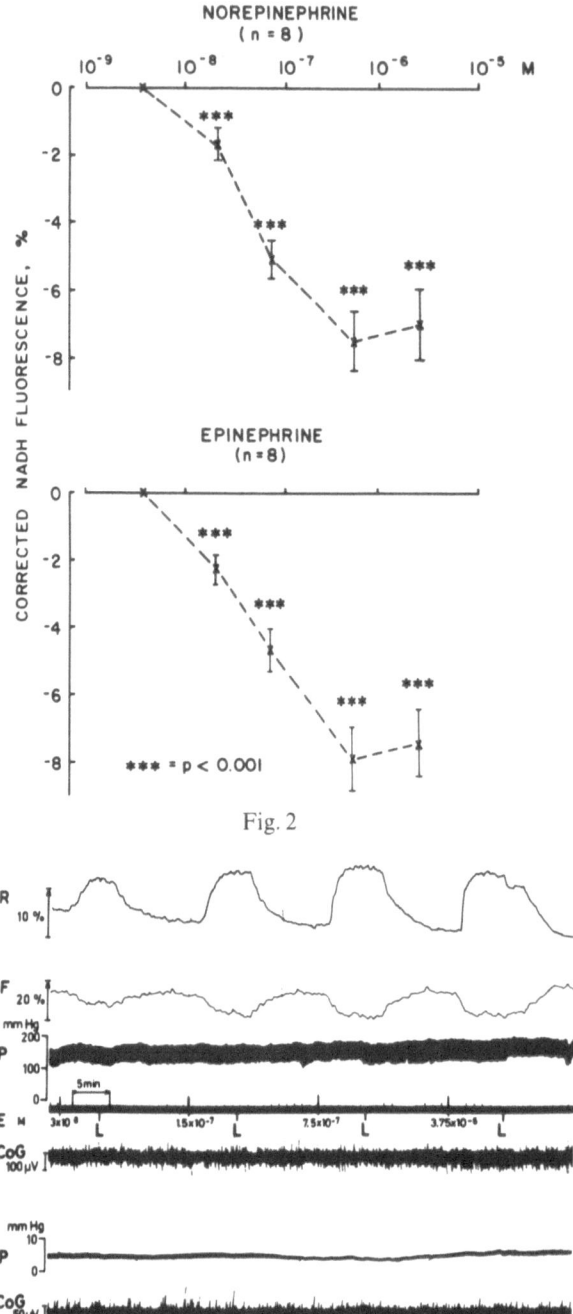

Fig. 2

Fig. 3. Effect of NE superfusion on cerebrocortical vascular volume (*R*) and redox state (*CF*) in a typical experiment. *L* shows the replacement of NE containing CSF with mock CSF. *ICP* intracranial pressure, *ECoG* at the bottom of the Fig. was recorded from the superfused brain cortex. An increase in *R* and a decrease in *CF* imply vasoconstriction and NADH oxidation, respectively

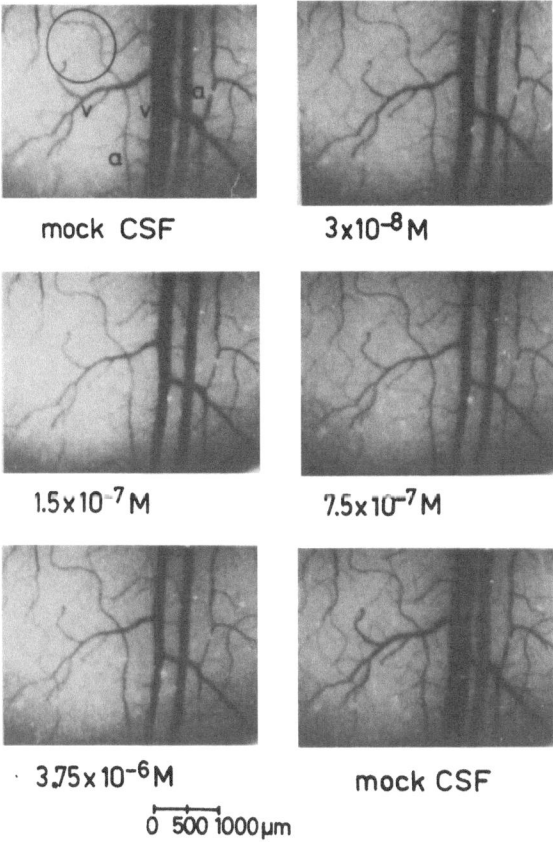

mock CSF 3×10^{-8} M

1.5×10^{-7} M 7.5×10^{-7} M

3.75×10^{-6} M   mock CSF

0   500 1000 µm

Fig. 4. Effect of norepinephrine (*NE*) superfusion on large pial arterial (*a*) and venous (*v*) vessels in a typical experiment

Discussion

   In the present study catecholamines, applied topically to the surface of the brain cortex by superfusion, resulted in a uniform vascular response, *i.e.* arterial and venous pial vessels constricted and the total volume of small intracortical vessels decreased. Our results agree with the data obtained by others[3,8,9] who injected NE by micropipettes into the perivascular space of pial vessels. Concerning the greater sensitivity of large pial veins compared to arteries in response to NE, our results are also consistent with the recent findings of Auer *et al.*[1], Edvinsson *et al.*[2] and Ulrich *et al.*[8]. Assuming that the NE concentration in the extracellular space in close contact with the sympathetic nerve terminals is in the order of 10^{-6} M[6], on the basis of our data and others' from literature[3,8,9], a significant role of the sympathetic nervous system in the regulation of cerebral vascular resistance seems to have been established.

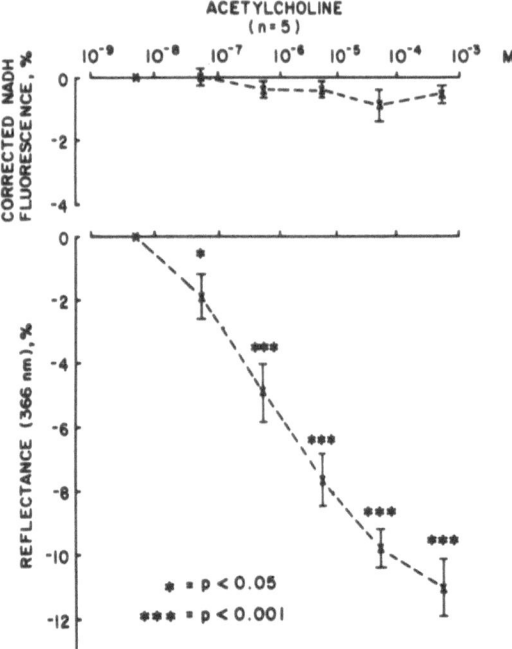

Fig. 5. Effect of acetylcholine superfusion on cerebrocortical NAD/NADH redox state and vascular volume. Abbreviations and marks of the figure are the same as in Figs. 1 and 2

Theoretically, catecholamine-induced NADH oxidation could be attributed either to the relative lack of substrate, or to increased mitochondrial oxygen consumption. Since catecholamines stimulate glucose metabolism, this NADH oxidation is interpreted as being due to increased tissue respiration.

Acetylcholine had the opposite effect on the diameter of pial vessels and the volume of small intracortical vessels as compared with catecholamines. Since cerebral perfusion pressure remained constant during Ach superfusion, CBF and cerebrocortical tissue oxygen supply should have been increased. If cerebrocortical microregions, whose oxygen tension in normoxic animals is close to zero[5], were biochemically hypoxic, as it was suggested by Rosenthal et al.[7], NADH oxidation should have occurred during Ach superfusion. Because the cerebrocortical NAD/NADH redox state was not influenced by any concentration of Ach, it is concluded that the cerebral cortex of normoxic animals is biochemically not hypoxic.

References

1. Auer, L. M., Johansson, B. B., Kuschinsky, W., Edvinsson, L.: Sympatho-adrenergic activity of cat pial veins. J. Cerebr. Blood Flow Metabol. *1*, suppl. 1 (1981), 311—312.
2. Dóra, E., Kovách, A. G. B.: Effect of acute arterial hypo- and hypertension on cerebrocortical NAD/NADH redox state and vascular volume. J. Cerebr. Blood Flow Metabol. *2* (1982), 209—219.

3. Edvinsson, L., McCulloh, J., Uddman, R.: Sympathetic and peptidergic mechanisms in cerebral veins. J. Cerebr. Blood Flow Metabol. *1*, suppl. 1 (1981), 327—328.
4. Kovách, A. G. B., Dóra, E., Szedlacsek, S., Koller, A.: Effect of the organic calcium antagonist D-600 on cerebrocortical vascular and redox responses evoked by adenosine, epilepsy, and anoxia. J. Cerebr. Blood Flow Metabol., *3* (1983), 51—61.
5. Metzger, H., Heuber, S.: Local oxygen tension and spike activity of the cerebral grey matter of the rat and its response to short intervals of O_2 deficiency or CO_2 excess. Pflügers Arch. *370* (1977), 201—209.
6. Owman, C., Edvinsson, L.: Histochemical and pharmacological approach to the investigation of neurotransmitters, with particular regard to the cerebrovascular bed. In: Cerebral Vascular Smooth Muscle and its Control, Ciba Foundation Symposium, Vol. 56, new series (Owman, C., Edvinsson, L., eds.), pp. 275—304. Amsterdam: Elsevier. 1978.
7. Rosenthal, M., LaManna, J. C., Jöbsis, F. F., Lavasseur, J. E., Kontos, H. A., Patterson, J. L.: Effects of respiratory gases on cytochrome a in intact cerebral cortex: Is there a critical pO_2? Brain Res. *108* (1976), 143—154.
8. Ulrich, K., Auer, L. M., Kuschinsky, W.: Cat pial venoconstriction by topical microapplication of norepinephrine. J. Cerebr. Blood Flow Metabol. *2* (1982), 109—111.
9. Wahl, M., Kuschinsky, W., Bosse, O., Olesen, J., Lassen, N. A., Ingvar, D. H., Michaelis, J., Thurau, K.: Effect of l-norepinephrine on the diameter of pial arterioles in the cat. Circ. Res. *31* (1972), 248—256.
10. Wahl, M., Kuschinsky, W.: The dilatatory action of adenosine on pial arteries of cats and its inhibition by theophylline. Pflügers Arch. *362* (1976), 55—59.

Author's address: E. Dóra, M.D., Cerebrovascular Research Center, Department of Neurology, University of Pennsylvania School of Medicine, Johnson Pavilion (G2), 36th & Hamilton Walk, Philadelphia, PA 19104, U.S.A.

Cerebral Blood Flow, Blood Volume
and Intracranial Pressure

Department of Histology, University of Lund, Lund, Sweden

Sympathetic Influence on Intracranial Pressure Through Alterations in Cerebrospinal Fluid Production and Cerebral Blood Volume

CH. OWMAN and M. LINDVALL

Summary

The mammalian choroid plexus receives a well-developed supply of adrenergic sympathetic nerves originating in the superior cervical ganglia, as shown by fluorescence histochemistry. Electron microscopy has demonstrated that the sympathetic nerve terminals directly innervate small arterioles in the plexus, as well as its secretory epithelium. With Pappenheimer's radioactive inulin dilution technique it has been established that the local sympathetics exert a strong inhibitory influence on the rate of CSF production, particularly prominent during hydrocephalus. This action is mediated both by a direct effect on the secretory cells and through alteration in the blood perfusion of the choroid plexus.

Also the pial veins, which hold a considerable portion of the intracranial blood volume, are innervated by sympathetic nerves from the same ganglia. Stimulation induces reductions, and ganglionectomy causes acute elevations, in the cerebral blood volume.

The sympathetic control of these two portions of the cerebrovascular bed—the choroid plexuses and the capacitance vessels—is of particular interest in view of the fact that intracranial pressure is primarily regulated through the CSF dynamics and cerebral blood volume.

Intraventricular pressure has been measured chronically in non-anaesthetized rabbits following various types of surgical intereference with the sympathetics in the neck. In agreement with the above-mentioned observations that bilateral postganglionic sympathectomy increases CSF formation and cerebral blood volume to approximately equal degrees, the denervation also leads to a marked elevation of intracranial pressure. If the CSF outflow pathways are blocked by previous cisternal kaolin administration, sympathectomy is no longer compatible with life, whereas non-denervated animals survive through the recruitment of sufficient compensatory mechanisms. The intracranial sympathetic system thus appears to be an important control mechanism in the intracranial pressure dynamics.

Keywords: CSF production; cerebral blood volume; sympathetic influence.

Introduction

Since the brain is enclosed within the rigid compartment of the skull, the intracranial pressure is primarily regulated through the cerebrospinal fluid

(CSF) dynamics and the cerebral blood volume. Most of this volume, probably some 70%, resides on the venous side, in the capacitance vessels. The entire circulatory system of the brain receives a well-developed sympathetic innervation, which is able to influence the calibre of both arteries and veins[10,20], and thereby the cerebral blood volume[3].

The choroid plexus can be considered to be a very specialized part of the cerebrovascular bed, equipped with a secretory epithelium and responsible for the major portion of the CSF production. The fluid is formed by an active mechanism at a rate that varies considerably from one species to another, but which is fairly constant when expressed as a fraction of the total CSF volume or on the basis of plexus weight. Also the choroid plexus receives an extensive sympathetic nerve supply, which has been found to inhibit CSF production[17].

The cerebrovascular sympathetic innervation originates almost exclusively from the superior cervical ganglia[6], which are easily accessible for stimulation and denervation experiments. On the basis of this knowledge, studies have been carried out to show that the cranial sympathetic nerves markedly influence intracranial pressure[8], probably through their effects on CSF production and cerebral blood volume. The main results from these investigations will be described and discussed below.

Adrenergic Innervation of the Choroid Plexus

The production of CSF is an energy-requiring process involving important functions of carbonic anhydrase and an ouabain-sensitive sodium-potassium activated ATPase. Approximately half a per cent of the total CSF volume is replaced by newly formed fluid every minute. Although the cellular mechanisms underlying the formation of CSF are now fairly well understood, our knowledge of the various control functions regulating the CSF circulation is surprisingly poor, particularly in view of the serious clinical conditions that may result from an impaired circulation of CSF. A wide variety of drugs and experimental conditions have been shown to influence the normal rate of formation. Nevertheless, pharmacological treatment of hydrocephalus has, in comparison with surgical drainage of CSF, been of little success.

Investigations on the possible involvement of local neurogenic mechanisms in the control of CSF formation from the choroid plexus have to a large extent been hampered by the widely held opinion that the choroid plexus is a non-innervated structure. This is primarily because the early morphological investigations were often difficult to interpret, since the histological staining methods used did not visualize nerve fibres with a sufficiently high degree of selectivity. It was not until the introduction of modern neurohistochemical techniques that it became possible to prove that the choroid plexus is indeed supplied with nerves and to define their identity, origin, and distribution in detail. A rational basis now exists for applying biochemical, pharmacological and physiological approaches in attempts to elucidate the role of neural mechanisms in the regulation of CSF formation.

The Falck-Hillarp histofluorence technique has been applied to the study of the plexus nerve supply in a number of animal species[5,13]. There are considerable differences both with respect to the nerve supply of the plexus in the individual ventricles and to the total plexus innervation in different species. The general pattern of distribution of the adrenergic fibres is, however, quite similar in all species: the nerve terminals form networks around small vessels and in the plexus tufts, with isolated fibres and nerve terminals running in close relation to the epithelium. The plexuses of the pig and cat are most well-supplied with adrenergic nerves; the rabbit, monkey, guinea pig, rat and hamster show an intermediary density; whilst the cow and mouse have the fewest nerve fibres.

The histochemical findings have been correlated with bioenzymatic measurements of the noradrenaline content in the plexus tissue[13]. All examined species showing the presence of adrenergic nerve terminals in their plexuses also contain measurable amounts of noradrenaline, the transmitter level varying between 0.10 and 0.73 ng/mg tissue protein. Quantitative determinations have been made of the amine content in the choroid plexuses of the different ventricles in the rabbit[16]. In this animal the plexus of the third ventricle contains 1.46 ng/mg protein, that of the lateral ventricles 0.27 ng/mg, and the plexus of the fourth ventricle 0.15 ng/mg. The values are in good agreement with the fluorescence histochemical findings, showing the innervation density to be in the order third ventricle plexus > lateral plexuses > fourth ventricle plexus.

In histofluorescence studies it has been established that bilateral excision of the superior cervical ganglion abolishes the adrenergic nerves in all plexuses, as shown in the cat and rat[5].

The Sympathetic Neuro-Effector System in the Choroid Plexus

Adrenergic nerve terminals are characterized by the presence of dense-cored synaptic vesicles, though it is not always possible to distinguish them from non-adrenergic terminals on this basis by fixation with glutaraldehyde-osmium tetroxide[11]. In an electron microscopic study of the lateral plexuses from cat and rabbit[2], the adrenergic nature of the nerve terminals has been confirmed by the appearance of electron-dense synaptic vesicles after the injection of the "false adrenergic transmitter", 5-hydroxydopamine, which is selectively taken up and accumulated in the synaptic vesicles of adrenergic nerves[24].

The choroid plexus contains an approximately equal number of terminals equipped with electron-dense and electron-lucent synaptic vesicles, all 30–60 nm in diameter[2]. The absence of 5-hydroxydopamine accumulation in the latter type of terminals indicates that they are non-adrenergic, probably cholinergic[14].

Both types of nerves come as close as 20 nm to the membrane of the epithelial cells, where they are located either contiguous to the base of the cell or cellular processes, between adjacent cells, or within a cellular invagination[2]. Similar close relationships have been described in other secretory tissues[23]. The distance between the nerve endings and the smooth muscle cells of the arterioles in the plexus tissue is less than 100 nm, which is the distance found in vascular beds known to receive a functioning autonomic innervation[1].

Thus, the adrenergic axons present in the choroid plexus come close enough to the smooth muscle cells of small arterioles as well as the secretory epithelium to suggest that both components in the plexus receive a true, functioning innervation.

Sympathetic Influence on CSF Production

It is evident from the previous description that prerequisites exist for a sympathetic nervous influence both on the secretory epithelium and on the blood flow through the plexus tissue, with a consequent net effect on the rate of CSF formation. In order to obtain a direct measure of the sympathetically mediated action on the bulk CSF formation, the ventriculocisternal perfusion technique of Pappenheimer *et al.*[22] has been applied to rabbits, utilizing ^{14}C-labelled inulin[9].

In experiments on rabbits the production rate of CSF has been determined during electrical stimulation of the sympathetic trunks in the neck[15]. Stimulation during 1–2 hours reduces the rate of CSF formation by a mean of 32% compared to the control situtation before stimulation. After cessation of stimulation, the production rate returns approximately halfway towards normal during a 1-hour observation period. Bilateral excision of the superior cervical ganglia[15], which produces extensive sympathetic denervation of the choroid plexuses, results a week later in a 33% increase in bulk CSF production as compared to an unoperated control group. The results are in agreement with the finding of enhanced carbonic anhydrase activity of the plexus following sympathectomy[2].

In view of our observation that the sympathetic nerve terminals in the choroid plexus have a close relation both to the secretory epithelium and to the smooth muscle cells of the arterioles, indicating a true innervation of both components[2] the effects of sympathetic nerve stimulation or denervation could have resulted from a direct action on the secretory cells, changes in plexus blood flow through a vascular response, or a combination of both. Against the background of the results presented above, there is reason to believe that the sympathetic action on the rate of CSF production is a combined secretory and vascular effect in the choroid plexus; the former predominating.

Sympathetic Influence on CSF Production During Hydrocephalus

Experimental obstructive hydrocephalus can be induced by the intracisternal injection of hydrated aluminium silicate (kaolin). This chemically inert substance causes severe inflammatory changes in the arachnoid membrane, subarachnoid fibrosis, and blockade of the outlets from the fourth ventricle when injected into the cisterna magna. The intracranial pressure increases secondary to the hydrocephalic process leading to dilation of the cerebral ventricles. We have studied the effects of hydrocephalus on the sympathetic modulation of CSF formation in preliminary experiments using rabbits injected with 0.5 ml of a sterilized 30% kaolin suspension into the cisterna magna.

After 3 days the cisternal kaolin injection results in about a tenfold elevation of the ventricular fluid pressure measured by a cannula placed in one of the lateral ventricles. Approximately 10% of the animals do, in fact, not survive this degree of intracranial hypertension. The pressure tends to normalize within a week, but it is still sufficiently increased to maintain significant dilation of the ventricles[21], measured as the septum-to-caudate distance in coronal sections of the brain. Even though the intracranial pressure was thus not very much elevated under this condition of low-pressre hydrocephalus (the mean level was 1.5 mm Hg), it was sufficient to cause marked reductions in brain catecholamine levels[21].

During the acute stage of hydrocephalus, the CSF production rate was found to be reduced to about one fourth of normal. In this series of experiments the production was measured by ventricle-to-ventricle perfusion of ^{14}C-inulin, kaolin being deposited into the cisterna magna where the CSF outflow has usually been collected in the measurements on untreated rabbits as previously described.

Bilateral stimulation of the sympathetic trunk in the neck in the course of 1 hour produced only an 18% reduction (compared with 32% in untreated animals), in the bulk formation of CSF, and there was only a slight return in the production rate during the 1-hour period following cessation of the nerve stimulation. It seems that the sympathetic nerves of these animals are already markedly activated, in order to inhibit the CSF production as part of an acute defense mechanism to cope with the blockade of the outflow pathways. In earlier experiments we have shown[7] that the cranial sympathetic nervous system is of critical importance during the acute, kaolin-induced hydrocephalus in the sense that sympathetic denervation markedly and rapidly aggravates the intracranial hypertension to a level that is no longer compatible with life.

After 3 weeks following kaolin-induced hydrocephalus, which is now combined with a near-normal level of intracranial pressure, the CSF production rate is still reduced by approximately 75% as compared with untreated controls. However, sympathetic nerve stimulation at this stage causes a decrement in the production rate that is at least of the same magnitude as that found in the untreated animals. Moreover, the rate normalizes by as much as 70% during the ensuing period of 1 hour following the stimulation. This agrees with the establishment of more long-term types of compensatory mechanisms involving not only reduced CSF formation from the impaired choroid plexuses[19], but also an increased CSF absorption via normal pathways which are not involved in the kaolin blockade, as well as the formation of alternate pathways for its outflow. It should be noted that, in the pathological conditions of longstanding hydrocephalus, any increase in the otherwise minimal entry of radioactive inulin from the perfused ventricles into the brain in the Pappenheimer technique will tend to overestimate the production rate compared with the control animals not having hydrocephalus. In the actual stimulation experiment, on the other hand, the animals serve as their own control.

Sympathetic Influence on Cerebral Blood Volume and Intracranial Pressure

Approximately 70% of the blood volume resides in the capacitance vessels, *i.e.* in the venous section of the vascular bed. An intricate control of cerebral blood volume is necessary in view of the special situation existing for the brain, it being rigidly enclosed in the skull. In fact, cerebral blood volume and the CSF circulation constitute the primary factors involved in the regulation of intracranial pressure. Anatomical and functional requirements have been shown to exist for a sympathetic neurogenic control not only of the choroid plexus responsible for the major part of the CSF formation, but also of the cerebrovascular bed. Thus, the pial arteries and arterioles, intracerebral arterioles, as well as the venous system of the brain receive sympatheic fibres demonstrable by fluorescence histochemistry, and the nerve terminals fulfill ultrastructural and pharmacological criteria for functional innervation of the vascular smooth muscles[20]. Also these sympathetic nerves originate in the superior cervical ganglia.

As a functional approach to the possible influence of the sympathetic nerves on intracranial pressure, changes in ventricular fluid pressure have been correlated with changes in cerebral blood volume at various stages after pre- or postganglionic sympathetic denervation. The ventricular fluid pressure was measured in conscious rabbits via a small cannula implanted into the left ventricle of the brain. The CSF pressure values given below represent the mean ventricular fluid pressure during a 10 hour period, starting 5 hours after implantation of the cannula. Cerebral blood volume was determined in mice by the dilution of an intravenously injected non-diffusible tracer, ^{131}I-labelled human serum albumin. Preganglionic denervation was performed by transection of the cervical sympathetic trunks below the superior cervical ganglia, and postganglionic denervation was produced by excising these ganglia. In a separate study, noradrenaline was measured chemically in the pial vascular system of rabbits at various stages after the two types of denervations[6].

After postganglionic sympathectomy, noradrenaline rapidly leaks from the degenerating nerves, and its concentration falls to values indistinguishable from zero in sympathetically innervated structures, such as the brain vascular system. This leakage of noradrenaline is accompanied by stimulation of the adrenergic receptors[18], and the stimulation ceases only when the degenerating nerves have lost their transmitter. However, since the sympathetic nerves constitute an important mechanism for inactivation of noradrenaline in the region of the receptors, the loss of the nerves means that the denervated receptors are exposed to higher local concentrations of noradrenaline. The activation of the receptors by circulating noradrenaline becomes pronounced when the so-called denervation supersensitivity[12] is established.

Within one day after postganglionic sympathectomy, during the noradrenaline leakage, the ventricular fluid pressure becomes significantly reduced below control levels whereas later on, after 3–6 days when the degenerating sympathetic nerves have lost their transmitter, the ventricular fluid pressure is higher than in the controls. As soon as denervation supersensitivity of the adrenergic receptors to circulating noradrenaline has been established, the ventricular fluid pressure returns to levels similar to those of the controls at about 2 weeks after the operation[8]. The patterns is similar, but the changes become exaggerated, if the CSF outflow pathways have been blocked by cisternal kaolin 2 days prior to the ventricular fluid pressure measurement[8]. The pressure also becomes elevated if the intracranial adrenergic nerves have been depleted of their transmitter by the intraventricular injection of reserpine[4].

During preganglionic sympathetic denervation the postganglionic adrenergic neuron remains intact, but its activity is reduced. Accordingly, noradrenaline shows a tendency to accumulate in the postganglionic neurons[6]. During this period of reduced activity, the ventricular fluid pressure is seen to be significantly increased, as can be observed both in normal and kaolin-treated animals[8].

The patterns of changes in ventricular fluid pressure after post- or preganglionic sympathetic denervation[8] and the sequence of corresponding changes in cerebral blood volume[3] closely resemble one another. The measurements of the cerebral blood volume were, however, performed in smaller animals (mice), in which the time scale after the denervations is shorter.

Concluding Remarks

The cerebrovascular system is extensively supplied with sympathetic nerves whose stimulation leads the vasoconstriction[20]. The sympathetic vasoconstrictor fibres of the brain also supply the capacitance vessels, which contain approximately 70% of the blood volume. We have shown that the cranial sympathetic nerves exert a pronounced influence on cerebral blood volume[3].

The choroid plexus can be considered as a very specialized region of the cerebrovascular bed, responsible for the major part of the CSF production. Also this structure, both its vasculature and its secretory epithelium, receives a well developed sympathetic innervation[17]. The sympathetic innervation of all these regions of the brain circulation originates almost exclusively from the superior cervical ganglia. The sympathetic fibres have an efficient inhibitory role in the CSF formation, through their action both on the blood vessels and directly on the secretory epithelium of the choroid plexus. This sympathetic inhibition can also be demonstrated under hydrocephalic conditions, particularly in low-pressure hydrocephalus.

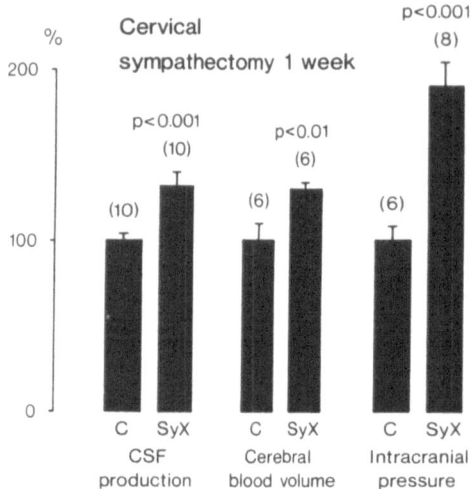

Fig. 1. Summary of data from cranial sympathetic denervation which leads to almost equal degrees of increase in CSF formation and cerebral blood volume, both important factors in the regulation of intracranial pressure, which increases markedly following the sympathectomy

The intracranial pressure can be monitored over several hours in conscious rabbits by recording from a cannula implanted in one of the lateral ventricles. Postganglionic sympathetic denervation tends to produce an elevation of the intracranial pressure, an effect that becomes exaggerated to a lethal degree if the CSF outflow is blocked by intracisternal administration of kaolin[8].

The cerebral blood volume and the CSF circulation constitute the two major mechanisms responsible for the regulation of intracranial pressure. The present series of experiments strongly suggest that the cranial sympathetic nerves represent an important protective system in the control of intracranial pressure, through their effects on these two regulatory mechanisms. This is particularly well demonstrable in experiments utilizing postganglionic denervation, as illustrated in Fig. 1, leading to an approximately equal increase in cerebral blood volume volume and rate of CSF formation, both resulting in a pronounced elevation of intracranial pressure.

Acknowledgements

Supported by grants from the Swedish Medical Research Council (No. 04X-732).

References

1. Burnstock, G.: Structure of smooth muscle and its innervation. In: Smooth Muscle (Bülbring, E., et al., eds.), pp. 1—69. London: E. Arnold Ltd. 1970.
2. Edvinsson, L., Håkanson, R., Lindvall, M., Owman, Ch., Svensson, K.-G.: Ultrastructural and biochemical evidence for a sympathetic neural influence on the choroid plexus. Exp. Neurol. 48 (1975a), 241—251.

3. Edvinsson, L., Nielsen, K. C., Owman, Ch., West, K. A.: Sympathetic adrenergic influence on brain vessels as studied by changes in cerebral blood volume of mice. Europ. Neurol. *6* (1971/72), 193—202.

4. Edvinsson, L., Nielsen, K. C., Owman, Ch., West, K. A.: Intracranial pressure in conscious rabbits after intraventricular reserpine. J. Neurosurg. *40* (1974), 743—746.

5. Edvinsson, L., Nielsen, K. C., Owman, Ch., West, K. A.: Adrenergic innervation of the mammalian choroid plexus. Amer. J. Anat. *139* (1974a), 299—308.

6. Edvinsson, L., Owman, Ch., Rosengren, E., West, K. A.: Concentration of noradrenaline in pial vessels, choroid plexus, and iris during two weeks after sympathetic ganglionectomy or decentralization. Acta Physiol. Scand. *85* (1972), 201—206.

7. Edvinsson, L., Owman, Ch., West, K. A.: Modification of kaolin-induced intracranial hypertension at various time-periods after superior cervical sympathectomy in rabbits. Acta Physiol. Scand. *83* (1971), 51—59.

8. Edvinsson, L., Owman, Ch., West, K. A.: Influence of sympathetic denervation on intracranial pressure. In: Intracranial Pressure II (Lundberg, N., *et al.*, eds.), pp. 453—459. Berlin-Heidelberg-New York: Springer. 1975 b.

9. Heisey, S. R., Held, D., Pappenheimer, J. R.: Bulk flow and diffusion in the cerebrospinal fluid system of the goat. Amer. J. Physiol. *203* (1962), 775—781.

10. Heistad, D. D., Marcus, M. L.: Cerebral blood flow. Effect of nerves and neurotransmitters. Developments in Neuroscience, Vol. 14. New York-Amsterdam-Oxford: Elsevier/North Holland. 1982.

11. Hökfelt, T.: Electronmicroscopic studies on peripheral and central monoamine neurons. Stockholm: Ivar Haeggström. 1968.

12. Langer, S. Z., Draskoczy, P. R., Trendelenburg, U.: Time course of the development of supersensitivity to various amines in the nictitating membrane of the pithed cat after denervation or decentralization. J. Pharmacol. Exp. Ther. *157* (1967), 255—273.

13. Lindvall, M.: Fluorescence histochemical study on regional differences in the sympathetic nerve supply of the choroid plexus from various laboratory animals. Cell Tiss. Res. *198* (1979), 261—267.

14. Lindvall, M., Edvinsson, L., Owman, Ch.: Histochemical study on regional differences in the cholinergic nerve supply of the choroid plexus from various laboratory animals. Exp. Neurol. *55* (1977), 152—159.

15. Lindvall, M., Edvinsson, L., Owman, Ch.: Sympathetic nervous control of cerebrospinal fluid production from the choroid plexus. Science *201* (1978), 176—178.

16. Lindvall, M., Owman, Ch.: Early development of noradrenaline-containing sympathetic nerves in the choroid plexus system of the rabbit. Cell Tiss. Res. *192* (1978), 195—203.

17. Lindvall, M., Owman, Ch.: Autonomic nerves in the mammalian choroid plexus and their influence on the formation of cerebrospinal fluid. J. Cerebr. Blood Flow Metabol. *1* (1981), 245—266.

18. Lundberg, D.: Some aspects of the pharmacology of the degeneration contraction of rat periorbital smooth muscle after sympathetic denervation. Acta Univ. Uppsal. *79* (1970), 1—10.

19. Martin, A. E., Wald, A., Hochwald, G. M., Malhan, C.: Kaolin-induced hydrocephalus impairs CSF secretion by the choroid plexus. Neurology *28* (1978), 945—949.

20. Owman, Ch., Edvinsson, L.: Neurogenic control of the brain circulation. Proceedings of the International Symposium held at the Wenner-Gren Center, Stockholm June 22—24, 1977. Oxford: Pergamon Press.

21. Owman, Ch., Edvinsson, L., Rosengren, E., Svendgaard, N., West, K. A.: Effects of intracranial hypertension, low-pressure hydrocephalus and subsequent ventriculo-peritoneal shunting on monoamine neurons in rabbit brain. In: Intracranial Pressure II (Lundberg, N., *et al.*, eds.), pp. 183—188. Berlin-Heidelberg-New York: Springer. 1975.

22. Pappenheimer, J. R., Heisey, S. R., Jordan, E. F., Downer, J.: Perfusion of the cerebral ventricular system in unanesthetized goats. Amer. J. Physiol. *203* (1962), 763—774.
23. Scott, B. L., Pease, D. C.: Electron microscopy of the salivary and lacrimal glands of the rat. Amer. J. Anat. *104* (1959), 115—161.
24. Tranzer, J. P., Thoenen, H.: Electronmicroscopic localization of 5-hydroxydopamine (3,4,5-trihydroxy-phenyl-ethylamine), a new "false" sympathetic transmitter. Experientia *23* (1967), 743—745.

Authors' address: Prof. Ch. Owman, Ph.D., M.D., Department of Histology, University of Lund, S-22362 Lund, Sweden.

[1] Departments of Surgery (Neurosurgery), Crittenton Hospital, Rochester, Michigan, U.S.A., [2] Department of Physics, Oakland University, Rochester, Michigan, U.S.A.

CSF and Venous Pulse Waves; a Look at Myogenic Autoregulation

H. D. Portnoy[2], M. Chopp[1], C. Branch[1], and M. Shannon[1]

Summary

Arterial CSF pulses were recorded in dogs during induced systemic hypertension and hypotension, hypercapnia and alterations in CSF pressure. Fourier analysis was performed on the pulses and transfer functions obtained. The transfer functions indicated that two types of transmission of the pulse wave occurred: linear and non-linear.

To explain the data, a model for control of cerebral blood flow through precapillary resistance vessels, particularly the arterioles, is presented. The resistance vessels are modelled as a Starling resistor in which the transmural pressure of the vessels is balanced against the vasomotor tone (defined as the active tension of the smooth muscle of the wall at the unstretched radius of the muscle). Non-linear transmission, as demonstrated by preferential attenuation of the fundamental frequency of the pulse, occurs when the transmural pressure is balanced by the vasomotor tone regardless of the change in arterial pressure, *i.e.,* when myogenic autoregulation is intact. Linear transmission, as shown by an approximately uniform transfer of all the harmonics of the pulse, occurs when the transmural pressure exceeds vasomotor tone, such as during marked arterial hypertension or when the arteriolar smooth muscle is inhibited, *i.e.,* when myogenic autoregulation is impaired.

Keywords: Cerebral vasomotor tone; CSF pulse waves; myogenic autoregulation; venous pulse waves.

Introduction

The cerebrovascular bed can be modelled as two Starling resistors in series which represent, first, the cerebral resistance vessels and, second, the cerebral veins and parasagittal intradural venous channels (lateral lacunae). A Starling resistor is a collapsible tube encased in a fluid filled rigid container; and, in addition, the pressure on the outside of the tube, the external pressure, must be greater than the pressure at the outlet of the tube, the outflow pressure. Under these circumstances, it can be shown that as long as flow is maintained, the pressure at the inlet of the tube, the inflow pressure, will be maintained nearly equal to the external pressure[2]. Over a wide range of flows, the cerebral veins and lateral lacunae meet the criteria of a Starling resistor, namely, the CSF pressure is greater than the sagittal sinus pressure, and the venous pressure approximates the CSF pressure[3, 9]. Of more significance, is that the resistance vessels, particularly the arterioles, can also be demonstrated to be Starling resistors and, as such, give rise to myogenic autoregulation[5, 6]. This paper demonstrates that pulse waves traversing the resistance vessels are modified in a specific way when myogenic autoregulation is intact and the pulse waves modified in another way when myogenic autoregulation is impaired[6]. These pulse waves transmitted into the veins and then into the CSF can be recorded and utilized to determine the ongoing state of myogenic autoregulation.

Materials and Methods

Fifty-three dogs divided into eight groups were studied. The dogs were anaesthetized with methohexital and pancuronium bromide and mechanically ventilated with 25% O_2 in 75% N_2. Lidocaine 1% was infiltrated at all the operative sites and the animals placed in a head holder in the sphinx position. The right lateral cerebral ventricle, right lingual artery and sagittal sinus were cannulated. The CSF (Pcsf), systemic arterial (Ps) and sagittal sinus (Pss) pressures were recorded on FM tape and an optical recorder and a train of ten individual pulse waves from each of the pressures analyzed, using a single pulse Fast Fourier Transform computer programme developed in this laboratory. The power amplitude and the amplitude transfer function spectra were derived between the fundamental frequency and 15 Hz. The transfer function spectra (XFRa) were calculated between the Pcsf and Ps (XFRa—Pcsf/Ps) and between Pss and Pcsf, though only the former will be considered in this report.

Group 1: In 13 animals, the response to hypercapnia and intracranial hypertension was evaluated. This data has been previously published[6].

Group 2: Systemic arterial hypertension was induced by the infusion of norepinephrine ($0.1–3.0 \, \text{mcg} \cdot \text{kg}^{-1} \cdot \text{min}^{-1}$) in five animals and dopamine ($5–90 \, \text{mcg} \cdot \text{kg}^{-1} \cdot \text{min}^{-1}$) in five animals.

Group 3: Systemic arterial hypotension was induced by the infusion of sodium nitroprusside or nitroglycerin ($1–24 \, \text{mcg} \cdot \text{kg}^{-1} \cdot \text{min}^{-1}$) in five animals, each.

Group 4: The same as group 3, except that the methohexital was discontinued one hour prior to the infusion of the hypotensive agent.

Group 5: In five animals, progressive hypovolemic hypotension was produced by sequential removal of $2 \, \text{ml} \cdot \text{kg}^{-1}$ aliquots of blood.

Results

Two types of transmission of the pulse waves were found: linear and non-linear. Linear transmission was recognized by an XFRa spectrum which was flat, *i.e.,* all harmonics of the wave were transferred to approximately the same degree. Non-linear transmission was identified by an XFRa spectrum which demonstrated preferential attenuation of the lower frequencies, particularly the fundamental frequency[6].

Control condition (all groups): The XFRa—Pcsf/Ps demonstrated non-linear transmission.

Group 1: See reference [6].

Group 2: With a progressive increase in Ps, the XFRa spectra were similar to the control until approximately 150 mm Hg above which the spectra became more linear (Fig. 3).

Group 3: With increasing infusion rate, both vasodilators produced systemic hypotension. There was minimal change in the XFRa spectra as compared with the control.

Group 4: There was less decrease in Ps with the same rates of infusion of the vasodilators than seen in group 3. As the infusion rate of sodium nitroprusside reached $8\,\text{mcg}\cdot\text{kg}^{-1}\cdot\text{min}^{-1}$, the XFRa spectra began to become more linear.

Group 5: As the Ps reached 40–60 mm Hg, the XRFa spectra started to change from non-linear to linear (Fig. 4).

Discussion

The Arteriole as a Starling Resistor

Arteriolar smooth muscle tension can be divided into active tension developed by active constriction of the smooth muscle fibres and elastic tension developed by stretching of the fibres. The arterioles can be pictured as a Starling resistor in which the inflow pressure is balanced by the active tension of the smooth muscle at the unstretched radius of the fibres, plus the pressure of the fluid surrounding the vessel[5, 6] (Fig. 1). We have defined the active tension at the unstretched radius of the fibres as "vasomotor tone" since it encompasses both the concept of smooth muscle tension and specific radius. From Laplace's law (Fig. 1), vasomotor tone, as defined above, represents a pressure in direct opposition to the arteriolar transmural pressure. Since transmural pressure equals internal pressure minus external pressure, in this arteriolar Starling resistor, transmural pressure is balanced against vasomotor tone. If transmural pressure exceeds vasomotor tone, then elastic tension is developed and the system does not operate as a Starling resistor since the vessel is not collapsible. If vasomotor tone exceeds transmural pressure, the vessel closes. Only when the transmural pressure is balanced by vasomotor tone does the system work as a Starling resistor.

Myogenic Autoregulation

It has been demonstrated in a Starling resistor that if the inflow pressure is coupled to the external pressure, such that the inflow pressure determines the external pressure, flow through the tube remains constant if there is no change in upstream or downstream resistances[8]. If vasomotor tone exactly matches transmural pressure, then flow through the arteriole will remain constant with changes in transmural pressure, that is, with variations in systemic arterial pressure[6, 7]. The mechanism is best understood by visualizing a cross-section of the arteriole. The transmural pressure acts to distend the vessel whilst the smooth muscle fibres act to close the vessel.

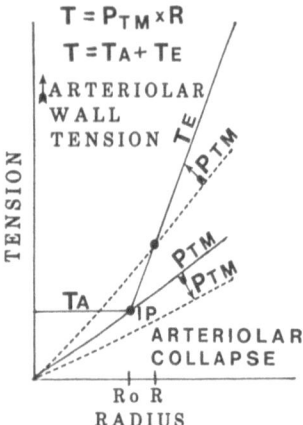

Fig. 1 A. According to Laplace's law, total tension (T) is equal to the transmural pressure (P_{TM}) multiplied by the radius. T is composed of active tension (T_A) due to the contraction of the smooth muscle fibres and elastic tension (T_E) due to stretching of the fibres. The unstretched radius of the vessel is R_0; the stretched radius, R. Vasomotor tone, defined as T_A/R_0, is a pressure in opposition to P_{TM} when the tension produced is derived from muscle tone alone. The point of balance between P_{TM} and VMT is the instability point (IP) since, above this point the vessel is open and the tension rapidly increases, and below this point there is arteriolar collapse

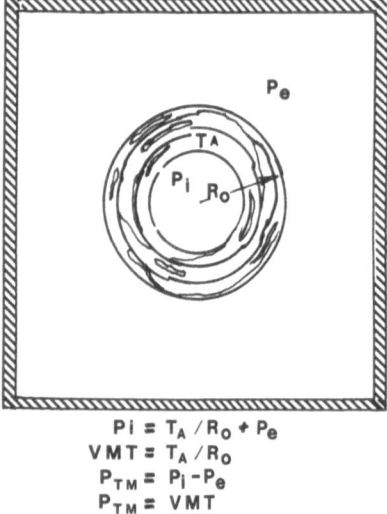

$$P_i = T_A/R_0 + P_e$$
$$VMT = T_A/R_0$$
$$P_{TM} = P_i - P_e$$
$$P_{TM} = VMT$$

Fig. 1 B. The arteriole acts as a Starling resistor only when the inflow or internal pressure (P_i) is balanced against the pressure external to the vessel (P_e) and the pressure represented by T_A/R_0 or VMT. Since $P_i - P_e$ equals P_{TM}, P_{TM} is balanced only by vasomotor tone

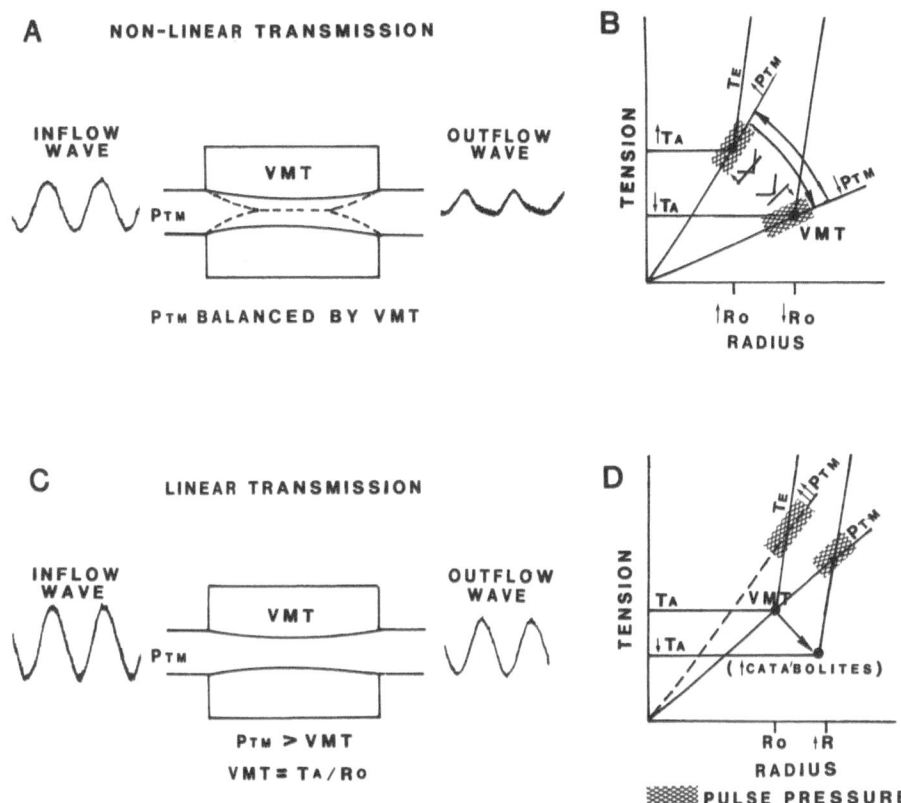

Fig. 2. *Intact myogenic autoregulation:* A. Starling resistor analog. Non-linear pulse wave transmission when systolic $P_{TM} > VMT >$ diastolic P_{TM}. The tube opens and closes during each pulse which results in a rectified outflow wave. Illustrated sine waves (2 Hz) were recorded using a Starling resistor under the described conditions. B. Laplacian analog. Within the range of myogenic autoregulation, any alteration in P_{TM} will be balanced by VMT at the instability point (*IP*). If P_{TM} rises, P_{TM} transiently exceeds VMT which signals an increase in VMT and restoration of regulation. If P_{TM} falls, VMT transiently exceeds P_{TM} which signals for a decrease in VMT until regulation is restored. A pulsatile P_{TM} (cross-hatched area) causes the arterioles to open and close, resulting in non-linear pulse transmission. *Impaired myogenic autoregulation:* C. Starling resistor analog. When the pressure external the tube (*VMT*) is less than the minimum pressure of the inflow wave (diastolic P_{TM}), the tube never closes and there is linear transmission of the pulse. D. Laplacian analog. If the diastolic P_{TM} is greater than VMT, the vessel never closes and there is linear transmission of the pulsatile P_{TM}. This will occur when the systemic arterial pressure is very high and the resultant P_{TM} ($\uparrow\uparrow P_{TM}$) cannot be matched by VMT, or when the smooth muscle is inhibited so that VMT cannot balance the existing P_{TM}

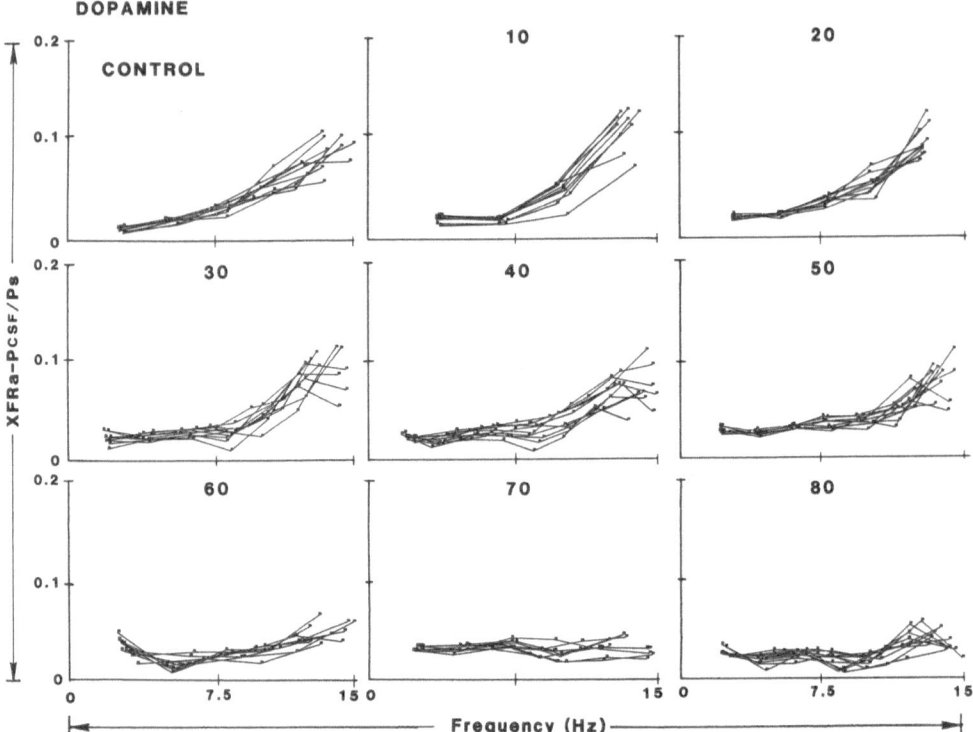

Fig. 3. Changes in the transfer function spectra between CSF and systemic arterial pressure (XFRa—Pcsf/Ps) with progressive increase in the infusion rate of dopamine. The number above each transfer function spectrum represents the rate in $mcg \cdot kg^{-1} \cdot min^{-1}$. The control demonstrates non-linear transmission with suppression of the fundamental frequency. With increasing systemic hypertension, the fundamental frequency increases until the spectra become relatively flat, indicating linear transmission

Between these two opposing forces, lie the pacemaker cells. These cells form the innermost layer of the smooth muscle fibres and generate the phasic electrical potential which determines the tension of the smooth muscle[4]. When the transmural pressure exceeds the vasomotor tone, the pacemaker cells undergo elastic tension. This is the signal to increase muscle fibre tension and thus increase vasomotor tone. When transmural pressure falls below vasomotor tone, the arteriole tends to collapse and this is the signal for a decrease in vasomotor tone. A servomechanism is thus produced which attempts to balance the vasomotor tone against the transmural pressure. When this balance is attained, cerebral blood flow is regulated at a constant level and myogenic autoregulation is intact (Fig. 2).

Under normal circumstances, this servomechanism is continuously seeking the transmural pressure. When the transmural pressure is constant, the muscle fibres first contract until the vasomotor tone exceeds the transmural pressure, then the fibres relax until the transmural pressure

HYPOVOL. HYPOTENSION

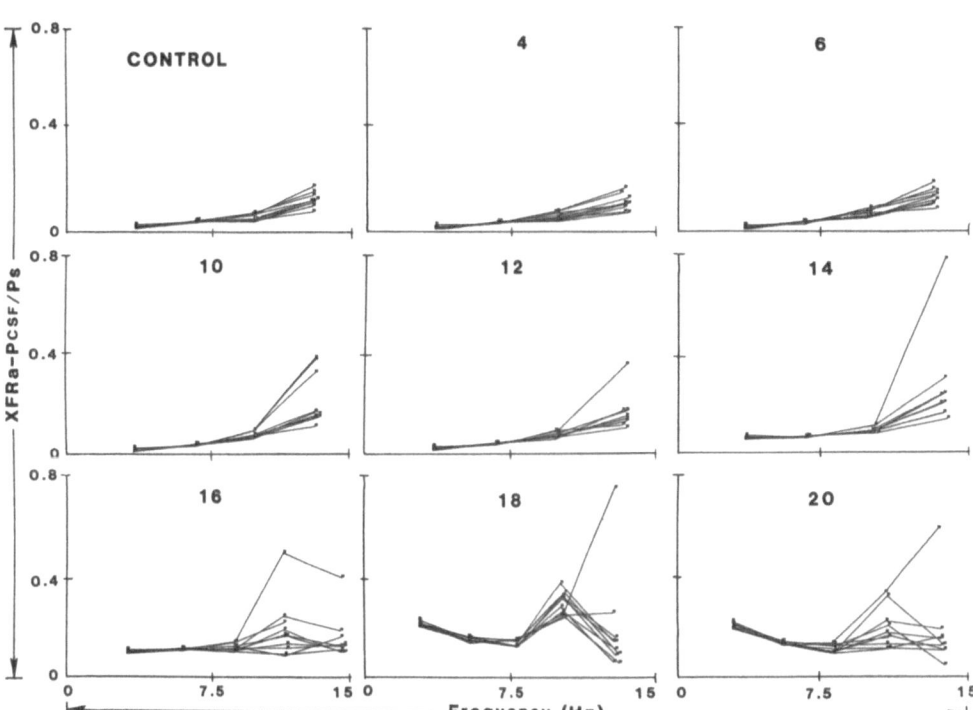

Fig. 4. Alterations in XFRa—Pcsf/Ps accompanying progressive withdrawal of blood (hypovolemia). Number above each transfer function spectrum represents blood withdrawn in ml · kg⁻¹. Note the change from non-linear to linear transmission following the removal of 14–16 ml · kg⁻¹ of blood (Ps approx. 60 mm Hg)

exceeds the vasomotor tone, and then the cycle repeats itself. This mechanism undoubtedly underlies the 5–6-minute oscillations of arteriolar diameter demonstrated by Auer[1].

Myogenic autoregulation becomes impaired when this finely balanced mechanism breaks down[6, 7]. This occurs under two circumstances. The first is when the smooth muscle can no longer contract in response to a rising transmural pressure (Fig. 3). The second is when the muscle fibres are inhibited and cannot respond properly to a given transmural pressure, such as during hypercapnia, in the presence of an arteriolar vasodilator, during marked systemic hypotension, or, during an increase in intracranial pressure[6, 7]. In both circumstances, the transmural pressure exceeds the vasomotor tone (Fig. 2).

The CSF/Venous Pulsewave as an Indicator of Myogenic Autoregulation

The cardiac pulse passing through the arterioles is altered in a specific manner depending on whether the system is autoregulating or not. If vasomotor tone balances the mean transmural pressure, *i.e.,* myogenic autoregulation is intact, then the pulse pressure is greater than the vasomotor tone during part of the pulse and below the vasomotor tone during the remainder of the pulse. Under these circumstances, transmission of the pulse wave is non-linear and the fundamental frequency of the pulse is fragmented into higher harmonics (Fig. 2).

When myogenic autoregulation is impaired, *i.e.,* when transmural pressure exceeds vasomotor tone, the arteriole does not collapse during any part of the pulsewave cycle. Transmission through the arterioles is linear and there is no fragmentation of the fundamental frequency (Fig. 2).

These altered wave forms are transmitted through the capillaries to the venous bed and then transmitted through the walls of the veins to the CSF. Simultaneous measurements of CSF and bridging vein pulsations have shown them to be nearly identical (unpublished data). Such a transfer of pulse waves through the wall of a collapsible tube has also been demonstrated in a Starling resistor (unpublished data).

The presence of linear or non-linear transmission of the arterial pulse into the venous bed can be determined by measuring the systemic arterial pressure and CSF pressure; the latter representing the venous pressure. The recorded waves can be analyzed using Fourier analysis and the CSF pressure/systemic arterial pressure transfer function determined (XFRa-Pcsf/Ps). When the transfer function is non-linear as shown by suppression of the fundamental frequency, myogenic autoregulation is intact. As autoregulation becomes impaired, the transfer function spectrum becomes linear and the transfer function of the fundamental frequency becomes progressively larger. An example of transfer function alterations with intracranial hypertension and hypercapnia has been previously reported. Other examples of this transformation are presented in Figs. 3 and 4. In Fig. 3, the change from non-linear to linear transmission is seen during marked systemic hypertension, and in Fig. 4, the alteration is seen during hypovolemic hypotension.

References

1. Auer, L. M., Gallhofer, B.: Rhythmic activity of cat pial vessels in vivo. Eur. Neurol. *20* (1981), 448—468.
2. Brower, R. W., Noordergraf, A.: Theory of steady flow in collapsible tubes and veins. In: Cardiovascular Systemic Dynamics (Baan, J., Noordergraff, A., Raines, J., eds.), pp. 256—265. Cambridge, Mass.: MIT Press. 1978.
3. Holt, J. P.: Flow through collapsible tubes and through in situ veins. IEEE Trans. Biomed. Eng. *16* (1969), 274—283.
4. Mellander, S., Johansson, B.: Control of resistance, exchange and capacitance functions in the peripheral circulation. Pharm. Rev. *20* (1968), 117—196.
5. Permutt, S., Riley, R. L.: Hemodynamics of collapsible vessels with tone: The vascular waterfall. J. Appl. Physiol. *18* (1963), 924—932.
6. Portnoy, H. D., Chopp, M., Branch, C., *et al.*: Cerebrospinal fluid pulse wave-form as an indicator of cerebral autoregulation. J. Neurosurg. *56* (1982), 666—678.

7. Portnoy, H. D., Chopp, M., Branch, C., *et al.*: CSF pulse wave, ICP and autoregulation. In: Intracranial Pressure V, in press.
8. Rodbard, S.: Autoregulation in encapsulated, passive, soft-walled vessels. Amer. Heart J. (1963), 924—932.
9. Shulman, K.: Small artery and vein pressures in the subarachnoid space of the dog. J. Surg. Res. *5* (1965), 666—678.

Author's address: H. D. Portnoy, M.D., Oakland Neurological Clinic, P.C., 1431 Woodward Avenue, Bloomfield Hills, MI 48013, U.S.A.

Departments of Anesthesiology/Critical Care Medicine and Environmental Health Sciences, The Johns Hopkins Medical Institutions, Baltimore, Maryland, U.S.A.

Effects of Cerebral Venous and Cerebrospinal Fluid Pressure on Cerebral Blood Flow

E. M. WAGNER and R. J. TRAYSTMAN

Summary

Regional cerebral blood flow responses to elevated jugular venous or cerebrospinal fluid pressure (Pcsf) were measured in thirty-nine anaesthetized ventilated dogs. When jugular venous pressure was elevated, arterial blood pressure and Pcsf were maintained constant at control values. When Pcsf was elevated, arterial and jugular venous pressure were maintained constant at control values. Cerebral perfusion pressure was calculated as the difference between arterial pressure and Pcsf, or jugular venous pressure, whichever downstream pressure was the highest. With an elevation in Pcsf or jugular venous pressure, the cerebral circulation autoregulated and no alterations in cerebral blood flow occurred as long as perfusion pressure remained above 60 mm Hg. When cerebral perfusion pressure decreased to values below 60 mm Hg, cerebral blood flow decreased significantly. We conclude that the brain autoregulates its blood flow to changes in perfusion pressure. In addition, our results suggest that the dominant mechanism of cerebral autoregulation is metabolic, not myogenic.

Keywords: Cerebral autoregulation; cerebrospinal fluid pressure; cerebral venous pressure; cerebral perfusion pressure; myogenic; metabolic.

Introduction

Blood flow through the cerebral vascular bed is influenced by mean arterial inflow pressure, venous outflow pressure, and the pressure surrounding cerebral vessels, cerebrospinal fluid pressure (Pcsf). Like other vascular beds, the cerebral circulation has an inherent capacity to maintain its blood flow constant over a wide range of perfusion pressure[9]. Although autoregulation is well established for the cerebral circulation[11, 18] its mechanism is poorly understood. Metabolic[4], myogenic[2], and neurogenic[8], theories have been suggested to explain the mechanism responsible for cerebral autoregulation. Inherent to this discussion is to properly define cerebral perfusion pressure. Autoregulatory responses have been demonstrated for changes in arterial inflow pressure. However, the cerebral blood flow (CBF) response to changes in perfusion pressure, mediated through increases in downstream venous pressure and/or increases in Pcsf, have not been systematically assessed. Elevations in the cerebral venous pressure to 20 mm Hg have shown small increases[7], decreases[1, 3], and no changes in CBF[14]. Several studies have shown decreases in CBF with elevation in Pcsf[5, 6, 10, 13, 20]. However, the absolute value of Pcsf, which ultimately resulted in a decrease of CBF was variable, as was the magnitude of the decrease in CBF. When increases in Pcsf were translated into perfusion pressure changes (*i.e.,* arterial pressure minus Pcsf), typical cerebral autoregulatory curves were obtained[13, 19]. The specific goal of this study was to examine the CBF response to changes in perfusion pressure mediated through alterations in cerebral venous pressure and Pcsf.

Methods

All experiments were performed on adult mongrel dogs of either sex (15–20 kg). Animals were anaesthetized with sodium pentobarbital (30 mg/kg, iv.) and mechanically ventilated through a tracheal cannula to insure constant, normal, blood gas and pH values. The preparation was designed to allow alterations in jugular venous pressure independent of changes in systemic arterial pressure. Hence, in these open chested animals, the superior and inferior vena cavae (below the kidney) were cannulated and drained into a reservoir. The inferior vena cava was included to eliminate any anastomotic drainage via the spinal veins. Blood was pumped from the reservoir into the right atrium. Thus arterial pressure could be maintained at any level by altering the speed of the pump which controlled venous return to the heart. Venous pressure was measured via a catheter placed in a jugular vein. Pcsf, measured in the lateral ventricle, could be manipulated and drained via a cisternal puncture connected to a volume reservoir containing mock CSF solution[22]. CBF was measured with radio-labelled microspheres (15 ± 3 µ diameter) using the reference sample method[12]. Three experimental protocols were studied.

Group 1: Jugular Venous Pressure Elevation

Total and regional CBF was measured in 16 animals with an elevated jugular venous pressure. Venous pressure was altered by raising the outflow reservoir draining the superior vena cava and lower portion of the inferior vena cava. Arterial pressure and Pcsf (close to atmospheric) were maintained constant for two CBF measurements; the first made at a control venous pressure (close to atmospheric pressure) and the second at an experimentally elevated jugular venous pressure (10 to 35 mm Hg). Perfusion pressure for this group of animals was defined as arterial pressure minus jugular venous pressure. This group of animals was further divided into two groups based on whether the perfusion pressure was greater than, or less than 60 mm Hg.

Group 2: Jugular Venous Pressure Elevation with Accompanying CSF Pressure Increases

Jugular venous pressure elevations were repeated in this series of 12 animals, however Pcsf was allowed to increase concomitantly with venous pressure elevation. Arterial pressure was maintained constant for both control and experimental blood flow measurements.

Perfusion pressure in this series of experiments, where Pcsf was always slightly less than jugular venous pressure, was defined as mean arterial pressure minus jugular venous pressure. Animals in this group were further subdivided, based on whether perfusion pressure was greater than, or less than 60 mm Hg.

Group 3: Pcsf Alterations

The CBF response to Pcsf alterations was studied in this series of 11 animals while maintaining arterial pressure and jugular venous pressure (close to atmospheric pressure) constant. Cerebral perfusion pressure was defined as arterial pressure minus Pcsf. Animals were further subdivided by their experimental perfusion pressure values.

Results

The results from the three groups of experiments are presented in Table 1 and absolute control CBF values are presented in Table 2. All values are presented as the mean ± SEM.

Group 1: Jugular Venous Pressure Elevation

The effect of alterations in jugular venous pressure on the CBF was determined under conditions where both the arterial pressure and the Pcsf

Table 1. *Percent Change in Regional Blood Flow*

Perfusion Pressure (mm Hg)	Group 1		Group 2		Group 3	
	< 60	> 60	< 60	> 60	< 60	> 60
Mean Hemisphere	—33.4*	—6.3	—29.5*	—0.3	—26.7*	4.0
Spinal Cord	—36.7*	—5.4	—31.0*	—9.0	—26.0*	—6.0
Cerebellum	—36.1*	—12.0*	—28.0*	—5.0	—32.8*	6.0
Medulla	—34.3*	—12.8*	—24.6*	—10.0	—26.9*	—2.6
Pons	—37.6*	—7.4*	—27.2*	—4.1	—23.8*	—2.1
Midbrain	—35.5*	—12.9*	—27.8*	—3.6	—26.6*	0.8
Diencephalon	—33.8*	—9.4*	—25.3*	—0.7	—27.8*	4.0
Caudate	—35.3*	—6.8	—25.8*	—9.9	—24.3*	—3.4
Hippocampus	—39.9*	—7.5	—32.7*	—4.3	—35.3*	12.3
Parahippocampal gyrus	—39.9*	—7.1	—31.3*	—7.6	—32.6*	1.3
Occipital lobe	—34.4*	—2.8	—26.8*	3.3	—28.1*	0.2
Temporal lobe	—34.0*	—4.3	—30.8*	—0.8	—26.7*	5.3
Parietal lobe	—32.4*	—3.4	—28.7*	0.7	—24.1*	1.3
Frontal lobe	—31.8*	—4.9	—29.6*	0.4	—24.8*	1.4

The change in flow from its control value is presented as the percent change.
* = $p < 0.5$.
Group 1 = Jugular venous pressure elevation.
Group 2 = Jugular venous pressure elevation with Pcsf increases.
Group 3 = Pcsf alterations.

Table 2. *Control Regional Blood Flow*

Flow (ml/min/100 g)	Group 1	Group 2	Group 3
Mean hemisphere	29.8 ± 1.3	32.0 ± 1.7	30.5 ± 1.6
Spinal cord	17.0 ± 1.2	21.7 ± 1.2	19.8 ± 1.5
Cerebellum	34.3 ± 1.2	39.7 ± 2.8	38.5 ± 1.8
Medulla	25.3 ± 1.5	29.8 ± 1.9	27.1 ± 1.9
Pons	20.2 ± 1.2	22.5 ± 1.5	20.5 ± 1.5
Midbrain	32.8 ± 1.5	36.7 ± 2.5	32.7 ± 2.4
Diencephalon	27.4 ± 1.2	30.4 ± 1.7	27.4 ± 1.5
Caudate	55.2 ± 4.2	57.6 ± 3.5	46.4 ± 3.7
Hippocampus	29.8 ± 1.8	29.7 ± 1.6	31.2 ± 2.3
Parahippocampal gyrus	27.4 ± 1.3	28.4 ± 1.5	29.4 ± 1.6
Occipital lobe	31.5 ± 1.4	32.6 ± 1.8	31.2 ± 1.6
Temporal lobe	29.3 ± 1.4	31.8 ± 1.9	30.7 ± 1.8
Parietal lobe	29.3 ± 1.2	30.4 ± 1.7	30.1 ± 2.0
Frontal lobe	29.2 ± 1.3	30.9 ± 1.6	30.0 ± 2.0

Absolute control flows are presented for mean hemispheric and regional blood flow in mo/min/100 g.
Group 1 = Jugular venous pressure elevation.
Group 2 = Jugular venous pressure elevation with Pcsf increases.
Group 3 = Pcsf alterations.

were maintained constant. In the sub group of animals with an experimental perfusion pressure greater than 60 mm Hg, the jugular venous pressure was elevated to an average of 21.6 ± 1.9 mm Hg, with a resultant perfusion pressure of 77.2 ± 3.3 mm Hg. In the subgroup of experiments where perfusion pressure was reduced to levels less than 60 mm Hg, the average jugular venous pressure elevation was 29.0 ± 1.2 mm Hg with a resultant perfusion pressure of 47.6 ± 1.1 mm Hg. Table 1 shows that mean hemispheric blood flow was not altered with venous pressure elevation, provided the perfusion pressure remained above 60 mm Hg. At perfusion pressures below 60 mm Hg, mean hemispheric blood flow decreased by 33.4% from the control flow of 29.8 ± 1.3 ml/min/100 g. No region of the brain showed a change in blood flow with jugular venous pressure elevation until the perfusion pressure was reduced to levels less than 60 mm Hg with the following exceptions: the cerebellum, medulla, pons, midbrain and diencephalon. These regions showed a small but significant decrease in blood flow with a perfusion pressure greater than 60 mm Hg.

Group 2: Jugular Venous Pressure Elevation with Accompanying CSF Pressure Increases

In this series of animals, the effect of alterations of jugular venous pressure on CBF was determined under conditions where arterial pressure was held constant, but Pcsf was allowed to increase concomitantly with venous pressure elevations. In those experiments with a perfusion pressure above 60 mm Hg, jugular venous pressure was elevated to 18.1 ± 1.7 mm Hg and Pcsf to 17.2 ± 2.3 mm Hg with a resultant perfusion pressure of 84.8 ± 4.7 mm Hg. In the group with perfusion pressure below 60 mm Hg, jugular venous pressure elevation averaged 30.8 ± 2.3 mm Hg and Pcsf was 28.8 ± 1.7 mm Hg with a resultant perfusion pressure of 48.8 ± 2.2 mm Hg. No change was observed in the mean hemispheric or any regional cerebral blood flow until perfusion pressure was reduced to levels below 60 mm Hg. Mean hemispheric blood flow was reduced by 29.5% from a control flow of 32.0 ± 1.7 ml/min/100 g at perfusion pressures below 60 mm Hg. Regional CBF decreased between 25 and 33%.

Group 3: Pcsf Alterations

The effect of altering Pcsf on CBF was determined under conditions where arterial and jugular venous pressure were maintained constant. In the subgroup with a perfusion pressure greater than 60 mm Hg, Pcsf was elevated to 22.0 ± 4.7 mm Hg with a resultant perfusion pressure of 70.4 ± 3.1 mm Hg. In the subgroup with a perfusion pressure less than 60 mm Hg, Pcsf was elevated to an average of 29.0 ± 2.4 mm Hg, with a resultant perfusion pressure of 45.7 ± 1.2 mm Hg. No change in mean hemispheric or any regional cerebral blood flow was seen until the perfusion pressure was reduced to less than 60 mm Hg. Mean hemispheric blood flow was reduced by 26.7% from a control flow of 30.5 ± 1.6 ml/min/100 g at perfusion pressures below 60 mm Hg. Regional CBF decreased between 24 and 35%.

Discussion

CBF is influenced by mean arterial pressure, venous outflow pressure, and the surrounding pressure of the cerebral vessels, Pcsf. Our results demonstrate that reductions in the perfusion pressure, mediated through increases in jugular venous and/or Pcsf, have little effect on CBF until the perfusion pressure is reduced to below 60 mm Hg. Comparable 25–35% reductions in mean hemispheric as well as regional CBF were obtained with a reduction in perfusion pressure to approximately 45 mm Hg. These results suggest that the brain regulates its blood flow to changes in the perfusion pressure gradient as opposed to the absolute inflow pressure. Alterations in downstream or surrounding pressure affect this perfusion pressure gradient and thereby influence flow. Our results are in accord with those reported by Raisis et al.[17] who showed that venous pressure elevation had no effect on CBF until very high venous pressures were attained (greater than 75 mm Hg). These high pressures effectively reduced perfusion pressure below the lower limit of cerebral autoregulation. Other results, showing decreases in CBF with small increases in venous pressure[2, 3], may have been obtained if the method for increasing venous pressure resulted in a diversion of flow from measured outflow tracts to anastomotic venous drainage channels[21]. In those studies reporting no change in CBF with venous pressure elevation, the reduction in perfusion pressure may not have been below the lower limit of autoregulation. The present study emphasizes the importance of the perfusion pressure gradient and minimizes the significance of the absolute magnitude of venous pressure. In addition, the results of our study eliminate the notion that a venous-arteriolar constrictor response is operative in the cerebral circulation.

Precisely how the surrounding pressure of a vascular system can affect flow has been described by Permutt and Riley[16] for the pulmonary circulation. They compared the pulmonary circulation with a Starling resistor and showed that when the pressure surrounding the vasculature exceeded the intravascular downstream pressure, the pressure gradient for blood flow is represented by arterial pressure minus the surrounding pressure. The intravascular downstream pressure is of no significance in terms of the control of blood flow in this situation. This situation is analogous to that of the cerebral vasculature. Cerebral venous pressure exceeds Pcsf until venous pressure abruptly drops in the superior sagittal sinus[20]. Nakagawa et al.[15] have shown that within the cerebral venous system there exists a point just proximal to the superior sagittal sinus which is fully exposed to Pcsf increases and exactly equals the surrounding pressure. Based on this analysis, Pcsf can provide the downstream pressure for flow through the cerebral vessels. Although previous investigators[5, 6, 10, 13, 20] have reported a variety of elevated Pcsf values which decrease CBF, Miller et al.[13] showed that when elevated Pcsf values were reported in terms of perfusion pressure, cerebral autoregulation was present. Our work supports this finding.

In our experiments, perfusion pressure was defined as the mean arterial pressure minus either Pcsf or jugular venous pressure, depending on which pressure was of greater magnitude. Although our protocol allowed for only a two point analysis with control and experimental CBF measurements, cerebral autoregulation was evident. Based on values reported in the literature for arterial inflow autoregulation[11, 18], we chose 60 mm Hg as the perfusion pressure above which cerebral vessels showed autoregulation, and below which cerebral autoregulation was compromised. Most regions of the brain responded in a similar manner to changes in perfusion pressure. However, when jugular venous pressure was increased, resulting in a perfusion pressure greater than 60 mm Hg, small but significant reductions (7–12%) in blood flow to the cerebellum, medulla, pons, midbrain and diencephalon were observed. The physiological significance of this finding is unclear, however the possibility exists that these lower regions of the brain respond differently to venous pressure elevation and have a higher autoregulatory limit than other brain regions. Yet in the second group of animals studied where jugular venous pressure was slightly greater than Pcsf, all regions of the brain responded in a similar manner.

We conclude that the brain autoregulates its blood flow to changes in perfusion pressure. The results of our experiments also provide important information concerning the mechanism responsible for cerebral autoregulation. At a perfusion pressure greater than 60 mm Hg, cerebral vessels autoregulate despite venous pressure elevations and an increase in vascular transmural pressure. If a myogenic mechanism for cerebral autoregulation were operative, one would expect that increases in vascular transmural pressure would result in vasoconstriction and a reduction in CBF. Since no change in mean hemispheric flow was seen in any of the three groups of animals until perfusion pressure was reduced to levels less than 60 mm Hg, we conclude that myogenic mechanisms are not primarily responsible for cerebral autoregulation. The results of our study are consistent with a metabolic theory for cerebral autoregulation.

References

1. Ekstrom-Jodal, B.: Effect of increased venous pressure on cerebral blood flow in dogs. Acta Physiol. Scand. Suppl. *350* (1970), 51—61.
2. Ekstrom-Jodal, B., Häggendahl, E., Nilsson, N. J.: On the relation between blood pressure and blood flow in the cerebral cortex of dogs. Acta Physiol. Scand. Suppl. *350* (1970), 29—49.
3. Emerson, T. E., Parker, J. L.: Effect of local increases of venous pressure on canine cerebral hemodynamics. In: Cerebral Circulation and Metabolism (Langfitt, T. W., McHenry, L., Reivich, M., Wollman, H., eds.), pp. 10—13. Berlin-Heidelberg-New York: Springer. 1975.
4. Fujishima, M.: The metabolic mechanism of cerebral blood flow autoregulation in dogs. Japan Heart J. *12* (1971), 376—382.
5. Greenfield, J. D., Tindall, G. T.: Effect of acute increase in intracranial pressure on blood flow in the internal carotid artery of man. J. Clin. Invest. *44* (1965), 1343—1351.
6. Häggendahl, E., Löfgren, J. Nilsson, N. J., Zwetnow, N. N.: Effects of varied cerebrospinal fluid pressure on cerebral blood flow in dogs. Acta Physiol. Scand. *79* (1970), 262—271.
7. Jacobson, I., Harper, A. M., McDowall, D. G.: Relationship between venous pressure and cortical blood flows. Nature *200* (1963), 173—175.
8. James, I. M., Millar, R. A., Purves, M. J.: Observations on the extrinsic neural control of cerebral blood flow in the baboon. Circ. Res. *25* (1964), 77—93.
9. Johnson, P. C.: Autoregulation of blood flow. Circ. Res. *15* (Suppl. 1) (1964), 2—9.
10. Johnston, I. H., Rowan, J. O., Harper, A. M., Jennet, W. B.: Raised intracranial pressure and cerebral blood flow. J. Neurol. Neurosurg. Psychiat. *35* (1972), 285—296.
11. Lassen, N. A.: Autoregulation of cerebral blood flow. Circ. Res. (Suppl. 1) (1964), 201—204.
12. Marcus, M. L., Heistad, D. D., Ehrhardt, J. C., Abboud, F. M.: Total and regional cerebral blood flow measurement with 7-10, 15, 25 and 50 μm microspheres. J. Appl. Physiol. *40* (4) (1976), 501—507.
13. Miller, J. D., Stanek, A., Langfitt, T. W.: Concepts of cerebral perfusion pressure and vascular compression during intracranial hypertension. In: Progress in Brain Research (Meyer, J. S., Schmede, J. P., eds.), pp. 411—432. Amsterdam: Elsevier. 1972.
14. Moyer, J. H., Miller, S. I., Snyder, H.: Effect of increased jugular pressure on cerebral hemodynamics. J. Appl. Physiol. *7* (1954), 245—247.
15. Nakagawa, Y., Tsuru, M., Yada, K.: Site and mechanism of compression of the venous system during experimental intracranial hypertension. J. Neurosurg. *41* (1974), 427—434.

16. Permutt, S., Riley, R.: Hemodynamics of collapsible vessels with tone: the vascular waterfall. J. Appl. Physiol. *18* (5) (1963), 924—932.
17. Raisis, J. E., Kindt, G. W., McGillicuddy, J. E., Gianotta, S. L.: The effects of primary elevation of cerebral venous pressure on cerebral hemodynamics and intracranial pressure. J. Surg. Res. *26* (1979), 101—107.
18. Rapela, C. E., Green, H. D.: Autoregulation of canine cerebral blood flow. Circ. Res. *15* (Suppl. 1) (1964), 205—211.
19. Sadoshima, S., Thames, M., Heistad, D. D.: Cerebral blood flow during elevation of intracranial pressure: role of sympathetic nerves. Am. J. Physiol. *241* (1981), H78—H84.
20. Shulman, K., Verdier, G. R.: Cerebral vascular resistance changes in response to cerebrospinal fluid pressure. Am. J. Physiol. *213* (5) (1967), 1084—1088.
21. Wagner, E. M., Traystman, R. J.: Cerebral venous outflow and arterial microsphere flow with venous pressure elevation. Am. J. Physiol. *244* (1983), H505—H512.
22. Wei, E. P., Raper, A. J., Kontos, H. A., Patterson, J. L.: Determinants of response of pial arteries to norepinephrine and sympathetic nerve stimulation. Stroke *6* (1975), 654—659.

Authors' addresses: Elizabeth M. Wagner, Ph.D., Department of Environmental Health Sciences, The Johns Hopkins University, School of Hygiene and Public Health, 615 North Wolfe Street, Baltimore, MD 21205, U.S.A., R. J. Traystman, Ph.D. (reprints), Director, Anesthesiology/Critical Care Medicine, Research Laboratories, Department of Anesthesiology/Critical Care Medicine, The Johns Hopkins Hospital, Blalock 1408, 600 North Wolfe Street, Baltimore, MD 21205, U.S.A.

[1] Department of Neurosurgery, Fujita-Gakuen Medical University, Kutsukake, Toyoake, Aichi 470–11, Japan, [2] Department of Neurosurgery, School of Medicine, Keio University, 35 Shinanomachi, Shinjuku, Tokyo, 160 Japan

The Changes in the Cerebral Microcirculation During Increased Intracranial Pressure

H. Sano[1], T. Kawase[2], and S. Toya[2]

Summary

Cerebral microcirculatory dynamics were observed during intracranial hypertension by fluorescein serial angiography via a 15×10 mm glass-covered skull window in 23 mongrel dogs. An epidural balloon was inflated contralaterally at a steady rate of 0.1 ml/min. Serial photographs gave information on pial vessels larger than 20 μ in diameter, and the following variables were estimated in the pial arteries and veins of 200 μ in diameter; arterial exposure period (AEP), venous exposure period (VEP), arterio-venous exposure ratio (AVER = VEP/AEP) and regional circulation time (RCT). Subdural pressure, systemic arterial pressure, respiration and EEG were monitored to recognize the state of ICP.

During the early stage of increased ICP, venous stasis occurred without any change in blood flow. At the later stage of raised ICP, however, venous blood was squeezed out by the increased ICP. There was arterial blood stasis resulting in cerebral ischaemia. These changes occurred at about 50 mm Hg of ICP.

Keywords: Intracranial pressure; cerebral circulation; fluorescein.

Increased intracranial pressure is one of the important problems, and there have been many papers[2-7] about the relationship between ICP and cerebral blood circulation since Cushing[1] reported a pressor response to increasing ICP. Fluorescein serial angiography can offer information on pial vessels in the arterial, capillary and venous phases separately. So fluorescein serial angiography was performed whilst increasing the intracranial pressure (ICP) in order to study the dynamics of the cerebral microcirculation.

Material and Methods

The study was carried out in the following way on 23 mongrel dogs weighing 9–22 kg under ketalar and pentobarbital intravenous anaesthesia: An epidural rubber balloon was inflated contralaterally at a steady rate of 0.1 ml/min to elevate the intracranial pressure (ICP), while monitoring systemic arterial pressure (SAP), respiration, EEG, CBF, and ICP.

SAP was continuously measured by a catheter placed into the abdominal aorta through the femoral artery, and using a pressure transducer (MPU-0.5, Nippon Koden). Respiration was recorded by a thermister (MTR-ITA Nippon Koden) which was put at the tip of the tracheotomy tube during the spontaneous respiration of room air. EEG was monitored from the right frontoparietal region. CBF was recorded from the right frontal region using the plate type element by heat clearance method (shincorder model TE 201). ICP was monitored using a pressure transducer (model MPU-0.5, Nippon Koden) which was connected to a small rubber balloon (2 mm in diameter) inserted into the subdural space. Every small balloon was checked for its reliability up to a pressure of 1,000 mm H$_2$O. All those monitoring systems were connected to a Polygraph (Nippon Koden RM-150) and they were recorded continuously.

Cerebral microcirculation dynamics was observed through a glass covered skull window (15 × 10 mm approximately) by fluorescein serial angiography, which was done at each stage of increased ICP. 5 ml of 0.5% fluorescein sodium solution was injected through a catheter placed in the right common carotid artery, and series of photographs were taken 2 or 3 times per second using a motor driven camera equipped with a telephoto lense and fluorescein filters. These serial photographs were able to give information on pial vessels larger than 20 μm diameter. The following variables were estimated in the pial arteries and veins of 200 μm diameter; Arterial Exposure period (AEP) is defined by the duration of the appearance of fluorescein into and disappearance from an arteriole with a diameter of 200 μm. The venous exposure period (VEP) is defined by the duration of the fluorescein between its appearance into and disappearance from 200 μm venule. The regional circulation time (RCT) is the time between the fluorescein flowing into the arteriole and its flowing into venule in the same territory. The arterio-venous exposure ratio (AVER) = VEP/AEP can be considered as an indicator of the arterio-venous flow ratio, and has the advantage of not being influenced by the injection time or pressure of the fluorescein solution.

The stages in the levels of ICP are shown below:

Stage I: stage without any change in SAP, respiration or CBF.
Stage II: CBF decreases slightly without any change in SAP.
Stage III: stage with increased SAP.
Stage IV: SAP decreases and CBF and respiration are arrested.

Results

The control value of AEP was 1.0–2.8 sec (mean 1.8 sec), VEP was 4.5–7.7 sec (mean 6.1 sec), RCT was 2.5 ± 0.5 sec and AVER was 3.6 ± 1.0. At stage I (10–20 mm Hg of ICP) AVER increased slightly because of slightly prolonged VEP without any change of RCT or AEP (Figs. 5–7). At stage II (20–50 mm Hg of ICP) RCT was prolonged slightly and AVER increased markedly to 5.6 ± 1.5 due to the prolonged VEP, which implies venous stasis. The pial veins were dilated and extravasation of fluorescein-dye was observed (Figs. 1 and 5–7).

At stage III (50–100 mm Hg of ICP) RCT was markedly prolonged to 16 ± 4 sec. AVER was suddenly reduced to approximately 2 due to the

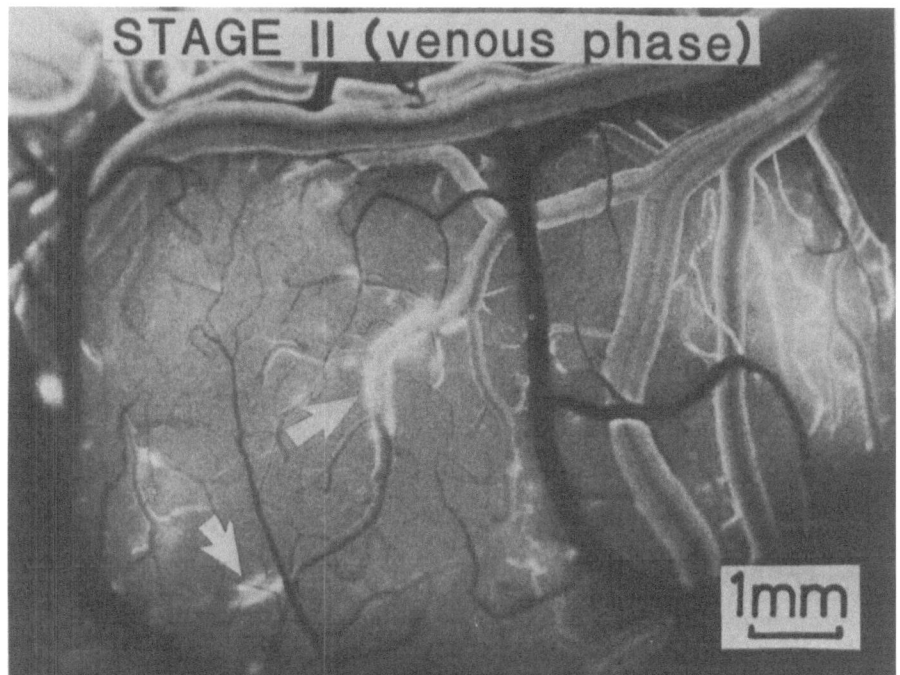

Fig. 1. Venous phase of fluorescein serial angiogram showed dilatation of pial veins and extravasation of fluorescein dye (white arrow) at stage II

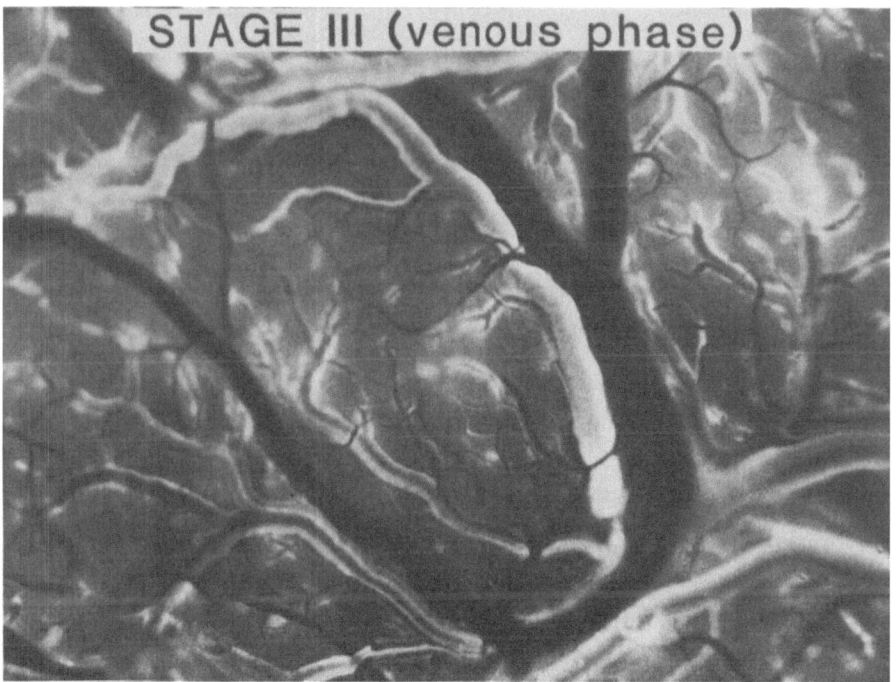

Fig. 2. Venous phase of fluorescein serial angiogram at stage III revealed extravasation of fluorescein dye

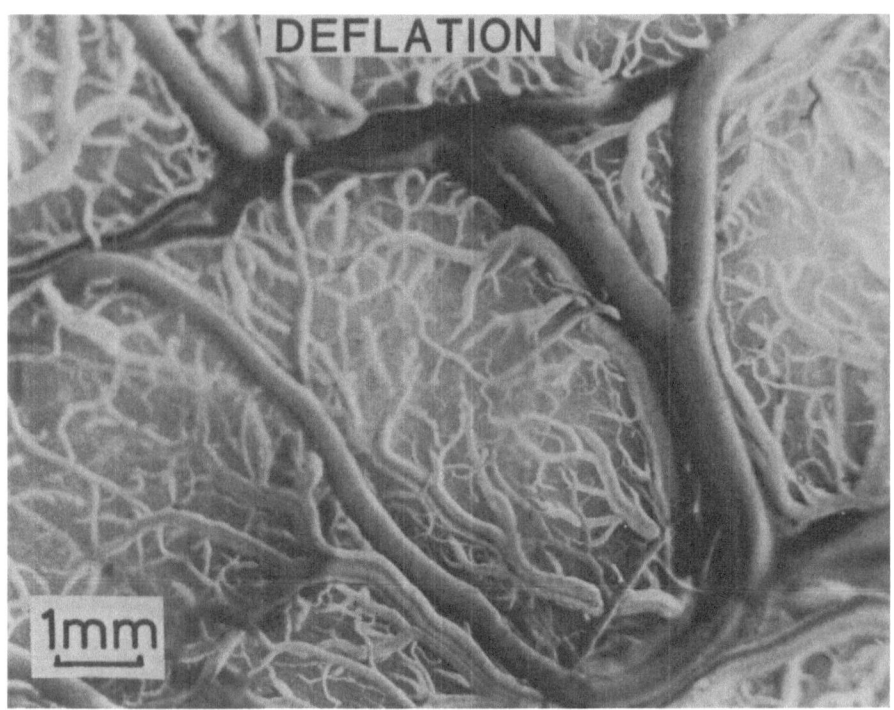

Fig. 3. Capillary phase of fluorescein serial angiogram immediately after decreased ICP from early stage IV showed hyperaemia, early venous filling with laminar flow and extravasation were observed

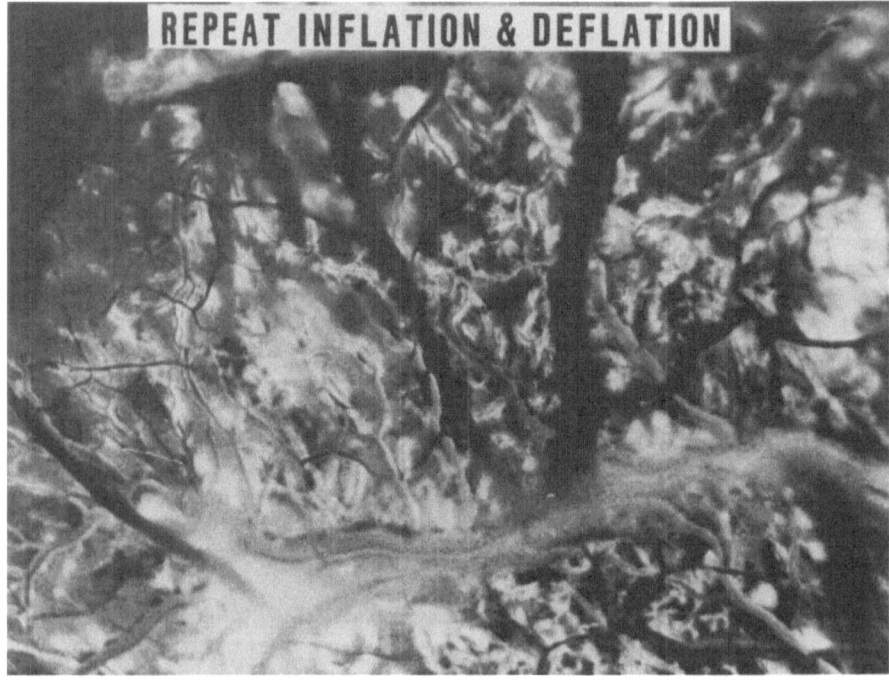

Fig. 4. Venous phase of fluorescein serial angiogram after repeated inflation and deflation of balloon for ICP showed marked extravasation not only of fluorescein dye but also of blood cells

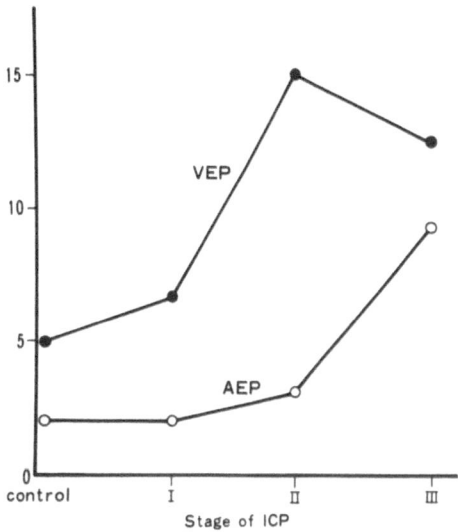

Fig. 5. Relation of AEP and VEP to the stage of ICP

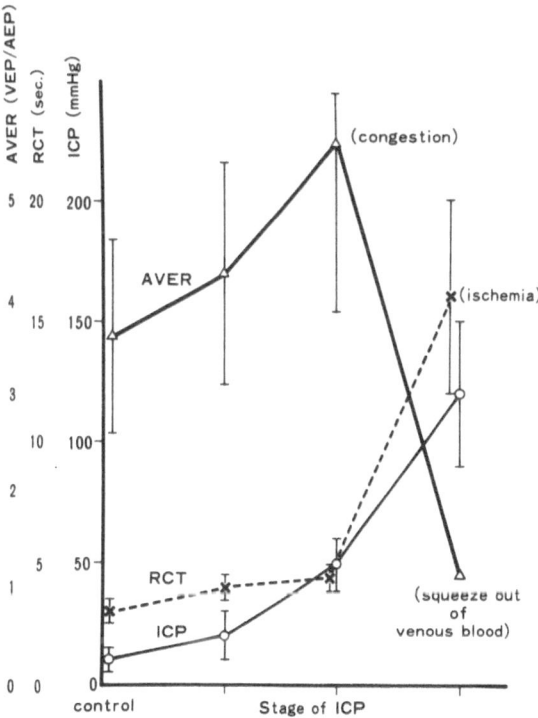

Fig. 6. Relation of ICP, RCT, and AVER to the stage of ICP

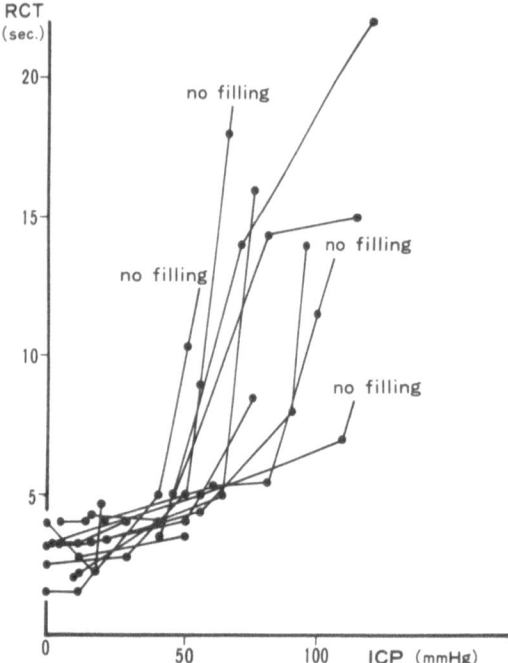

Fig. 7. Relation between RCT and ICP. RCT was immediately prolonged at the critical level of ICP of 50 mm Hg

prolonged AEP. VEP was shorter than in stage II, because of the squeezing out of venous blood (Figs. 2 and 5–7). At stage IV the pial vessels were not filled and the EEG was flattened. After decreasing the ICP early venous filling with a marked extravasation of fluorescein-dye was observed (Fig. 3).

After repeated inflation and deflation of the balloon in order to test the effect of increasing ICP, there was a marked extravasation not only of fluorescein-dye but also of blood cells (Fig. 4). Fig. 5 shows the relationship between AEP and VEP at the various stages of ICP. VEP increased slightly during stage I and markedly during stage II, but decreased to a value near to AEP when stage III has been reached. AEP was not changed until stage II but it increased when stage III was reached. Fig. 6 shows the relationship between ICP, RCT and AVER according to the stages of ICP. RCT was slightly prolonged during stage I and II but suddenly it was markedly prolonged in stage III. AVER gradually increased until stage II but then it suddenly decreased to a level around 1 in stage III. Fig. 7 shows the relationship between RCT and ICP. RCT was suddenly prolonged at the critical level of ICP of 50 mm Hg.

Discussion

There have been many reports about ICP and CBF, since Cushing[1] published his report in 1902. By observation of pial vessels through a glass-covered skull window during increasing ICP Wolf et al.[6] found that venules dilated primarly and then became narrow. This resulted in cessation of blood flow because of arterial compression. Wolf et al.[7] reported again that CBF would be reduced by 55 mm Hg when there is an increase of 55 mm Hg in ICP.

Kety et al.[2] reported that CBF was reduced at an ICP of 450 mm H_2O in human beings.

Langfitt et al.[3] reported that CBF was suddenly reduced by a rapidly increasing ICP, however, CBF decreased only slightly until the ICP increased by between 35 and 50 mm Hg when it was increased slowly.

The present authors' data shows that a critical level of reduced CBF was seen with an ICP increased to 50 mm Hg. The critical level of the other reports was slightly different probably because of the speed and method of increasing the ICP.

Ohwada et al.[5] reported that unless the pressure was elevated to an extremely high level (900–2,000 mm H_2O), the dye appears in the cortical arteries immediately. No circulatory change was seen when the ICP was below 400 mm H_2O. Therefore, inflow into the cerebral arterial system is not influenced by a moderately increased ICP but is mainly dependent on the pressure difference between the intracranial pressure and systemic blood pressure.

The present authors' data show that venular stasis started without a prolongation of RCT at an ICP of 20 mm Hg. At an ICP level of between 30 and 40 mm Hg venous stasis became marked and CBF decreased. Over 50 mm Hg capillary and arteriolar stasis occurred with a rapidly prolonged RCT. These results agree well with the fact that capillary pressure is about 50 mm Hg. Namely, at first an increased ICP causes venous stasis without a concomitant reduction in CBF because of autoregulation. CBF starts to decrease because of a damaged autoregulation caused by an elevated ICP. When ICP increases up to the level of capillary pressure, capillary and arteriolar stasis will occur. As the greater part of the vascular bed is made up of capillaries and venules, CBF and RCT are markedly prolonged.

Conclusions

During the early stages of an increased ICP, venous stasis occurs without even a minimal change in CBF. At the late stage of elevated ICP, venous blood is squeezed out by the increased ICP, and the stasis of blood may cause cerebral ischaemia. The critical level of change in ICP is found to be about 50 mm Hg.

References

1. Cushing, H.: Some experimental and clinical observations concerning states of increased intracranial tension. Am. J. Med. Sci. 124 (1902), 375—400.
2. Kety, S. S., Shenkin, H. A., Schmit, C. F.: The effect of increased intracranial pressure of cerebral circulatory function in man. J. Clin. Invest. 27 (1948), 493—499.

3. Langfitt, T. W., Kassell, N. F., Weinstein, J. D.: Cerebral Blood Flow with Intracranial Hypertension. Neurology *15* (1965), 761—773.
4. Lundberg, N.: Continuous recording and control of ventricular fluid pressure in neurosurgical practice. Acta Psychiat. Neurol. Scand., Supplement *149* (1960), 36.
5. Ohwada, T., Kuramae, T., Ueno, K., Abe, H., Kashiwaba, T., Yada, K.: Cerebral circulation and intracranial hypertension. Brain and Nerve *22* (1970), 247—255.
6. Wolf, H. G., Forbes, H. S.: The cerebral circulation observation of pial circulation during changes in intracranial pressure. Arch. Neurol. Psychiat. *20* (1928), 1035—1047.
7. Wolf, H. G., Blugmant, H. L.: The cerebral circulation, the effect of normal and increased intracranial cerebrospinal fluid pressure on the velocity of intracranial blood flow. Arch. Neurol. Psychiat. *21* (1929), 795—804.

Author's address: H. Sano, M.D., Department of Neurosurgery, Fujita-Gakuen Medical University, Toyoake, Aichi 470-11, Japan.

Macpherson Laboratory, White Memorial Medical Center, Los Angeles, California, U.S.A.

The Influence of Sympathetic Extracranial Venoconstriction on Canine Cerebral Venous, Sigmoid Sinus, and Mean Arterial Pressures

W. J. PEARCE and J. A. BEVAN

Summary

In the present study, we have examined the effects of sympathetically mediated extracranial venoconstriction on cerebral venous, sigmoid sinus, and mean arterial pressures in dogs. When the normal drainage pattern of cerebral venous effluent blood remained intact, sympathetic stimulation of the left superior cervical ganglion produced no significant effect on cerebral venous or mean arterial pressures, but decreased sigmoid sinus pressure at low frequencies of constant current stimulation. When the entire cerebral venous effluent blood was diverted through the extracranial veins, sympathetic stimulation resulted in significant increases in both cerebral venous and mean arterial pressures, but did not affect sigmoid sinus pressure. All effects of sympathetic stimulation were blocked by 2 mg/kg intravenous phentolamine.

These results demonstrate that in the dog, the veins draining the cerebral venous effluent blood constrict markedly during sympathetic stimulation, in vivo. In the intact dog, this venoconstriction diverts cerebral venous drainage away from extracranial veins and toward the sigmoid/vertebral sinus route of venous return. However, when the sigmoid/vertebral sinus route is blocked, extracranial venoconstriction has a pronounced effect on cerebral venous pressure. The data also suggest that a reversible systemic hypertensive response can result from moderate increases in cerebral venous pressure.

Keywords: Cerebral venous pressure; extracranial venoconstriction; sympathetic influence.

Introduction

In many species, including man, a significant portion of the cerebral venous effluent is carried by the extracranial veins. Thus, changes in the vascular tone of extracranial veins may affect the intracranial and cerebral venous pressures, cerebral blood volume, or may influence the vascular route through which the cerebral effluent blood returns to the heart. Previous studies in our laboratory have demonstrated that the extracranial veins, which drain the canine cerebral venous effluent blood, receive a dense, mixed adrenergic innervation, exhibit a norepinephrine sensitivity equivalent to that of other contractile veins, and demonstrate a

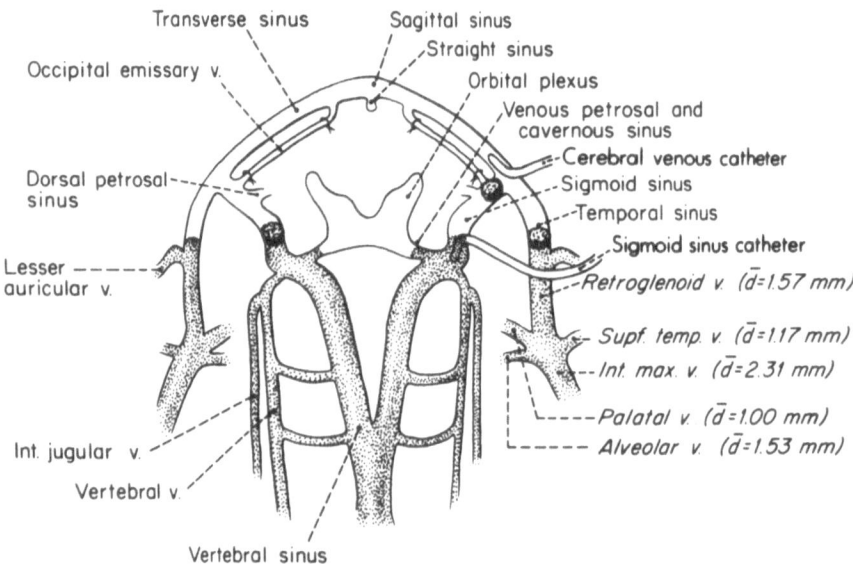

Fig. 1. All canine cerebral venous effluent blood entering the paired transverse sinuses drains either via the paired sigmoid, or the paired temporal sinuses. Effluent blood from the sigmoid sinuses empties into the vertebral sinuses whereas that from the temporal sinuses drains via the retroglenoid veins into the internal maxillary veins. Also shown in the above figure are the relative positions of the cerebral venous catheter, the sigmoid sinus catheter, and the occlusion sites. The stippled portions of the diagram indicate extracranial structures

phentolamine-sensitive constriction in response to transmural stimulation in vitro [16-18]. These extracranial retroglenoid veins extend approximately 20 mm from their cranial margins to their junction with the internal maxillary veins, and drain effluent blood from the intraosseus temporal sinuses (see Fig. 1). In the intact dog, these paired veins drain approximately 50% of the cerebral venous effluent blood, with the remaining portion returning to the heart via the intraosseus sigmoid and vertebral sinuses. The present experiments were conducted to examine the effect of sympathetically mediated extracranial venoconstriction on canine cerebral venous, sigmoid sinus, and mean arterial pressures.

Methods

A. Animal Preparation

Male dogs weighing between 25 and 30 kg were anaesthetized with 120 mg/kg intravenous alpha-chloralose (60 mg/ml in isotonic sodium tetraborate). A sufficient depth of anaesthesia and paralysis was maintained by continuous i.v. infusion (4.5 ml/kg/hr) of an isotonic solution containing 10.0 mg/ml alpha-chloralose, 3.36 mg/ml sodium bicarbonate, 1.0 mg/ml dextrose, 0.3 mg/ml potassium chloride, 5.11 mg/ml sodium chloride, and 0.22 mg/ml gallamine triethiodide. Following tracheal intubation, each animal was mechanically ventilated with 30% oxygen, 70% nitrogen at a rate and depth which maintained the end-tidal carbon dioxide concentration at 5.0%. Oesophageal temperature was maintained at 39 °C by a proportionally controlled heating pad. Arterial pressure was measured via a femoral arterial catheter connected to a Statham 23 BB pressure transducer.

The left superior cervical ganglion was exposed and two platinum stimulating electrodes were placed around the fibre bundle passing from the ganglion to the vagus nerve. Following electrode placement, the electrodes and nerve were covered with a sponge soaked in mineral oil. To prevent antidromic stimulation of the heart, all communications betwen the ganglion and the vagus nerve were severed. A constant current stimulator (Grass S 44/PSIU 6) was employed to deliver square-wave impulses of 5 milliseconds duration. The current of stimulation in each animal was that which produced a maximal pupillary dilatation during a 2 Hz test stimulation.

The right temporal sinus was opened with a dental drill and occluded downstream with two heparin soaked cotton pellets. A small polyethylene cannula (PE 90) was introduced into the sinus and advanced 2 cm toward the junction of the right and left transverse sinuses (see Fig. 1). The right temporal sinus cannula was then secured to the cranium and the opening in the sinus was sealed with bone wax. This cannula was connected to a Statham pressure transducer (P 23 V) for the measurement of cerebral venous pressure.

The right sigmoid sinus was opened and occluded upstream with cotton pellets. A PE 90 cannula was introduced into the sinus and advanced 2 cm toward the vertebral sinus. After this cannula was secured and the sinus was sealed with bone wax, it was connected to a Statham transducer (P 23 V) for the measurement of sigmoid sinus pressure. The left sigmoid sinus was also exposed, but was not opened or occluded until later in the experiments (see below).

B. Protocol

The experimental protocol consisted of two frequency response determinations, one performed prior to left sigmoid sinus occlusion and one performed after occlusion. Prior to left sigmoid occlusion, cerebral effluent blood drained via both the sigmoid and temporal sinuses. Following occlusion, all cerebral effluent blood drained exclusively through the left temporal sinus and thus through the left retroglenoid vein. The frequencies of stimulation employed were 1, 2, 4, 8, and 16 Hz administered in random order. Each period of stimulation lasted 60 seconds and nine minutes were allowed to elapse between consecutive stimulations. When both frequency response determinations had been made, 2 mg/kg phentolamine (1 mg/ml in isotonic saline) were infused intravenously. Thirty minutes later, the ganglion was stimulated for 60 seconds at 8 Hz.

C. Data Collection and Analysis

Mean arterial (MAP), cerebral venous (CVP), and sigmoid sinus pressure (SSP) values were digitized and stored once every two seconds during protocol execution with the aid of an on-line computer system. Following the completion of the experimental protocol, averages were taken for each variable for the 10-second interval occuring immediately prior to each stimulation and during the last 10 seconds of each stimulation. Paired differences between corresponding averages were then calculated for each frequency of stimulation for each variable both before and after left sigmoid sinus occlusion. For each variable and frequency, these absolute differences were averaged and plotted against stimulation frequency as shown in Figs. 2 and 3. Each population of paired differences was evaluated for statistical significance using a paired-t test. Throughout the text, values are given as the mean ± its standard error for 6 experiments. Unless otherwise stated, statistical significance implies a P value less than 0.05.

Results

A. Cerebral Venous Pressure (see Fig. 2)

At rest CVP was 11.9 ± 1.3 mm Hg. Following each stimulation, CVP returned to its control value prior to restimulation. During pre-occlusion

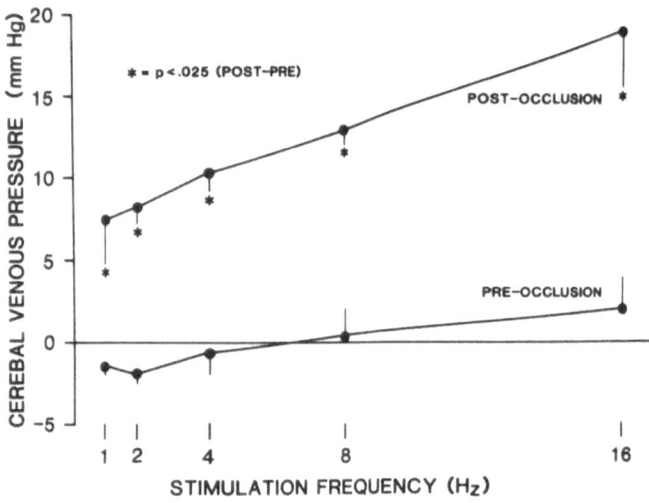

Fig. 2. Above are shown the absolute changes in cerebral venous pressure plotted against frequency of stimulation both before and after occlusion of the left sigmoid sinus. The asterisks indicate statistical significance at the P < 0.025 level for the paired differences between pre- and post-occlusion responses

stimulation, CVP exhibited a slight tendency to decrease at stimulation frequencies below 4 Hz and tended to increase somewhat during stimulation at 16 Hz, although none of these changes were statistically significant. During post-occlusion stimulations, CVP increased significantly at all frequencies of stimulation. The magnitudes of these increases ranged from 7.3 ± 2.4 mm Hg at 1 Hz to 18.9 ± 3.3 mm Hg at 16 Hz and were directly proportional to the frequency of stimulation. Following the administration of intravenous phentolamine, sympathetic stimulation had no measurable effect on CVP.

B. Sigmoid Sinus Pressure (SSP) (see Fig. 3)

At rest, SSP was 10.2 ± 1.1 mm Hg. Following each stimulation, SSP returned to its resting value prior to restimulation. During pre-occlusion stimulations at 1 and 2 Hz, SSP decreased significantly by 2.4 ± 0.6 mm Hg and 2.2 ± 0.7 mm Hg respectively. SSP also tended to decrease somewhat at 4 and 8 Hz, but these changes were not significant. The small increase in SSP observed at 16 Hz was not significant. During post-occlusion stimulations, SSP tended to increase in proportion to the frequency of stimulation; however, none of these increases were significant. Although the stimulation-induced changes in SSP were consistently more positive after occlusion than before, none of these differences were significant. Following the administration of intravenous phentolamine, sympathetic stimulation had no measurable effect on SSP.

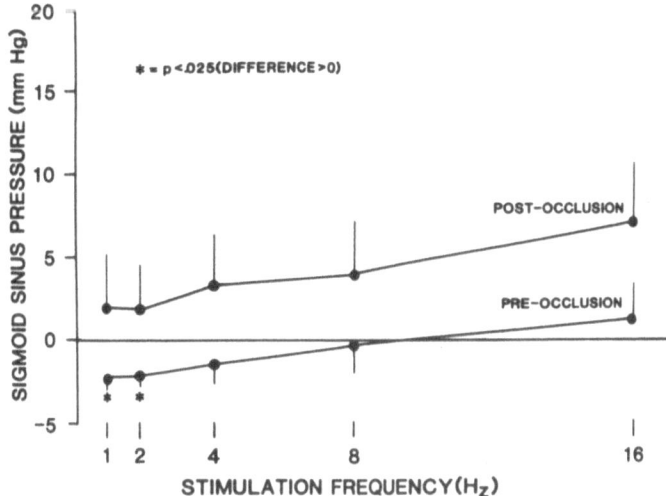

Fig. 3. Above, the absolute changes in sigmoid sinus pressure are plotted against stimulation frequency both before and after left sigmoid sinus occlusion. The asterisks indicate statistical significance at the P < 0.025 level for the paired differences between pre-stimulation and stimulation conditions

C. Mean Arterial Pressure (MAP)

At rest, MAP averaged 112.2 ± 9.4 mm Hg. Following each stimulation, MAP returned to control values prior to restimulation. Pre-occlusion stimulations had no significant effect on MAP. During post-occlusion stimulations at 1 and 2 Hz, MAP did not change significantly. However, during post-occlusion stimulations at 4, 8, and 16 Hz, MAP exhibited significant increases of 16.9 ± 14.6 mm Hg, 26.0 ± 14.9 mm Hg, and 29.6 ± 17.0 mm Hg respectively. Following the administration of intravenous phentolamine, sympathetic stimulation had no effect on MAP.

Discussion

The resting values of cerebral venous pressure (CVP) obtained in the present study agree well with previously published values[3, 8, 9, 15, 19, 20]. That CVP was routinely higher than sigmoid sinus pressure (SSP) suggests that the sigmoid/vertebral sinus pathway normally drains slightly more than 50% of the cerebral effluent blood in the intact dog (see Fig. 1). Although sigmoid sinus occlusion alone had no significant effect on resting CVP, SSP, or MAP, it had a significant effect on the stimulation-induced responses of these variables.

Prior to sigmoid occlusion, CVP did not change significantly during sympathetic stimulation. After occlusion however, CVP increased markedly during stimulation in a frequency dependent manner. Indeed, the shape of the CVP frequency-response curve (see Fig. 2) closely parallels that obtained from in vitro studies of sympathetically mediated retroglenoid vein contractility[16-18]. Because the retroglenoid veins constitute the only cerebral venous outflow path after occlusion (see Fig. 1), and have exhibited sympathetically mediated vasoconstriction in vitro, we attribute the post-occlusion, stimulation-induced increases in CVP to sympathetic vasoconstriction of the canine retroglenoid veins. Consistent with this conclusion is the observation that phentolamine blocked the stimulation-induced increase in CVP observed after sigmoid sinus occlusion. Although cerebral venous and intracranial pressures are increased under conditions of elevated cerebral blood flow[1, 11, 12, 14], it is highly improbable that cerebral blood flow increased during sympathetic stimulation in the present experiments[2, 7]. Stimulation-induced increases in MAP may also be excluded as a mechanism responsible for the observed post-occlusion increases in CVP since CVP increased significantly at low frequencies of stimulation (1 and 2 Hz) without any concomitant change in MAP.

Prior to sigmoid occlusion, SSP decreased significantly during stimulation at 1 and 2 Hz, a response attributable to sympathetically mediated reduction of blood flow to the cranial muscular beds which drain into the sigmoid/vertebral venous system. At increasing frequencies of pre-occlusion stimulation, however, SSP tended to increase. We attribute this latter response to an overall increase in the volume of cerebral effluent draining via the sigmoid/vertebral system, which in turn would be expected during sympathetically mediated retroglenoid venoconstriction; retroglenoid venoconstriction diverted effluent away from the temporal sinus and toward the sigmoid sinus prior to occlusion.

Following occlusion, SSP tended to increase with increasing frequencies of stimulation, although none of these changes were statistically significant. Thus, it remains possible that the outflow resistance in the sigmoid/vertebral sinus pathway increases slightly during sympathetic stimulation. This possibility seems unlikely, however, since the sigmoid/vetebral pathway travels primarily through bony channels of fixed diameter. Therefore, we consider the post-occlusion, stimulation-induced increases in SSP to be secondary to the concomitant increases in MAP which occured only after occlusion.

Prior to occlusion, stimulation had no effect on MAP. After occlusion, however, MAP increased significantly in a frequency dependent manner at 4, 8, and 16 Hz. These stimulation-induced increases cannot be due to antidromic stimulation of the heart since all fibres passing from the site of stimulation toward the heart had been cut. In addition, stimulus parameters and electrode location were identical before and after occlusion. Therefore, we conclude that the post-occlusion, stimulation-induced increases in MAP were in some way secondary to sigmoid sinus occlusion.

Although the systemic hypertensive response to increased intracranial pressure is commonly thought to be only a pre-terminal event[4, 5, 12], several studies have reported graded and reversible increases in MAP following experimentally induced intracranial hypertension[6, 9, 10, 19]. Because cerebral venous and thus intracranial pressures were significantly elevated during post-occlusion, but not pre-occlusion stimulations, and because MAP did not change during pre-occlusion stimulations, we interpret the post-occlusion, stimulation-induced increases in MAP as a reversible systemic hypertensive response to raised intracranial pressure. The elevated intracranial pressure, in turn, was secondary to post-occlusion retroglenoid venoconstriction.

In view of the present results, it is doubtful that retroglenoid venoconstriction in the intact dog has a significant effect on cerebral venous or intracranial pressures, since the retroglenoid pathway is normally in confluence with the sigmoid-vertebral pathway. Instead, retroglenoid venoconstriction probably acts to increase the fraction of cerebral venous effluent blood draining via the sigmoid-vertebral sinus pathway. Only in situations where the sigmoid-vertebral pathway is blocked should retroglenoid venoconstriction have a pronounced effect on cerebral venous, intracranial, and possibly arterial pressures.

In man, only a small fraction of the cerebral venous effluent blood drains via the vertebral plexus. Instead, the sigmoid and temporal sinuses constitute a single outflow tract which drains exclusively into the internal jugular veins. Thus, if human internal jugular veins have the capacity to constrict as do the canine retroglenoid veins, such constriction may influence the control of intracranial pressure, cerebral venous pressure, and cerebral blood volume in humans. In the light of such a mechanism, increases in human cerebral blood volume have been reported to accompany intracranial hypertension in selected cases[13], a finding consistent with internal jugular venoconstriction. Although the increased CBV and cerebral venous pressure occuring during intracranical hypertension have been attributed to the mechanical compression of intracranial veins[15], constriction of the internal jugular veins could also be involved, particularly since preliminary studies of monkey internal jugular veins suggest that these vessels receive a dense adrenergic innervation[18]. The contractile capacity of these vessels, however, awaits examination.

References

1. Candia, G. J., Heros, R. C., Lavyne, M. H., Zervas, N. T., Nelson, C. N.: Effect of intravenous sodium nitroprusside on cerebral blood flow and intracranial pressure. Neurosurg. *3* (1978), 50—53.

2. Edvinsson, L., MacKenzie, E. T.: Amine mechanisms in the cerebral circulation. Pharm. Rev. *28* (1976), 275—353.

3. Ekstrom-Jodal, B.: On the relationship between blood pressure and blood flow in the canine brain with particular regard to the mechanism responsible for cerebral blood flow autoregulation. Acta Physiol. Scand. Suppl. *350* (1970), 1—8.

4. Evans, J. P., Espey, F. F., Kristoff, F. V., Kimbell, F. D., Ryder, H. W.: Experimental and clinical observations on rising intracranial pressure. Arch. Surg. *63* (1951), 107—114.

5. Fitch, W., McDowall, D. G., Keaney, N. P., Pickerodt, V. W. A.: Systemic vascular responses to increased intracranial pressure. J. Neurol. Neurosurg. Psychiat. *40* (1977), 843—852.

6. Grubb, R. L., Raichle, M. E., Phelps, M. E., Phelps, M. E., Ratcheson, R. A.: Effects of increased intracrainial pressure on cerebral blood volume, blood flow, and oxygen utilization in monkeys. J. Neurosurg. *43* (1975), 385—398.

7. Heistad, D. D., Marcus, M. L., Gross, P. M.: Effects of sympathetic nerves on cerebral vessels in dog, cat, and monkey. Am. J. Physiol. *235* (1978), H544—H552.

8. Jacobson, I., Harper, A. M., McDowall, D. G.: Relationship between venous pressure and cortical blood flow. Nature *4902* (1963), 173—175.

9. Johnston, I. H., Rowan, J. O., Park, D. M., Rennie, M. J.: Raised intracranial pressure and cerebral blood flow. 5. Effects of episodic intracranial pressure waves in primates. J. Neurol. Neurosurg. Psychiat. *38* (1975), 1076—1082.

10. Leech, P., Miller, J. D.: Intracranial volume-pressure relationships during experimental brain compression in primates. 1. Pressure responses to changes in ventricular volume. J. Neurol. Neurosurg. Psychiat. *37* (1974), 1093—1098.

11. Leech, P., Miller, J. D.: Intracranial volume-pressure relationships during experimental brain compression in primates. 2. Effect of induced changes in systemic arterial pressure and cerebral blood flow. J. Neurol. Neurosurg. Psychiat. *37* (1974), 1099—1104.

12. Manz, H. J.: Pathophysiology and pathology of elevated intracranial pressure. In: Pathobiology Annual (Ioachim, H. L., ed.), pp. 359—381. New York: Raven Press. 1979.

13. Mathew, N. T., Meyer, J. S., Ott, E. O.: Increased cerebral blood volume in benign intracranial hypertension. Neurol. *25* (1975), 646—649.

14. Miller, J. D.: Volume and pressure in the craniospinal axis. Clin. Neurosurg. *22* (1975), 76—105.

15. Nakagawa, Y., Tsuru, M., Yada, K.: Site and mechanism for the compression of the venous system during experiemental intracranial hypertension. J. Neurosurg. *41* (1974), 427—434.

16. Pearce, W. J., Bevan, J. A.: Sympathetic stimulation, cerebral blood flow, an the role of extra-cerebral vasoconstriction. In: Advances in Physiological Sciences, Vol. IX: Cardiovascular Physiology: Neural Control Mechanisms (Kovach, A. G. G., *et al.,* eds.), pp. 157—166. New York: Pergamon Press. 1981.

17. Pearce, W. J., Bevan, J. A.: The possible influence of extracerebral venoconstriction on cerebral hemodynamics. J. Cerebr. Blood Flow Metabol. 1, Suppl. 1 (1981), 325—326.

18. Pearce, W. J., Bevan, J. A.: Neurogenic vasoconstriction of extracranial veins draining canine cerebral venous effluent. In: Cerebral Blood Flow-Effects of Nerves and Neurotransmitters (Heistad, D. W., *et al.,* eds.), pp. 87—94. New York: Elsevier. 1982.

19. Raisis, J. E., Kindt, G. W., McGillicuddy, J. E., Giannotta, S. L.: The effects of primary elevation of cerebral venous pressure on cerebral hemodynamics and intracranial pressure. J. Surg. Res. *26* (1979), 101—107.
20. Symon, L., Crockard, H. A., Juhasz, J., Branston, N. M.: The effect of intracranial hypertension on cerebrovascular resistance—an experimental study. Acta neurochir. (Wien) *35* (1976), 221—232.

Authors' address: W. J. Pearce, Ph.D., Director, Macpherson Laboratory, White Memorial Medical Center, 1720 Brooklyn Avenue, Los Angeles, CA 90033, U.S.A.

Experimental Research Department, Semmelweis Medical University, Department of Neurology, The University of Alabama in Birmingham

Integrated Microvessel Diameter and Microregional Blood Content as Determined by Cerebrocortical Video Reflectometry

A. Eke

Summary

Quantitative television reflectometry of the exposed brain cortex with digital processing of its reflectance image has been applied for the measurement of arterial, venous diameter and parenchymal blood content integrated over a selected region. A map of the region's blood content has also been constructed from several thousand individually determined values of tissue optical density (OD), which linearly correlate with the tissue blood content[2,3]. Image processing of these maps has separated the low density pixels within parenchymal areas from those of high density within vascular areas, and allowed the integration of the parenchymal blood content by means of the low density pixels. The difference between the number of high-density pixels in maps with and without arterial bolus perfusion—when temporarily no blood remains in the arteries—was found to be proportional to the integrated arterial diameter, whereas the high-density pixel-count during the arterial haemodilution measured the integrated venous diameter.

Keywords: Pial vessels; regional control of CBF; television reflectometry; image processing.

Introduction

Increasingly accurate determination of vessel diameter in the complex network of pial arteries and veins has long been of interest, not only because of the significance of these parameters in characterizing the supplying and draining functions of the network, but also because of their direct influence on the intraparenchymal flow. From the pioneering work of Bloch[1] and Wiederhielm[4], the television camera has been more and more frequently used in the measurement of vessel diameter by analyzing the voltage versus the distance of raster lines as the electron beam sweeps the image of an illuminated vessel. As with any of these "image splitting" methods, however, measurements could only be made in selected sites within a vascular network.

In a complex network of vessels, such as that over a region of the brain cortex, single or even several measurements of vessel diameter are insufficient for the quantitative analysis of vascular responses since, in view of their possible contribution to the control of regional blood flow, an integrated diameter of the whole regional network is more representative than a single diameter measured at a preselected site.

Analysis of the reflectance images of the brain cortex by the microflow mapping method of Eke[2] revealed that the measurement of integrated vessel diameters and parenchymal blood content can be done by computerized processing of reflectance images recorded before and during arterial bolus perfusion. A brief description of the techniques follows.

Description of the Method

The brain cortex is exposed by a cranial window for television reflectometry[2, 3]. Television images of the brain cortex are collected on video tape prior to and during bolus perfusion, induced by the intracarotid injection of an isotonic dextran-saline solution. The present method performs its analysis on two frames from this reflectometric record (Fig. 1): the one recorded prior to and that at the peak of the arterial transit of the haemodiluted bolus, when no blood remains in the arteries for a time longer than the TV framing rate ($\frac{1}{30}$ sec). Having each frame digitized at 8 bit resolution for 10,000 picture elements (pixels), a map of the region's blood content is constructed from the individual values of tissue optical density[2, 3]. A map like this is shown in Fig. 2. Digital processing of these maps can separate pixels within low-density parenchymal areas from those within high-density vascular areas, assigning 0 (white) and 1 (black) respectively to the corresponding pixels in the thresholded

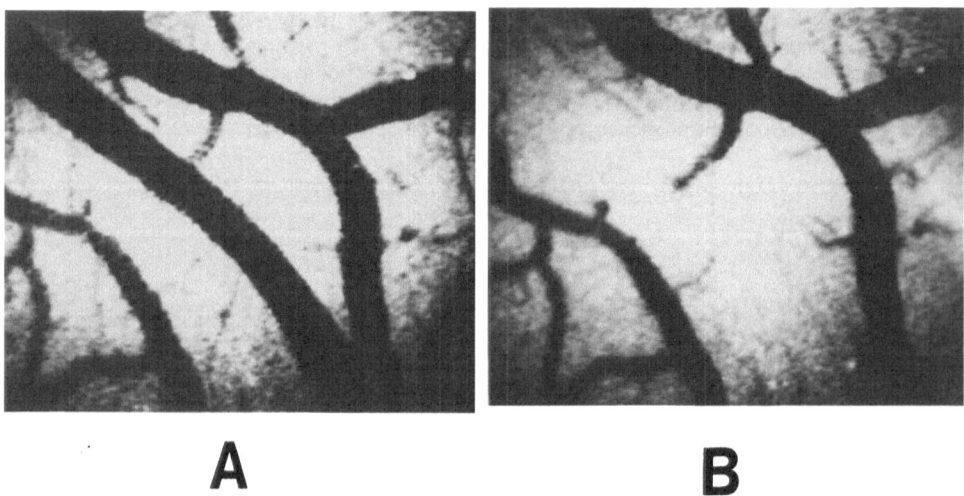

Fig. 1. Television images of 6 mm² of the brain cortex of an anaesthetized cat before (A) and at the peak of the arterial transit of a haemodiluted bolus (B) induced by an intracarotid injection of isotonic dextran-saline solution. Note that on image-B, the blood is completely eliminated by the bolus from the artery and that the parenchymal and venous blood content was not yet affected at all

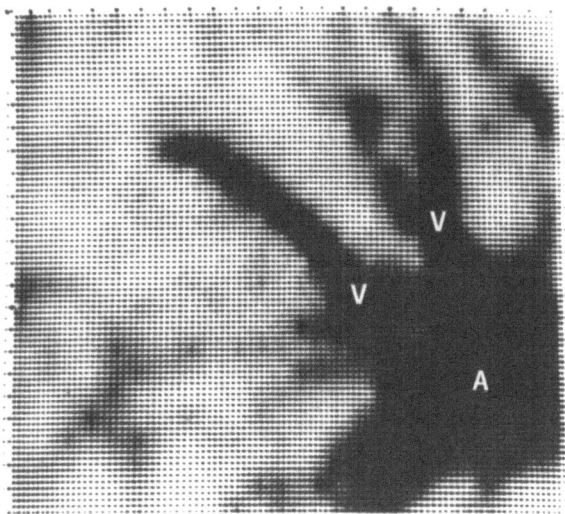

Fig. 2. Microregional blood content mapped over 9 mm³ of the brain cortex, as determined by the optical density of the tissue. It was calculated from digitized reflectance images and displayed by an Apple II microcomputer system. *A* artery, *V* veins. Light areas represent the parenchyma in which the blood content is somewhat heterogeneous

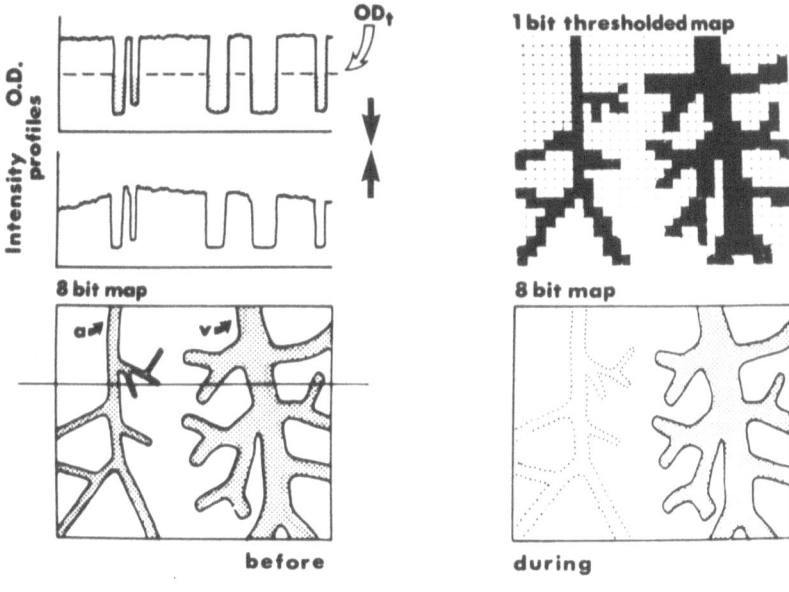

Fig. 3. A scheme of the analysis of the optical density (blood content) maps, such as shown in Fig. 2, for integrated arterial (*a*) and venous (*v*) diameters and parenchymal blood content. Two OD-maps are schematically represented on the lower panels. They are obtained by 8 bit conversion of the TV-frames of the reflectance images. Note, that the high density of the artery is completely diminished during bolus perfusion (see also Fig. 1) allowing a selective measurement of the integrated venous diameter. Intensity and optical density profiles of the reflectance, and optical density maps are shown in the upper left panel. (Note the direction of increase indicated by black arrows!) The map on the upper right separates the high-density vascular areas (black) from the low-density parenchymal areas (white) by having the 8 bit OD-map thresholded on an OD-level (*ODt*) selected well above that of the parenchymal areas but also much below that of the visible pial vessels. The integrated arterial and venous diameter equals the respective pixels counts of this map, whilst the integrated parenchymal blood content is a sum of the low-density OD-data

maps of 1 bit resolution (see Fig. 3). The difference between the number of high-density pixels in the frames with and without arterial haemodilution is proportional to the integrated arterial diameter, whereas the high-density pixel-count during the arterial haemodilution measures the integrated venous diameter. The integrated parenchymal blood content equals the sum of the low-density pixel data on the map recorded before the arterial bolus perfusion.

Conclusions

The integrated measurement of arterial and venous diameters, and parenchymal blood content may contribute a more adequate description of the significant factors in the regional control of CBF. Since integrated and distributed values for vascular diameters and parenchymal blood content are simultaneously provided by these reflectometric techniques for the same cerebrocortical area, this may lead to a better understanding of the interdependence of regional versus sub- or microregional controls of CBF described by integrated and distributed parameters, respectively.

References

1. Bloch, E. H.: A method for studying the dynamics of transcapillary transfer quantitatively at the microscopic level in situ in living organs. Angiology *14* (1963), 97—106.
2. Eke, A.: Reflectometric mapping of microregional blood flow and blood volume in the brain cortex. J. Cerebr. Blood Flow Metabol. *2* (1982), 41—53.
3. Eke, A., Hutiray, G., Kovach, A. G. B.: Induced hemodilution detected by reflectometry for measuring microregional blood flow and blood volume in cat brain cortex. Amer. J. Physiol. *236* (1979), H759—H768.
4. Wiederhielm, C. A.: Continuous recording of arteriolar dimensions with television microscope. J. appl. Physiol. *18* (1963), 1041—1042.

Author's address: Dr. A. Eke, Assistent Professor, Experimental Research Department, Semmelweis Medical University, Ulloi ut 78/a, H-1082 Budapest VIII, Hungary.

Pharmacological Effects

Department of Neurosurgery, Graz University, Austria, [1]Department of Surgical Neurology, Research Institute for Brain and Blood Vessels, Akita, Japan

Pial Venous Reaction and Cerebral Blood Volume During Methohexital Anaesthesia

K. Haselsberger, I. Sayama[1], and L. M. Auer

Summary

In a study on adult cats, the reactions of pial veins and arteries through a cranial window, as well as cerebral blood volume, blood pressure and intracranial pressure were observed during a 60-minute period of Methohexital anaesthesia, an ultrashort-acting barbiturate. While MAP and pial arteries showed no significant reaction, pial veins constricted significantly by 20%.

It is concluded that methohexital might exert a direct venoconstrictory effect, thus reducing the cerebral blood volume. This observation appears of clinical interest in situations of increased ICP and cerebral blood volume such as in the acute stage after severe head injury.

Keywords: Methohexital; pial venoconstriction; cats.

Introduction

Barbiturates have obtained attention as potential protectors of the ischaemic brain[6, 9, 11, 13, 16, 17, 19, 21]. They have therefore become of clinical interest for the induction of iatrogenic coma, especially in patients with severe head injury[8, 12, 14, 15]. Besides their direct lowering effect on the blood pressure[7], cerebral blood flow and volume are reduced via metabolic depression[1, 7, 17, 18]. Methohexital, a short-acting barbiturate, has so far been little investigated regarding its effect on cerebral blood flow as well as direct cerebrovascular response. In the baboon, methohexital seemed to reduce CBF in the healthy brain areas and increase it in the ischaemic regions; vasoconstriction in the healthy areas and a steal phenomenon have been suggested, but the types of vessels involved could not be specified[7]. In healthy man, a 45% CBF-reduction was observed following the i.v. bolus-injection of 1 mg/kg Methohexital[12a].

Since cerebral venous reactions are of importance with respect to their possible role as regulators of cerebral blood volume, the present study was undertaken.

Materials and Methods

Investigations were performed in 6 adult cats of either sex. The animals were initially anaesthetized with $30\,mgkg^{-1}$ sodium-pentobarbital i.v. Following endotracheal intubation, one femoral vein and artery were cannulated for continuous monitoring of mean arterial pressure using a Statham P 23 dB pressure transducer and a Hellige 1214 electromanometer, for blood gas analysis (AVL gas 937 C gas check) and intravenous administration of drugs. Animals were then immobilized with $60\,\mu g/kg^{-1}$ pancuronium-bromide and ventilated with a $3:1$ mixture of $N_2O:O_2$ using a Loosco baby respirator. $PaCO_2$ and PaO_2 were kept at normal levels during the experiments, mean 29.7 ± 2 mm Hg and 109.7 ± 3.5 mm Hg, respectively. With the animal in the sphinx position and the head fixed into a stereotaxic holder, a closed cranial window was made in the left parietal region as previously described in detail[2]. Pial venous and arterial diameter variations were continuously registered with a TV multichannel videoangiometer[4]. Cerebral blood volume variations were continuously recorded with a semiquantitative photometric technique by focussing a photomultiplier on a small cortical area free of pial vessels[3]. Intracranial pressure was measured via a plastic cannula in the cisterna magna, a Statham P 23 dB pressure transducer and a HSE electromanometer. After changing the respirator gas mixture from $N_2O:O_2$ to room-air, Methohexital anaesthesia was induced: therefore, one initial bolus of $1\,mg/kg^{-1}$ was slowly injected intravenously, followed by a 60-minute continuous infusion of $0.1\,mg/kg^{-1}\,min^{-1}$ Methohexital. Statistical analysis was performed using variance analysis (F-test, Duncan-test).

Results

The mean arterial pressure (MAP), 117.5 ± 2.8 SEM mm Hg on average under resting conditions, tended to initially decrease by about 10%, returning to the resting level within 20 minutes, and then there was a slowly progressing decrease during the remaining 40 minutes; the wide variation between individual animals never allowing the mean MAP to reach a significant change from the resting level (Fig. 1). Intracranial pressure showed no significant change, absolute mean values remaining between 1.5 and 2.5 mm Hg.

Pial veins with resting diameters between 74 and 190 μm uniformly constricted following the induction of Methohexital anaesthesia. The individual maximal constriction as a percentage of the resting calibre ranged from 13.2 to 40.9% and was 25.9% on average. The mean venoconstriction with time is depicted in Fig. 2, where constriction is shown to reach its maximum after 30 minutes at a level of -20.8%, becoming insignificantly less during further anaesthesia and returning towards the resting level following cessation of the Methohexital infusion.

By contrast, the reactions of the pial arteries were variable, individual maximal reactions ranging between 25.0% dilatation and 21.1% constriction, with an averaged maximal dilatation of 8.5% during the 60 minutes observation period; the mean maximal constriction was 0.7%. Average reactions at intervals of 5 minutes expressed the wide variation of individual reactions, all different mean values from the resting level being statistically insignificant.

Concomitant with venoconstriction, cerebral blood volume could eventually be seen to decrease (Fig. 3).

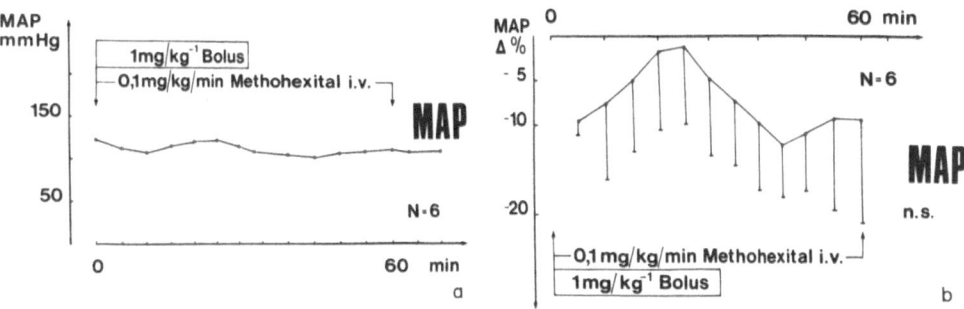

Fig. 1. Mean values of MAP (mean arterial pressure), a) absolute values in mm Hg, b) mean percentage changes from resting level ± SEM. Due to the wide variation of individual data, reductions of about 10% of the resting were not statistically significant

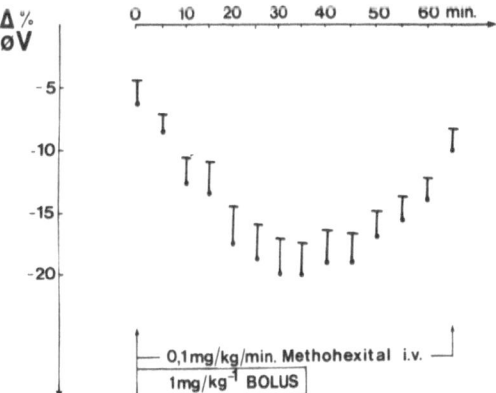

Fig. 2. Mean percent venoconstriction during the 60 minutes of Methohexital anaesthesia. After 30 minutes, a 20.8% reduction in the resting calibres is noted, statistically significant at the 1% level

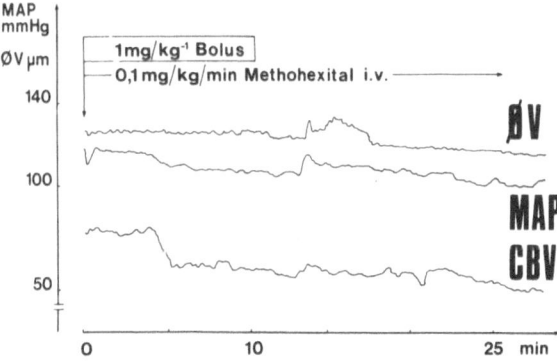

Fig. 3. Traces from an individual experiment, showing a pial vein's (∅V) (absolute diameter in μm) slowly decreasing during Methohexital infusion. Cerebral blood volume (CBV), semiquantitatively estimated, also decreases. MAP (mean arterial pressure in mmHg)

Discussion

The present experiments indicate a possible direct venoconstrictor effect of Methohexital: a significant reduction in venous calibre in a situation of unchanged resistance vessels and an insignificant 10% reduction of MAP suggest a reduction in the cerebral blood volume on the venous side that is independant from the possible and wellknown depressant effect on the cerebral metabolism. This is not in full accordance with the opinion of Branston *et al.*[7], though differences might be explained by different doses used. It could be suggested that a direct vascular effect of Methohexital has been found in the present study due to the relatively low dosage-rate of infusion. This direct vascular effect might be masked with higher doses due to the marked metabolic reduction.

It would be of special interest to investigate the effect of Methohexital-anaesthesia in neurosurgical intensive care patients with special respect to its potential role in the reduction of an elevated intracranial pressure via its effect on the cerebral blood volume. The latter has been shown to be elevated rather than decreased in experimental situations of increased ICP such as CSF-infusion and water intoxication[5] as well as in the acute stage after severe head injury. Studies with simultaneous measurement of cerebral blood flow, brain metabolism and cerebral blood volume in patients during Methohexital anaesthesia should be performed in order to elucidate potential benefits from this ultrashort-acting barbiturate.

Acknowledgement

We are grateful to Mr. H. Leiner for technical assistance and to Miss S. Auer for the statistical calculations.

This work was supported by the Austrian *Fonds zur Förderung der wissenschaftlichen Forschung* (Project No. 4368).

References

1. Astrup, J., Nordström, C. H., Rehncrona, S.: Rate of rise in extracellular potassium in the ischemic rat brain and the effect of pre-ischemic metabolic rate: Evidence for a specific effect of phenobarbitone. In: Cerebral Function, Metabolism and Circulation (Ingvar, D. H., *et al.*, eds.), pp. 7.8—7.9. Copenhagen: Munksgaard. 1977.
2. Auer, L. M.: The pathogenesis of hypertensive encephalopathy. Acta neurochir. (Wien) Suppl. *27* (1978), 1—111.
3. Auer, L. M.: A method for continuous monitoring of pial vessel diameter changes and its value for dynamic studies of the regulation of cerebral circulation. A preliminary report. Pflügers Archiv *373* (1978), 195—198.
4. Auer, L. M., Haydn, F.: Multichannel videoangiometry for continuous measurement of pial microvessels. Acta Neurol. Scand. *60*, Suppl. 72 (1979), 208—209.
5. Auer, L. M., Sayama, I., Johansson, B. B., Leber, K.: Pial venous reaction to sympathetic stimulation during elevated ICP. In: The Cerebral Veins (Auer, L. M., Loew, F., eds.), pp. 143—153. Wien-New York: Springer. 1983.
6. Bleyaert, A. L., Nemoto, E. M., Safar, P., *et al.*: Thiopental amelioration of post-ischemic encephalopathy in monkeys. In: Cerebral Function, Metabolism and Circulation (Ingvar, D. H., *et al.*, eds.), pp. 7.4—7.5. Copenhagen: Munksgaard. 1977.

7. Branston, N. M., Hope, D. T., Symon, L.: Barbiturates in focal ischemia of primate cortex: effects on blood flow distribution, evoked potentials and extracellular potassium. Stroke *10* (1979), 647—653.

8. Bruce, D. A., Raphaely, R. A., Swedlow, D., Schut, L.: The effectiveness of iatrogenic barbiturate coma in controlling increased ICP in 61 children. In: Intracranial Pressure IV (Shulman, K., *et al.,* eds.), pp. 630—632. Berlin-Heidelberg-New York: Springer. 1980.

9. Corkill, G., Sivaligam, S., Reitan, J. A., *et al.*: Dose dependency of the post-insult protective effect of pentobarbital in the canine experimental stroke model. Stroke *9* (1978), 10—12.

10. Crane, P. D., Braun, L. D., Cornford, E. M., *et al.*: Dose dependent reduction of glucose utilization by pentobarbital in rat brain. Stroke *9* (1978), 12—18.

11. Flamm, E. S., Demopoulos, H. B., Seligman, M. L., *et al.*: Possible molecular mechanisms of barbiturate-mediated protection in regional cerebral ischemia. In: Cerebral Function, Metabolism and Circulation (Ingvar, D. H., *et al.,* eds.), pp. 7.10—7.11. Copenhagen: Munksgaard. 1977.

12. Gaab, M. R., Bushe, K. A.: Die Behandlung der intrakraniellen Drucksteigerung. Intensivbehandlung *6* (1981), 34—52.

12 a. Herrschaft, H., Schmidt, H.: Der Einfluß von Methohexital-Natrium auf die globale und regionale Hirndurchblutung des Menschen. Anaesthesist *45* (1974), 340—344.

13. Hoff, J. T., Smith, A. L., Hankinson, H. L., *et al.*: Barbiturate protection from cerebral infarction in primates. Stroke *6* (1975), 28—33.

14. Marshall, L. F., *et al.*: The outcome with aggressive treatment in severe head injuries. Part II: Acute and chronic barbiturate administration in the management of head injury. J. Neurosurg. *50* (1979), 26—30.

15. Michenfelder, J. D., Theye, R. A.: Cerebral protection by thiopental during hypoxia. Anesthesiology *39* (1973), 510—517.

16. Michenfelder, J. D., Milde, J. H.: Cerebral protection by anaesthetics during ischaemia (a review). Resuscitation *4* (1975), 219—233.

17. Michenfelder, J. D., Milde, J. H., Sundt, T. M.: Cerebral protection by barbiturate anesthesia. Use after middle cerebral artery occlusion in Java monkeys. Arch. Neurol. *33* (1976), 345—350.

18. Pierce, E. C., Lambertsen, C. J., Deutsch, S., *et al.*: Cerebral circulation and metabolism during thiopental anesthesia and hyperventilation in man. J. Clin. Invest. *41* (1962), 1664—1671.

19. Smith, A. L., Hoff, J. T., Nielson, S., *et al.*: Barbiturate protection in acute focal cerebral ischemia. Stroke *5* (1974), 1—7.

20. Wright, R. L., Ames, A.: Measurement of maximal permissible cerebral ischemia and a study of its pharmacologic prolongation. J. Neurosurg. *21* (1974), 567—574.

21. Yatsu, F. M., Diamond, I., Graziano, C., *et al.*: Experimental brain ischemia: protection from irreversible damage with rapid acting barbiturate (Methohexital). Stroke *3* (1972), 726—732.

Author's address: Univ.-Prof. Dr. L. M. Auer, Department of Neurosurgery, Graz University, A-8036 Graz, Austria.

Department of Anesthesiology and Critical Care Medicine, The Johns Hopkins Medical Institutions, Baltimore, Maryland, U.S.A.

Effects of Nitroglycerin and Nitroprusside on Cerebral Veins

M. C. Rogers and R. J. Traystman

Summary

Cerebral haemodynamic effects of intravenous nitroglycerin (NTG) and nitroprusside (NPR) were compared in ten anaesthetized, paralyzed, ventilated dogs, at normal cerebrospinal fluid pressure (Pcsf). NTG (5, 25, 50 µg/kg) increased Pcsf to 181, 250, and 273% of its control value and decreased arterial blood pressure to 89, 80, and 78% of the control level, respectively. NPR (5, 25, 50 µg/kg) increased Pcsf to 142, 169, and 177% of the control value and decreased arterial blood pressure to 88, 72, and 69% of the control level. Haemorrhagic hypotension to arterial blood pressure levels (54% of control) lower than those produced with either NTG or NPR did not alter Pcsf in five dogs. Cerebral blood flow was unchanged with NTG, NPR, or haemorrhagic hypotension. For any given decrease in arterial blood pressure, NTG produced a greater increase in Pcsf than NPR, or haemorrhagic hypotension. We conclude that NTG has a greater effect on cerebral veins than NPR. The increase in cerebral volume responsible for the increased Pcsf with these agents occurs as a consequence of an increase in the cerebral venous volume.

Keywords: Cerebral venous vasodilatation; cerebrospinal fluid; cerebral blood flow; nitroglycerin; nitroprusside.

Introduction

While vasodilatory drugs such as nitroglycerin (NTG) and nitroprusside (NPR) are known to have different effects on the systemic arteries and veins, the concept that these drugs may have different effects on cerebral arteries and veins remains unexplored. In the peripheral vessels, NTG acts primarily on the venous capacitance vessels[19], whereas NPR acts primarily on arterioles[14]. While NTG has been shown to increase cerebrospinal fluid pressure (Pcsf)[11] and dilate cerebral blood vessels[6], the effects of NTG on the relationship between the cerebral blood flow (CBF) and Pcsf are largely unknown.

NPR has been proposed as a useful agent in the treatment of cerebral vasospasm following subarachnoid haemorrhage[3]. Although NPR is known to be a potent peripheral vasodilator by virtue of its depressant effects on smooth muscle[7], some investigators have demonstrated cerebral vasoconstriction with NPR[1, 13]. Others have shown cerebral vasodilatation with NPR[4], whereas still others have shown no effect of NPR on CBF[2, 5]. As with NTG, NPR elevates Pcsf[18], but the effects of NPR on the relationship between CBF and Pcsf are unclear.

We investigated whether NTG and NPR affected cerebral arteries or veins by evaluating the increase in Pcsf with systemic hypotension produced by: NTG, NPR, or haemorrhage.

Methods

Experiments were done on fifteen adult, mongrel dogs of either sex (20–24 kg) anaesthetized with sodium pentobarbital (30 mg/kg, i.v.). Supplemental doses of anaesthetic were administered as required. Heparin (500 units/kg, i.v.) was used as the anticoagulant, with additional doses given every 90 minutes. The animals were paralyzed with succinylcholine (40 mg) and ventilated with a positive pressure respirator connected to a tracheal cannula. Tidal volume and respiratory rate were adjusted to give an alveolar (end-expiratory) carbon dixoide tension of 4.0% as monitored on a CO_2 gas analyzer. Dissection to expose the femoral artery and vein, carotid arteries, and cranium was done with an electric cautery. Arterial blood pressure (iliac arterial pressure) was measured via a cannula advanced from the femoral artery. To prevent cooling, the dogs were covered with a plastic sheet and all surgical areas, where possible, were sutured. Rectal temperature was maintained around 38 °C throughout the experiment with the use of a thermostatically controlled pad. All pressures were measured with Statham P-23 strain gauges, and all data were recorded on a Gould-Brush recorder. Oxygen tension (PO_2), carbon dioxide tension (PCO_2) and pH were measured immediately after the samples were obtained. End-expiratory CO_2 was maintained constant throughout the experiment.

The technique used to measure cerebral venous blood outflow has been described previously[10, 17]. The confluence of the cerebral sinuses was cannulated and the lateral sinuses and occipital emissary veins were occluded with bone wax to prevent communication between the intracranial and extracranial venous circulations. From the confluence of the sinuses the blood then passed through a previously calibrated electromagnetic flow probe, before returning to the dog via the femoral vein. With this technique, approximately 50–70% of the mass of the brain is drained at the confluence of the sagittal and straight sinuses. The technique has been used successfully to examine cerebral autoregulation[10], sympathetic[17], adrenergic[8], hypoxic[9, 15, 16], and hypercapnic[17] effects on the cerebral circulation. In addition, the verification procedure for this technique has been described in detail elsewhere[16, 17]. Cerebral venous outflow pressure was measured upstream from the flow probe. This pressure merely measures the resistance to the flow of blood induced by the flow transducer, since the outflow was set at the level of the right atrium and all pressures were referred to this common zero reference plane. Cerebral perfusion pressure (CPP) was estimated as systemic arterial pressure minus Pcsf. Intracranial vascular resistance was calculated by dividing CPP by CBF.

Pcsf was measured continuously via a catheter placed in the lateral ventricle or cisterna magna. NTG (N = 5) and NPR (N = 5) were administered intravenously (5, 25, 50 μg/kg) in random order over 15 seconds. NTG and NPR were dissolved in saline so that the volume delivered always remained the same. Control injections were done with the vehicle (saline) only. Fifteen minutes were allowed between each dose of drug so that all haemodynamic parameters had returned to control values before subsequent doses were given. Alternatively, systemic hypotension (N = 5) was produced by the withdrawal of blood from large bore catheters placed in the femoral artery and vein, at a rate designed to mimic the degree and time course of hypotension produced by NTG and NPR.

Control versus experimental values were compared using the paired t test and significance levels were accepted at the 0.05 level.

Results

Administration of NTG (5, 25, 50 µg/kg) increased Pcsf to 181, 250, and 273% of its control value and decreased mean arterial blood pressure to 89, 80, and 78% of the control level (Table 1). NPR (5, 25, 50 µg/kg) increased Pcsf to 142, 169, and 177% of its control value and decreased the mean arterial blood pressure to 88, 72, and 69% of the control level (Table 2). Haemorrhagic hypotension to arterial blood pressure levels (54% of control) less than those produced with either NTG or NPR resulted in no change in Pcsf (Table 3). CBF was unchanged from control values with NTG, NPR or haemorrhagic hypotension (Tables 1–3). For any given decrease in mean arterial blood pressure, NTG produced a greater increase in Pcsf than NPR or haemorrhagic hypotension. With 50 µg/kg NTG, or NPR, the ratio of the increase in Pcsf to the decrease in arterial blood pressure was 0.395 for NTG and 0.233 for NPR. With haemorrhagic hypotension this ratio was — 0.069.

Discussion

Our data show that NTG and NPR increase Pcsf and decrease mean arterial blood pressure, thus reducing cerebral perfusion pressure markedly. In these studies, cerebral perfusion pressure was always above the classically accepted lower limit for autoregulation (60 mm Hg). Thus the cerebral vasculature was able to autoregulate its blood flow and no changes in CBF were observed. This reduction in arterial blood pressure resulting from NTG or NPR has been well described[19] as has the rise in Pcsf[11, 18]. However, the mechanism by which NTG and NPR produce a rise in Pcsf is unclear. Our study suggests that, whatever the mechanism, it appears to be independent of changes in CBF. Perhaps the simplest mechanism which might account for the rise in Pcsf with NTG and NPR is autoregulatory vasodilatation of cerebral resistance vessels (arterioles), which could increase cerebral blood volume with no alteration in CBF. We do not feel this mechanism is of importance here since a reduction in arterial blood pressure alone, which also evokes cerebral autoregulatory vasodilation, does not increase Pcsf (Table 3). In addition, this autoregulatory mechanism could not explain the fact that NTG was able to increase Pcsf to a greater extent than NPR for any reduction in arterial blood pressure.

While it is convenient to consider cerebral vascular resistance as occurring almost exclusively in cerebral arterioles, this may in fact not be the case. It has been shown that 46% of the pressure drop across the cerebral vasculature occurred between small pial arterioles (25–40 µ diameter) and

Table 1. *Haemodynamic Effects of 5, 25, and 50 µg/kg Nitroglycerin*

	Control	5 µg/kg	Control	25 µg/kg	Control	50 µg/kg
Arterial blood pressure (mmHg)	138 ± 5	123 ± 6*	148 ± 8	119 ± 7*	138 ± 7	109 ± 6*
Cerebrospinal fluid pressure (mmHg)	11 ± 3	20 ± 4*	12 ± 3	30 ± 6*	11 ± 3	30 ± 4*
Cerebral perfusion pressure (mmHg)	127 ± 7	103 ± 9*	136 ± 8	89 ± 7*	127 ± 9	79 ± 7*
Cerebral blood flow (ml/min)	25 ± 2	26 ± 3	24 ± 1	23 ± 2	24 ± 2	25 ± 5

Each value = Mean ± SEM; N = 5; * = < 0.05.

Table 2. *Haemodynamic Effects of 5, 25, and 50 µg/kg Nitroprusside*

	Control	5 µg/kg	Control	25 µg/kg	Control	50 µg/kg
Arterial blood pressure (mmHg)	139 ± 6	122 ± 9*	145 ± 9	105 ± 13*	140 ± 8	97 ± 14*
Cerebrospinal fluid pressure (mmHg)	12 ± 2	17 ± 4*	16 ± 4	27 ± 6*	13 ± 5	23 ± 8*
Cerebral perfusion pressure (mmHg)	127 ± 7	105 ± 6*	129 ± 10	78 ± 14*	127 ± 12	74 ± 18*
Cerebral blood flow (ml/min)	22 ± 1	23 ± 1	23 ± 1	21 ± 2	24 ± 3	24 ± 2

Each value = Mean ± SEM; N = 5; * = < 0.05.

Table 3. *Haemodynamic Effects of Haemorrhage*

	Control	Haemorrhage
Arterial blood pressure (mm Hg)	136 ± 5	74 ± 8*
Cerebrospinal fluid pressure (mm Hg)	12 ± 2	8 ± 4
Cerebral perfusion pressure (mm Hg)	124 ± 6	66 ± 6*
Cerebral blood flow (ml/min)	24 ± 2	23 ± 3

Each value = Mean ± SEM; N = 5; * = < 0.05.

the cerebral veins, *i.e.*, an anatomic area distal to the site of normal arteriolar resistance[12]. Furthermore, it is clear that the cerebral capillaries and veins contain the majority of the cerebral blood volume and changes in cerebral capacitance vessels must occur in order for significant changes in cerebral blood volume to occur. In addition, NTG has been shown to be a potent venous vasodilator in other peripheral vascular beds[19] and NPR has been shown to have effects on the vascular muscle of resistance and capacitance vessels[14]. Thus, it may be anticipated that NTG and NPR exert their major effect on cerebral capacitance vessels (veins and venules), reducing cerebral venous resistance, thus leading to an increase in cerebral blood volume and Pcsf. This mechanism also suggests that cerebral venous capacitance changes must be necessary for the increase in Pcsf to occur during a decrease in arterial blood pressure which allows cerebral perfusion pressure to remain within autoregulatory limits. Autoregulatory alterations in cerebral vascular resistance produced by haemorrhagic hypotension, for example, would result in changes in cerebral vascular resistance and cerebral blood volume. Thus, no change in Pcsf would occur. This mechanism is also consistent with our observations in several animals that increases in Pcsf with NTG and NPR administration may occur in the absence of any alteration in arterial blood pressure.

In summary, these results support the concept that NTG has a greater effect on cerebral veins than NPR. The increase in cerebral blood volume responsible for the increased Pcsf with these agents occurs due to an increase in cerebral venous volume. Since neither NTG nor NPR altered CBF, and the fact that Pcsf increased more with NTG than with NPR, suggests that NTG is a more potent cerebral venous vasodilator than NPR. Since blood pressure fell and CBF was unchanged with NTG or NPR, the cerebral arteries and arterioles dilated to keep CBF constant (cerebral auto-regulation). The arterial dilatation probably had no effect on Pcsf, since with haemorrhagic hypotension Pcsf was unchanged, yet cerebral vasodilation clearly occurred under those conditions.

We conclude that the concept that vasodilator drugs differ in their effects on arteries and veins should be extended to the cerebral vasculature and that the ability of vasodilator drugs to result in cerebral vasodilation should be considered in their action.

References

1. Brown, E. D., Hanlon, K., Crockard, K., Mullan S.: Effect of sodium nitroprusside on cerebral blood flow in conscious human beings. Surg. Neurol. 7 (1977), 67—70.

2. Griffiths, D. P. G., Cummins, B. H., Greenbaum, R., Griffith, H. B., Staddon, G. E., Wilkins, D. G., Zorab, J. S. M.: Cerebral blood flow and metabolism during hypotension induced with sodium nitroprusside. Brit. J. Anaesth. 46 (1974), 671—679.

3. Heros, R. C., Zervas, N. T., Lavyne, M. H., Pickren, K. S.: Reversal of experimental cerebral vasospasm by intravenous nitroprusside therapy. Surg. Neurol. 6 (1976), 227—229.

4. Ivankovich, A. D., Miletich, D. J., Albrecht, R. F., Zahed, B.: Sodium nitroprusside and cerebral blood flow in the anesthetized and unanesthetized goat. Anesthesiology 44 (1976), 21—26.

5. Keaney, N. P., McDowall, D. G., Turner, J. M., Lane, J. R., Okuda, Y.: The effects of profound hypotension induced with sodium nitroprusside on cerebral blood flow and metabolism in the baboon. Brit. J. Anaesth. 45 (1973), 639.

6. Lowe, R. F., Gilboe, D. D.: Canine cerebrovascular response to nitroglycerin, acetylcholine, 5-hydroxytryptamine, and angiotensin. Amer. J. Physiol. 225 (1973), 1333—1338.

7. Needleman, P., Jakschik, B., Johnson, E. M.: Sulfhydryl requirement for relaxation of vascular smooth muscle. J. Pharmacol. Exper. Ther. 187 (1973), 324—331.

8. O'Neill, J. T., Traystman, R. J.: Adrenergic receptors in the intra- and extracranial vasculature. In: Neurogenic Control of Brain Circulation (Owman, Ch., Edvinsson, L., eds.), pp. 245—260. Oxford and New York: Pergamon Press. 1977.

9. Pitt, B. R., Radford, E. P., Jr., Gurtner, G. H., Traystman, R. J.: Interaction of carbon monoxide and cyanide on cerebral circulation and metabolism. Arch. Environ. Hlth. 34 (1979), 354—359.

10. Rapela, C. E., Green, H. D.: Autoregulation of canine cerebral blood flow. Circ. Res. 15 (Suppl. 1) (1964), I 205—211.

11. Rogers, M. C., Hamburger, C., Owen, K., Epstein, M. H.: Nitroglycerin effects on intracranial pressure. Anesthesiology 51 (1979), 227—229.

12. Shapiro, H. M., Stromberg, D. D., Wiederhielm, C. A.: Dynamic pressure in the pial arterial microcirculation. Amer. J. Physiol. 221 (1971), 279—283.

13. Stoyka, W. W., Schutz, H.: The cerebral response to sodium nitroprusside and trimetaphan controlled hypotension. Canad. Anaesth. Soc. J. 22 (1975), 275—283.

14. Tinker, J. H., Michenfelder, J. D.: Sodium nitroprusside: Pharmacology, toxicology and therapeutics. Anesthesiology 45 (1976), 340—354.

15. Traystman, R. J., Fitzgerald, R. S.: Cerebrovascular response to hypoxia in baroreceptor and chemoreceptor denervated dogs. Amer. J. Physiol. 241 (1981), H 724—H 731.

16. Traystman, R. J., Fitzgerald, R. S., Loscutoff, S. M.: Cerebral circulatory responses to arterial hypoxia in normal and chemodenervated dogs. Circ. Res. 42 (1978), 649—657.

17. Traystman, R. J., Rapela, C. E.: Effect of sympathetic nerve stimulation on cerebral and cephalic blood flow in dogs. Circ. Res. 36 (1975), 620—630.

18. Turner, J. M., Powell, D., Gibson, R. M., McDowall, D. G.: Intracranial pressure changes in neurosurgical patients during hypotension induced with sodium nitroprusside or trimetaphan. Brit. J. Anaesth. 49 (1977), 419—424.

19. Warren, S. E., Francis, G. S.: Nitroglycerin and nitrate esters. Amer. J. Med. 65 (1978), 53—62.

Authors' address: M. C. Rogers, M.D., Professor and Chairman, Department of Anesthesiology and Critical Care Medicine, The Johns Hopkins Hospital, Blalock 1415, 600 North Wolfe Street, Baltimore, MD 21205, U.S.A., R. J. Traystman, Ph.D. (reprints), Director, Anesthesiology and Critical Care, Medicine Research Laboratories, Department of Anesthesiology and Critical Care Medicine, The Johns Hopkins Hospital, Blalock 1408 B, 600 North Wolfe Street, Baltimore, MD 21205, U.S.A.

Departments of Neurosurgery and Clinical Pharmacology, University Hospital, Lund, Sweden, and the Wellcome Surgical Research Institute, University of Glasgow, Scotland

Effects on Feline Cortical Veins of Topical Application of Nifedipine

L. Brandt, B. Ljunggren, G. Teasdale, and K.-E. Andersson

Summary

Cat cortical venules were exposed in vivo to the calcium-antagonistic drug nifedipine. The topical application of the drug invariably induced marked venular dilatations. The venular responses were less pronounced but much more long-lasting than nifedipine-induced arteriolar dilatations in vessels of comparable diameter. The results suggest that cerebral venules have a tone which is dependent on extracellular calcium.

Keywords: Pial veins; nifedipine; topical application.

Introduction

It is generally accepted that changes in the concentration of calcium-ions within the smooth muscle cell regulate several cellular precesses, including the final step in the excitation-contraction coupling. Several studies have shown that smooth muscle is dependent on an extracellular source of calcium to initiate and maintain contraction[7]. Intracellularly stored calcium is also involved but is of less importance[2, 8]. Many pharmacological agents blocking transmembrane calcium-flux are used clinically, especially in cardiovascular disorders. Most of the information on the effects of calcium-inhibiting drugs has been obtained from animal studies on the systemic circulation. In the last years information has also accumulated concerning the effects of calcium-antagonistic drugs on isolated *cerebral* vessels[2] and *cerebral* blood flow[1, 3]. However, hitherto such information has been focussed on the arterial part of the cerebral vasculature, evidently due to the previously held opinion that cerebral veins should lack contractile properties[6]. The present in vivo study was undertaken to evaluate venular responses to the topical application of nifedipine.

Methods

In anaesthetized cats of either sex, a left-sided craniectomy (3 × 2 cm) was performed. The dura was reflected and the exposed gyri were irrigated continuously with mineral oil at a temperature of 37 °C. Pial venular calibre was measured by a television image-splitting technique [5]. Perivenular microinjections were made by the use of micropipettes with a tip diameter of approximately 8 μm. A hydraulic syringe was used to inject approximately 2 μl of the test solution into the perivenular space.

Nifedipine, obtained from Bayer AG., dissolved in ethanol, polyethylene glycol, and water, was further diluted in a mock cerebrospinal fluid solution. Nifedipine was used in a concentration of 20.0 μg/ml. This concentration of the drug has previously been shown to induce pronounced arteriolar dilatation under equal conditions [1]. Effects of test solutions containing the ethanol and polyethylene glycol vehicle only and corresponding to the nifedipine solution were also studied. To prevent decomposition of nifedipine, the solutions were protected from light exposure except during perivenular microinjections. All experiments were performed under steady state conditions.

The resting diameters of the investigated venules were in the range 41 to 216 μm.

The venular responses were compared with responses from a previous series of experiments in which the effects of the topical application of nifedipine of the same concentration on arterioles of comparable size were investigated.

Results

Perivenular microapplication of nifedipine invariably induced a dilatatory response which ranged from 16.2 to 47.9% (mean value 26.1%) of resting venular calibre (Fig. 1). When compared to the effect of the vehicle solution alone, the effect of nifedipine was more pronounced and persistent. Maximal nifedipine-induced dilatatory responses were 26 ± 4% as compared to maximal vehicle-induced dilatatory responses which rose to 15 ± 2%. The venules returned to their pre-injection diameter within approximately 15 min after the vehicle injections, whereas the effect of nifedipine persisted unreduced for more than 20 min. The time course for the nifedipine-induced venular dilatations were also in a striking contrast to nifedipine-dilated arterioles of comparable size (Fig. 2).

Discussion

The results strongly suggest that cerebral venules have a tone which is dependent on extracellular calcium. The difference in the time course between vehicle-induced dilatations and nifedipine-induced responses confirm the effect of nifedipine per se. The dilatations can not be explained as a result of an increased blood flow secondary to dilatations of adjacent arterioles since the experiments on arterioles showed that nifedipine-dilated arterioles invariably returned to their resting calibre within 15 min. The reason for the marked difference in time course remains unexplained. The observation that the venular dilatatory responses were less pronounced than the arteriolar responses could probably be explained by less elasticity of the venular wall and/or a lower intravascular pressure.

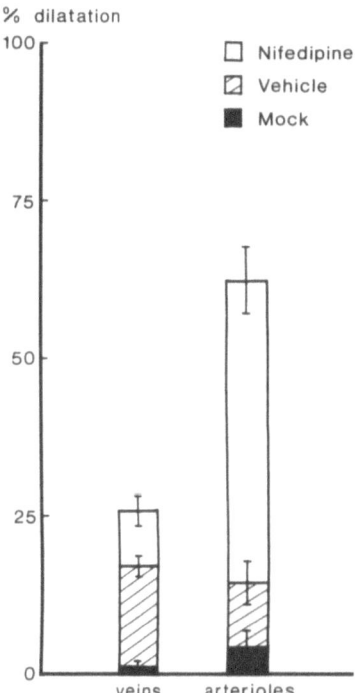

Fig. 1. Comparison between venular and arteriolar dilatatory responses to nifedipine (20 μg/ml), corresponding vehicle, and mock CSF, respectively. Bars indicate mean values ± SEM

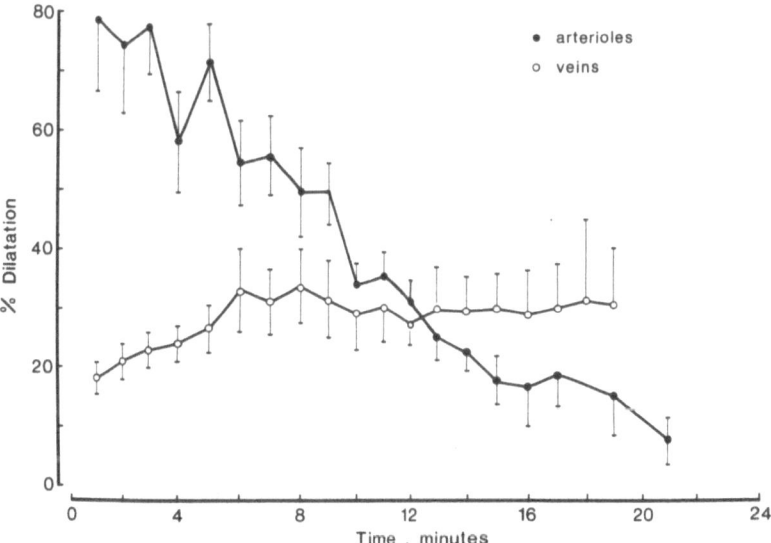

Fig. 2. Time course for the venular and arteriolar dilatations induced by nifedipine (20 μg/ml). Bars indicate mean values ± SEM

The predominant part of the intracranial blood volume is distributed in the venous part of the cerebral vasculature. Therefore the observed effects on cerebral venules of a calcium-antagonistic agent on cerebral venules must be taken into account when the possible use of calcium-antagonistic drugs in cerebrovascular disease is considered.

Acknowledgements

This study is dedicated to the Greta and Johan Kock Foundation and the Elsa Schmitz Foundation.

References

1. Brandt, L., Andersson, K.-E., Bengtsson, B., Edvinsson, L., Ljunggren, B., MacKenzie, E. T.: Effects of nifedipine on pial arteriolar calibre. An in vivo study. Surg. Neurol. *12* (1979), 349—352.
2. Brandt, L., Andersson, K.-E., Edvinsson, L., Ljunggren, B.: Effects of extracellular calcium and of calcium antagonists on the contractile responses of isolated human pial and mesenteric arteries. J. CBF and Metabolism *1* (1981), 339—347.
3. Harper, A. M., Craigen, L., Kazda, S.: Effect of the calcium antagonist, Nimodipine, on cerebral blood flow and metabolism in the primate. J. CBF and Metabolism *1* (1981), 349—356.
4. Fleckenstein, H.: Specific pharmacology of calcium in myocardium, cardiac pacemakers, and vascular smooth muscle. Ann. Rev. Pharmacol. Toxicol. *17* (1977), 149—166.
5. MacKenzie, E. T., Strandgaard, S., Graham, D. I., Jones, J. V., Harper, A. M., Farrar, J. K.: Effects of acutely induced hypertension in cats on pial arteriolar caliber, local cerebral blood flow and the blood brain barrier. Circ. Res. *39* (1976), 33—41.
6. Purves, M. J.: The physiology of the cerebral circulation. Cambridge, Great Britain: University Press. 1972.
7. Rüegg, J. C.: Smooth muscle tone. Physiol. Rev. *51* (1951), 201—248.
8. Small, J. V., Sobleszek, A.: Ca-regulation of mammalian smooth muscle actomyosin via a kinase-phosphatase-dependent phosphorylation and dephosphorylation of the 20,000 M light chain of myosin. Eur. J. Biochem. *76* (1977), 521—530.

Author's address: L. Brandt, M.D., Department of Neurosurgery, University Hospital, S-221 85 Lund, Sweden.

Department of Surgical Neurology, Research Institute for Brain and Blood Vessels, Akita, Japan, [1] Department of Neurosurgery, Graz University, Austria, [2] Department of Neurology, University of Lund, Sweden

Effect of Intravenous Dihydroergotamine on Cat Pial Veins and Intracranial Pressure

I. Sayama, L. M. Auer[1], and B. B. Johansson[2]

Summary

Diameter variations of pial vessels as well as mean arterial pressure (MAP), heart rate, body temperature and intracranial pressure (ICP) were continuously monitored before and after intravenous administration of $15 \mu g \cdot kg^{-1}$ dihydroergotamine (DHE) in six anaesthetized cats. Pial veins ($138.3 \pm 8.1 \mu m$; n = 36) showed no or minor calibre changes after injection whereas pial arteries ($105.0 \pm 7.0 \mu m$; n = 36) dilated in three cats and constricted in two. While MAP remained constant, heart rate decreased by $10.7 \pm 3.1\%$ from resting values. No significant change was observed in ICP. Thus, DHE does not constrict cerebral veins in anaesthetized cats.

Keywords: Dihydroergotamine; cat pial arteries and veins; intracranial pressure; heart rate.

Introduction

Dihydroergotamine (DHE) is a powerful constrictor of capacitance vessels in skeletal muscle and skin tissue thereby increasing the circulating blood volume[4, 5]. The drug has been reported to have little or no effect on cerebral blood flow in man and experimental animals[3]. Since no data have been published regarding cerebral capacitance vessels, the present study was designed to investigate whether DHE decreases blood volume and thus intracranial pressure by constricting cerebral veins.

Materials and Methods

Six cats with a body weight of 1.4–2.4 kg were anaesthetized with $30 \, mg \cdot kg^{-1}$ sodium-pentobarbital (Nembutal®), intubated endotracheally, immobilized with $60 \, \mu g \cdot kg^{-1}$ pancuroniumbromide (Pavulon®) and ventilated with a 3 : 1 mixture of $N_2O : O_2$ using a Loosco baby respirator. The aorta and vena cava inferior were cannulated with PVC catheters via both femoral arteries and one femoral vein. This allowed frequent blood gas measurements (AVL type 937 C gas check), monitoring of mean arterial pressure (MAP) (Statham P 23 dB pressure transducer and Hellige type 1214 electromanometer) and the administration of drugs. Body temperature was continuously monitored by a Philips rectal-thermosensor unit and maintained within the physiological range using a heating pad. After the animal's head was fixed into a stereotaxic frame in a sphinx position, a closed cranial window was made in the left parietal region[1]. The intracranial pressure (ICP) was registered via a plastic cannula in the cisterna magna, connected to a Statham P 23 dB transducer and an HSE electromanometer. Pial veins and arteries beneath the cranial window were observed and their diameter variations were continuously monitored as previously described using TV-angiometry[2] during resting conditions and for a period of 20 minutes following slow intravenous injection of 15 μg · kg[-1] DHE. Statistical analysis were performed with variance analyses (F-test, Duncan-test).

Blood gases were maintained at normal levels; the mean $PaCO_2$ was 34.4 ± 1.9 mm Hg initially and 36.9 ± 1.9 mm Hg at the end of the experiment, PaO_2 was 107.6 ± 5.7 mm Hg and 89.9 ± 2.3 mm Hg, respectively.

Results

Pial Veins and Arteries

Measurements were made on a total of 36 pial arterial portions and 36 venous portions. The mean resting diameter of the veins was 138.3 ± 8.1 μm (SEM), range 80–298 μm. Ten and 20 minutes after the injection of DHE the corresponding value was 139.0 ± 7.8 μm and 141.9 ± 7.6 μm, these minor non-significant changes were not visibly detectable. Pial arterial reactions (mean resting 105.0 ± 7.0 μm, range 52–172 μm) varied from animal to animal; constriction was observed in 2, dilatation in 3 and no change in 1 cat. The result was an insignificant mean dilatation from 105.0 ± 7.0 μm to 114,8 ± 7.8 μm after 10 min and 115.3 ± 8.2 μm after 20 min.

MAP, ICP, Heart Rate

MAP remained stable during the experiments (129.5 ± 2.8 mm Hg before and 127.5 ± 5.3 mm Hg 20 min after the injection of the drug). A slight not statistically significant increase in mean ICP was observed (4.1 ± 0.6 mm Hg before and 5.7 ± 1.0 mm Hg 20 min after DHE). However, in the three cats with pial arterial dilatation, ICP increased about 50%, and decreased accordingly when the arteries constricted. A decrease in heart rate was consistently noted with a mean decrease of 10.7 ± 3.1% as compared with the resting value (219.9 ± 19.2 · min[-1] before and 196.8 ± 20 · min[-1] 10 min after the injection of the drug).

Discussion

The potent constrictor effect of DHE on capacitance vessels in the skeletal muscle and the skin is considered to be mainly or entirely due to direct excitatory action on the vascular smooth muscle effectors[4]. According to the present results there is no corresponding effect on the cerebral veins and no consequent reduction in cerebral volume. Elevation of

ICP together with arterial dilatation and an ICP decrease with arterial constriction even suggests a change in the cerebral blood volume on the arterial side. Slight dilatory responses have earlier been noted in the resistance vessels of skeletal muscle, intestine and kidney and thought to be related to the central inhibitory action of DHE on autonomic nervous structures resulting in decreased sympathetic constrictor fibre discharges[4]. Dilatation of arteries, observed in other vascular beds after DHE, is completely abolished or even reversed after sympathectomy. The decrease in heart rate with no concomitant change in MAP is in agreement with several previous reports[3].

The lack of effect of DHE on pial veins cannot be attributed to poor penetration through the blood-brain barrier since the drug easily enters the brain[3]. DHE has only minor effects on the capacitance vessels of the intact intestine and kidney[4]. Thus it is evident that the results vary with the vascular beds studied.

In conclusion, there is no evidence that DHE constricts cerebral veins.

Acknowledgement

We are grateful to Mr. H. Leiner for technical assistance and to Miss S. Auer for the statistical calculations.

This work was supported by the Austrian *Fonds zur Förderung der wissenschaftlichen Forschung* (Project No. 4368).

References

1. Auer, L. M.: The pathogenesis of hypertensive encephalopathy. Acta neurochir. (Wien), Suppl. *27* (1978), 1—111.
2. Auer, L. M., Haydn, F.: Multichannel videoangiometry for continuous measurement of pial microvessels. Acta Neurol. Scand. *60*, Suppl. 72 (1979), 208—209.
3. Berde, B., Schild, H. O.: Ergot alcaloid and related compounds. Handbook of Experimental Pharmacology, Vol. 49. Berlin-Heidelberg-New York: Springer. 1978.
4. Mellander, S., Nordenfelt, I.: Comparative effects of dihydroergotamine and noradrenaline on resistance, exchange and capacitance functions in the peripheral circulation. Clin. Science *39* (1970), 183—201.
5. Nordenfelt, I., Mellander, S.: Haemodynamic effect of dihydroergotamine in orthostatic hypotension. Cardiology *61* (1976), 316—321.

Author's address: Prof. Dr. L. M. Auer, Department of Neurosurgery, Graz University, A-8036 Graz, Austria.

Department of Neurosurgery, Graz University, Austria, [1] Department of Surgical Neurology, Research Institute for Brain and Blood Vessels, Akita, Japan

In vivo Effect of Serotonin on Cat Pial Veins

K. Leber, L. M. Auer, and I. Sayama[1]

Summary

In a series of 6 adult cats, the reactions of pial veins to topically administered 10^{-4} M serotonin (5-HT) were investigated using the closed cranial window technique and multichannel-videoangiometry. Small pial veins dilated by $8.6 \pm 2\%$; large veins tended to remain unreactive or to constrict. The reaction pattern was thus similar to that of pial arteries, though less marked. Ketanserin, a 5-HT-2 blocker, tested against its solvent, in a randomized study on a further 12 cats, induced no significant vessel reactions. A potential constrictor effect might be masked by a dilatatory action of the solvent.

Keywords: Pial veins; serotonin; ketanserin.

Introduction

The monoamine serotonin, 5-hydroxytryptamine (5-HT), is known as a neurotransmitter in many pathways of the brain stem and cerebellum[13]. Moreover, various cerebroarterial effects have been described: constriction of larger arteries[5, 8, 14, 18, 19] was potentiated by the addition of blood[16], prostaglandin $F_{2\alpha}$ 15 and beta-lipoprotein[7]. Small pial arteries, by contrast, dilated. Apart from the somewhat controversial data[6] of cerebral arteries so far available, the influence of 5-HT on cerebral veins has never been studied. The present investigation on pial vessels was therefore performed with special regard to the interest in cerebral veins as a potential mediator for cerebral blood volume-regulation.

Material and Methods

The experiments were performed in two series: The first series was aimed at the investigation of the effects of topically administered 5-HT and the effect of an intravenously administered 5-HT blocker. In the second series, the effect of the blocker per se and its vehicle was to be tested.

Series 1: Experiments were performed in 6 cats of either sex with a body weight of 1.5–3 kg under sodium-pentobarbital and $N_2O:O_2$ (3:1) anaesthesia, endotracheal intubation and controlled ventilation after relaxation with $60 \mu g/kg$ pancuronium-bromide. PVC catheters were placed into both femoral arteries and one femoral vein for frequent blood gas measurements, continuous blood pressure recording and drug administration, respectively. The mean arterial pressure (MAP) was monitored via a Statham P 23 dB transducer, a Hellige 1214 electromanometer and a Rikadenki multichannel penwriter. Blood gas analyses were performed on an AVL type 937 C blood gas analyser. Body temperature was continuously monitored with a Philips rectal thermosensor unit and maintained between 37 and 38 °C using a heating pad. As previously described in detail [2] a closed cranial window was made in the left parietal region for the observation and continuous recording of pial venous and arterial diameter variations with a TV multichannel videoangiometer [3]. A subdural PVC cannula was placed underneath the cranial window for superfusion of pial vessels with a 10^{-4} solution of serotonin in mock CSF [9], administering a volume of 0.1 ml per minute with a small volume perfusion pump. A second cannula was placed into the cisterna magna in order to maintain the intracranial pressure at normal levels. After achieving a steady state of vessel reactions with 5-HT, the 5-HT-2-receptor blocker Ketanserin® * was intravenously injected (1 mg/kg) with the 5-HT superfusion still being maintained.

Series 2: In 12 cats, anaesthesia and ventilation were performed as described in series 1. Again, a closed cranial window was made and the intracranial pressure monitored via a cannula to the cisterna magna. In a normotensive, normocapnic and normoxic steady state, the 5-HT blocker Ketanserin or a placebo containing the vehicle of Ketanserin were injected intravenously in a dose of 1 mg/kg in a randomized manner. Pial venous and arterial reactions were again continuously registered using the TV angiometer technique. Statistical calculation was performed using the Wilcoxon test.

Results

Series 1: During the experiments, the $PaCO_2$ was maintained constant, 32.4 mm Hg on the average, PaO_2 106.1 mm Hg. Resting MAP was 127.1 ± 8.4 mm Hg remained unchanged during superfusion with 5-HT (131.3 ± 9.6 mm Hg). Following the intravenous injection of Ketanserin, however, the MAP fell significantly to 81.3 ± 8.2 mm Hg ($p < 0.01$).

During 5-HT superfusion pial veins with resting diameters between 75 and 200 μm (mean resting diameter = 137.3 ± 8.1 SEM μm) dilated by $7.3 \pm 2.6\%$ to 145.8 ± 8.4 μm ($p < 0.01$) (Fig. 1). The mean venous dilatation increased from 7.3% to $11.1 \pm 3.2\%$ between the resting state and 10 minutes after the injection of Ketanserin and to $10.2 \pm 3.4\%$ after 20 minutes following Ketanserin-injection. 5-HT dilated small veins up to 150 μm resting diameter significantly by $8.3 \pm 3.2\%$ ($p < 0.05$); the further dilatation due to intravenous Ketanserin by $12.4 \pm 4.2\%$ was statistically significant at the 1% level. There was no difference between the calibres at 10 and 20 minutes after Ketanserin injection (Figs. 1 and 2). Fig. 3 shows the resting diameters plotted against percentage changes due to the topically administered 5-HT, clearly indicating dilatation of small veins and a tendency of large ones to remain unreactive or even to constrict.

---

* Janssen, Beerse.

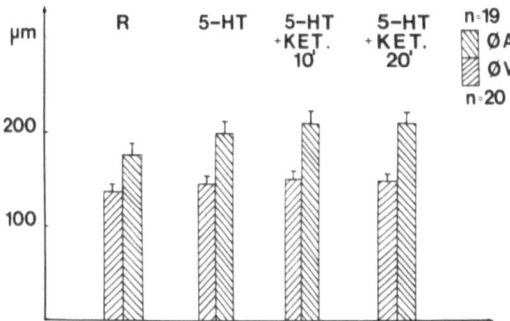

Fig. 1. Mean absolute diameters ± SEM of pial veins ($\varnothing V$) and arteries ($\varnothing A$) under resting conditions (R), during the perivascular administration of 5-HT and 10 and 20 minutes after the intravenous injection of Ketanserin

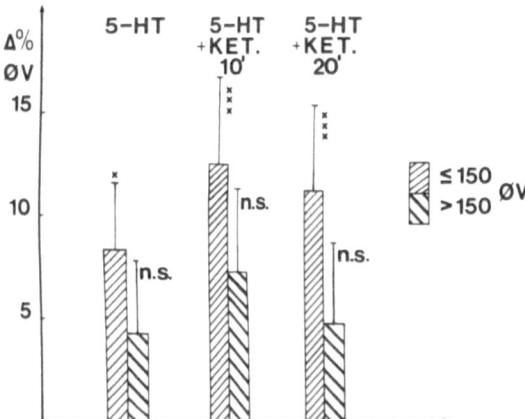

Fig. 2. Mean ± SEM percentage dilatation of pial veins ($\varnothing V$) with resting calibres larger and smaller than 150 μm

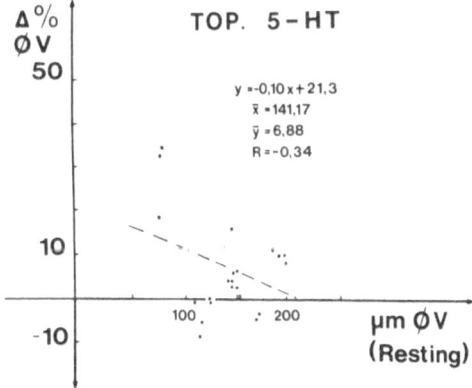

Fig. 3. Resting calibres of pial veins ($\varnothing V$) in μm plotted against percentage changes during the topical administration of 5-HT

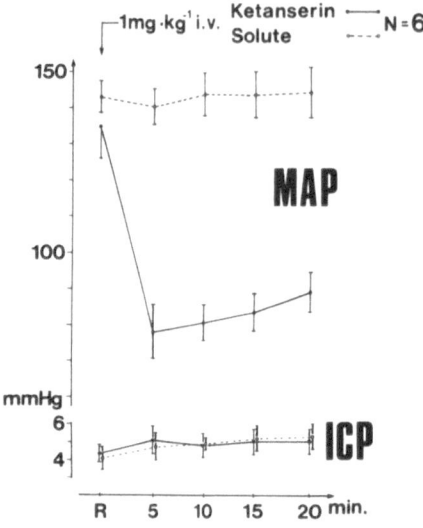

Fig. 4. Reactions of mean arterial pressure (± SEM) (MAP) and intracranial pressure (*ICP*) to the intravenous injection of Ketanserin or its vehicle. *R* resting state

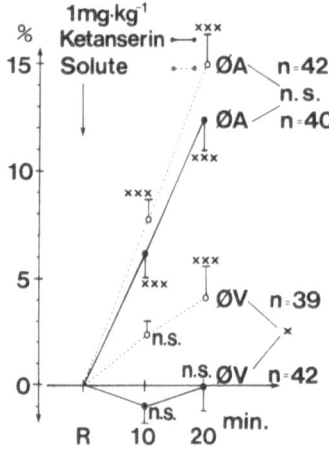

Fig. 5. Mean (± SEM) percentage changes in pial venous (∅*V*) and arterial (∅*A*) calibres. *R* resting state

As described elsewhere in detail[4] pial arteries with resting calibres between 67 and 259 μm (mean 176.2 ± 13.2 μm) were markedly dilated by 16.0 ± 4.4% with topically applied 5-HT (p < 0.01). Ketanserin induced further dilatation by 22.6 ± 3.3% after 10 minutes, which was almost unchanged after 20 minutes (Fig. 1).

Series 2: As indicated in Fig. 4, the MAP remained stable after the injection of a placebo containing the vehicle of Ketanserin, whereas it decreased significantly following the injection of Ketanserin. The intracranial pressure exhibited insignificant variations between 4 and 5 mm Hg in both groups. The heart rate remained stable in the placebo group, however, showing a marked decrease following the injection of placebo; wide variations in individual values being apparent in either group.

42 pial veins with resting diameters between 96 and 228 μm (mean 157.1 ± 6.5 μm) showed no significant variations following the injection of Ketanserin, whereas in the placebo-group, a 2.4 ± 0.6% and 4.1 ± 1.3% dilatation was noticed after 10 and 20 minutes, respectively, the difference between Ketanserin and placebo being significant at the 5% level. By contrast, pial arteries showed significant (p < 0.01) 6–8 and 12–15% dilatation after 10 and 20 minutes, respectively, in both groups (Fig. 5).

Discussion

In agreement with the previous report of MacKenzie et al.[10], small pial arteries dilated in this series and larger ones tended to constrict. A similar trend of vessel reaction was observed on the venous side in this study, although the reactions were far less marked than on the arterial side. The observation of small vessel dilatation in contrast to larger pial vessels which remained more or less unreactive or constricted and other authors' observations of large vessels which strongly constricted[1, 5, 12, 14, 17–19], suggests a dual effect of serotonin on cerebral vessels. Such a dual action of serotonin on cerebral vessels would explain the pathological circumstances of low blood flow and high blood volume as observed in patients during symptomatic cerebral vasospasm after subarachnoid haemorrhage[11].

The serotonin blocker Ketanserin exhibited no effect on the pial arteries; the degree of dilatation was identical with Ketanserin and its vehicle despite the fact that the blood pressure remained stable in the placebo group, whereas it fell markedly in the Ketanserin group. One could have expected more dilatation in the Ketanserin group by suggesting additional autoregulatory arterial dilatation. Masking of the slight constrictory effect by Ketanserin due to the blood pressure decrease cannot therefore be ruled out. Especially on the venous side, however, the significantly different behaviour between the Ketanserin group and the placebo group might indicate a masking effect of the vehicle, which inhibits the appearance of a Ketanserin-induced venoconstrictory effect. Further in vitro studies will, however, be necessary to verify this.

Acknowledgement

We gratefully acknowledge the technical assistance of Mr. H. Leiner and statistical calculations by Miss S. Auer.

This work was supported by the Austrian *Fonds zur Förderung der wissenschaftlichen Forschung* (Project No. 4368).

References

1. Allen, G. S., Henderson, L. M., Chou, S. N., French, L. A.: Cerebral arterial spasm. Part 2: In vitro contractile activity of serotonin in human serum and CSF on the canine basilar artery, and its blockage by methylsergide and phenoxybenzamine. J. Neurosurg. 40 (1974), 442—450.
2. Auer, L. M.: The pathogenesis of hypertensive encephalopathy. Acta neurochir. (Wien) Suppl. 27 (1978), 1—111.
3. Auer, L. M., Haydn, F.: Multichannel videoangiometry for continuous measurement of pial microvessels. Acta Neurol. Scand. 60, Suppl. 72 (1979), 208—209.
4. Auer, L. M., Leber, K., Sayama, I.: In vivo effect of serotonin on cat pial vessels. In preparation.
5. Edvinsson, L., Owman, Ch.: Pharmacological identification of adrenergic (alpha and beta), cholinergic (muscarinic and nicotinic), histaminergic (H 1 and H 2), and serotinergic receptors in isolated intra- and extracranial vessels. In: Blood Flow and Metabolism in the Brain (Harper, A. M., et al., eds.), pp. 1.18—1.24. Churchill Livingstone. 1975.
6. Edvinsson, L., MacKenzie, E. T.: Amine mechanisms in the cerebral circulation. Pharmacol. Rev. 28 (1977), 275—348.
7. Eidelman, B. H., Mendelow, A. D., McCalden, T. A., Bloom, D.: Lipoprotein potentiation of the cerebrovascular response to intra-arterial 5-hydroxytryptamine. Acta Neurol. Scand. 56, Suppl. 64 (1977), 78—79.
8. Harper, A. M., MacKenzie, E. T.: Cerebral circulatory and metabolic effects of 5-hydroxytryptamine in anaesthetized baboons. J. Physiol. 271 (1977), 721—733.
9. Kuschinsky, W., Wahl, M.: Local chemical and neurogenic regulation of cerebral vascular resistance. Physiol. Rev. 58 (1978), 656—689.
10. MacKenzie, E. T., Young, A. R., Stewart, M., Harper, A. M.: Effect of serotonin on cerebral function, metabolism and circulation. Acta Neurol. Scand. 56, Suppl. 64 (1977), 76—77.
11. Martin, W. R. W., Baker, R. P., Grubb, R. L., Raichle, M. E.: Cerebral blood volume, blood flow, and oxygen metabolism in cerebral ischemia and subarachnoid hemorrhage: An in vivo study using positron emission tomography. Acta neurochir. (Wien) 1984 (in press).
12. McCalden, T. A., Bevan, J. A.: The effect of calcium withdrawal and calcium antagonists on cerebrovascular tone and response to various agonists. In: Cerebral Blood Flow: effect of nerves and neurotransmitters (Heistad, D. D., et al., eds.), pp. 21—27. New York: Elsevier. 1982.
13. Nieuwenhuys, R., Voogd, J., van Huijzen, Chr.: The human cerebral nervous system, pp. 221—226. Berlin-Heidelberg-New York: Springer. 1981.
14. Rapela, C. E., Martin, J. B.: Reactivity of cerebral extra- and intraparenchymal vasculature to serotonin and vasodilator agents. In: Blood Flow and Metabolism in the Brain (Harper, A. M., et al., eds.), pp. 4.5—4.9. Churchill Livingstone. 1975.
15. Rosenblum, W. I.: Interaction of prostaglandins with biogenic amines applied to pial vessels. In: Blood Flow and Metabolism in the Brain (Harper, A. M., et al., eds.), pp. 4.14—4.15. Churchill Livingstone. 1975.
16. Svendgaard, N. A., Edvinsson, L., Owman, Ch.: Changes in sensitivity of cerebral vessels to noradrenaline and 5-hydroxytryptamine in the presence of subarachnoid blood. Acta Neurol. Scand. 56, Suppl. 64 (1977), 318—319.
17. Towart, R., Kazda, S.: Preferential vasodilator actions of the calcium antagonists nimodipine (Bay e 9736), nifedipine, and verapamil on contractions of cerebral vascular smooth muscle induced by neurotransmitter and vasoconstrictor substances. In: Cerebral Blood Flow: Effect of Nerves and Neurotransmitters (Heistad, D. D., et al., eds.), pp. 29—38. New York: Elsevier. 1982.

18. Welch, K. M. A., Helpern, J. A.: Do serotonin and noradrenalin neurons protect to cerebral capillaries? J. Cerebr. Blood Flow Metabol. *1*, Suppl. 1 (1981), S 371—S 372.
19. Young, A. R., MacKenzie, E. T.: In vivo effect of various drugs used in the therapy of cerebrovascular disease. J. Cerebr. Blood Flow Metabol. *1*, Suppl. 1 (1981), S 565—S 566.

Author's address: Prof. Dr. L. M. Auer, Department of Neurosurgery, Graz University, A-8036 Graz, Austria.

Cerebral Veins Under Various Pathological Circumstances

Department of Neurology, School of Medicine, Keio University, Tokyo, Japan

Vulnerability of Cerebral Venous Flow Following Middle Cerebral Arterial Occlusion in Cats

M. Kobari, F. Gotoh, M. Tomita, N. Tanahashi, and K. Tanaka

Summary

The haemodynamic changes in the cerebral venous system after middle cerebral arterial (MCA) occlusion were investigated in 23 cats. The regional cerebral blood volume and tissue carbon-black dilution curves were recorded on both parieto-temporal cortices by our photoelectric method. The inlet (arterial phase) curve and outlet (venous phase) curve were derived from each tissue carbon-black dilution curve and subjected to moment analysis, by means of which the mean transit time and gamma index were calculated.

After MCA occlusion, the mean transit time of the indicator became 1.9 times longer for the arterial phase of the blood flow and 2.9 times longer for the venous phase, whilst the gamma index increased from 0.35 to 0.44 for the arterial phase and remained constant for the venous phase.

These results suggest the rapid formation of the collateral circulation into the ischaemic tissue and the occurrence of venous outflow disturbances in the early phase of ischaemia following MCA occlusion.

Keywords: Occlusion of the middle cerebral artery; venous outflow disturbances; haemorheological changes; photoelectric method.

There have been numerous investigations on alterations in the cerebral circulation following occlusion of the middle cerebral artery (MCA), but few have considered the changes in the cerebral venous system. Using our photoelectric method[3], haemodynamic changes in the feline cerebral venous system in the early phase of ischaemia of the cerebral cortical tissue after MCA occlusion were investigated.

Materials and Methods

Twenty-three adult cats of both sexes weighing 1.9–4.0 kg were anaesthetized with 50 mg/kg body weight of alpha-chloralose and 500 mg/kg body weight of urethane. After tracheal intubation and immobilization with alcuronium chloride, respiration was controlled with a Harvard Respirator. The left femoral and left lingual arteries were catheterized for monitoring of the systemic arterial blood pressure and for intracarotid injection, respectively. A pair of photoelectric units described in detail elsewhere[3] were attached to both parieto-temporal cortices to monitor the regional cerebral blood volume in a thin layer of cerebral cortex supplied by the MCA. Tissue indicator dilution curves were recorded *in situ* in the same region after spike injection of 0.2 ml of 1/40 diluted carbon-black solution into the left carotid artery via the lingual artery. Cerebral ischaemia was produced by transorbital clipping of the left MCA at its origin. Tissue carbon-black dilution curves were obtained before and approximately five minutes after MCA occlusion.

The tissue carbon-black dilution curve, a cumulative distribution function, was first differentiated yielding two curves comprising the inlet curve and the outlet curve[5]. The error due to an "overlapping effect", in which some particles of carbon-black leave the tissue vasculature before their complete entrance into the tissue has been effected, was corrected for by extrapolation from the descending part of the outlet curve using the gamma density function[6]. The amount of carbon-black involved in the "overlapping effect" was then added to the inlet curve.

The mean transit time of the indicator and gamma index[2] were calculated by moment analysis[1] for inlet curves and outlet curves. The gamma index[2], defined by us, is a value that quantitatively describes the spatial dispersion of transit time of the indicator in the vasculature. Thus, changes in the gamma index indicate variations of multiplexity of the vasculature.

Results

Fig. 1 shows typical recordings of the cerebral blood volume and tissue carbon-black dilution curves before and during MCA occlusion. The postfixes i and c denote ipsilateral and contralateral to the occluded side, respectively. The small downward arrows indicate injection of carbon-black solution into the ipsilateral carotid artery. The ipsilateral cerebral blood volume decreased abruptly after occlusion of the MCA. The carbon-black dilution curve showed a marked prolongation immediately after MCA

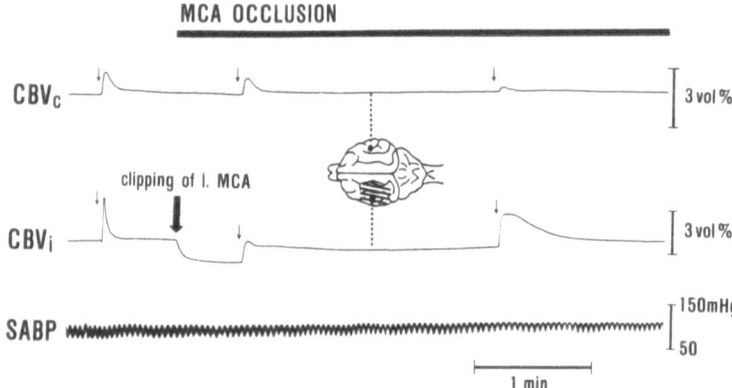

Fig. 1. Continuous recordings of contralateral (*CBVc*) and ipsilateral (*CBVi*) cerebral blood volume before and during MCA occlusion. Injection of carbon-black solution into the ipsilateral carotid artery is designated by small downward arrows, which are followed by the appearance of the tissue carbon-black dilution curves

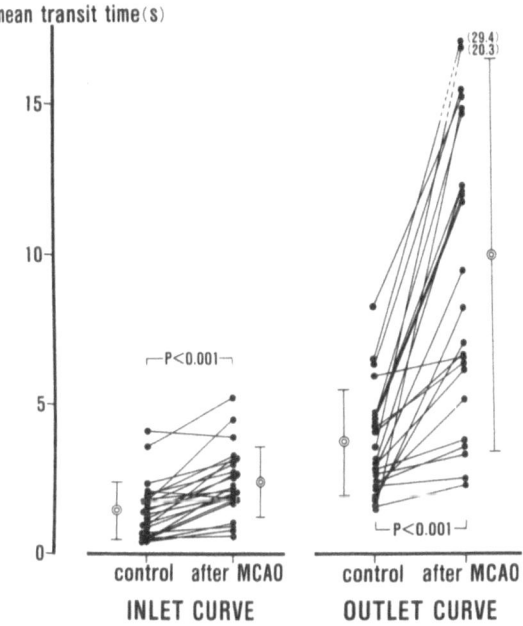

Fig. 2. The mean transit times for the inlet curves and the outlet curves before and during MCA occlusion. The mean and standard deviation are indicated by the double circles and vertical bars, respectively

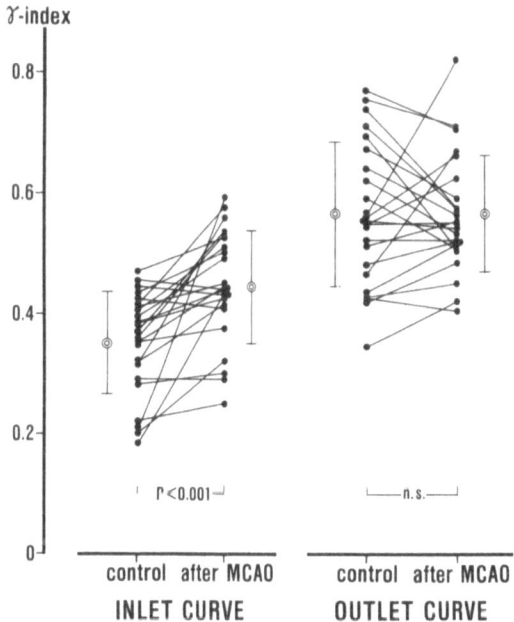

Fig. 3. The values of the gamma index for the inlet curves and the outlet curves before and during MCA occlusion

occlusion, but started to build up again in about two and a half minutes. These tissue carbon-black dilution curves, which were recorded before and during MCA occlusion, were subjected to the analysis mentioned above. No marked changes were observed in the cerebral blood volume or the carbon-black dilution curve in the contralateral side after MCA occlusion.

The mean transit times for the inlet curve and the outlet curve calculated from the carbon-black dilution curves before and during MCA occlusion are shown in Fig. 2. The mean transit time for the inlet curve was 1.4 ± 1.0 seconds before and 2.4 ± 1.2 seconds during cerebral ischaemia, whilst the mean transit time for the outlet curve was 3.7 ± 1.8 seconds before and 9.9 ± 6.5 seconds during MCA occlusion. The delays in the mean transit times after MCA occlusion were both statistically significant ($p < 0.001$), but the prolongation rates of the mean transit time for the inlet curve and the outlet curve were 1.9 ± 0.9 and 2.9 ± 1.9, respectively, which were significantly greater ($p < 0.01$) for the outlet curve than the inlet curve.

Fig. 3 summarizes the values of the gamma index for the inlet curve and the outlet curve, calculated from the carbon-black dilution curves before and during MCA occlusion. The gamma index for the inlet curve was 0.35 ± 0.08 before and 0.44 ± 0.09 during cerebral ischaemia, which represented a statistically significant increase ($p < 0.001$). The respective values for the outlet curve were 0.56 ± 0.12 and 0.57 ± 0.10, and the change was not statistically significant.

Discussion

The inlet curve and outlet curve obtained by differentiation of the tissue carbon-black dilution curve and corrected for the "overlapping effect" were regarded as reflecting the arterial phase and the venous phase of the blood flow, respectively[5]. The validity of this presumption is reported elsewhere[6]. The prolongation of the mean transit time for the arterial phase of the blood flow after MCA occlusion implies the development of a detour inflow route of the collateral circulation due to occlusion of the main cerebral artery, and the increase in the gamma index suggests its multiplexity. On the other hand, the marked prolongation of the mean transit time for the venous phase of the blood flow indicates that a more striking circulatory disturbance is occurring in the venous system after arterial occlusion. The venous flow disturbance cannot be attributed to morphological changes in the vasculature since the gamma index, which indicates the spatial dispersion of the transit time of the indicator, remained constant. One explanation for our results might be haemorheological changes in the venous blood flow, presumably due either to haemoconcentration by plasma or water leakage from venules, which should occur in a matter of seconds after interruption of the blood flow, or to the non-Newtonian property of the venous blood flow. The increase in local blood viscosity may lead to venous outflow disturbances and stasis of blood in the tissue.

Recently, we described "low perfusion hyperaemia" following MCA occlusion[4]. This phenomenon is characterized by the increase in cerebral blood volume or the paradoxical appearance of hyperaemia during the low cerebral blood flow stage after MCA occlusion. The mechanism of this phenomenon could be partly attributable to the above-mentioned disorder of the cerebral venous system.

In conclusion, disturbance of the venous outflow occurs in the early phase of ischaemia after MCA occlusion in cats, which might modify the course of the overall cerebral circulatory disturbance following cerebral ischaemia.

References

1. Tomita, M., Gotoh, F.: Moment analysis on indicator dilution curve (in Japanese). Igaku no Ayumi *90* (1974), 847—849.
2. Tomita, M., Gotoh, F.: Variation in gamma index (dimensionless dispersion) of hydrogen gas in a single human brain. In: Cerebral Circulation and Metabolism (Langfitt, T. W., *et al.,* eds.), pp. 135—137. Berlin-Heidelberg-New York: Springer. 1975.
3. Tomita, M., Gotoh, F., Sato, T., Amano, T., Tanahashi, N., Tanaka, K., Yamamoto, M.: Photoelectric method for estimating hemodynamic changes in regional cerebral tissue. Amer. J. Physiol. *235* (1978), H 56—H 63.
4. Tomita, M., Gotoh, F., Amano, T., Tanahashi, N., Tanaka, K.: "Low perfusion hyperemia" following middle cerebral arterial occlusion in cats of different age groups. Stroke *11* (1980), 629—636.
5. Tomita, M., Gotoh, F., Tanahashi, N., Tanaka, K., Kobari, M.: Photoelectric method for studying the intraparenchymal circulation of the brain. In: Basic Aspects of Microcirculation (Tsuchiya, M., *et al.,* eds.), pp. 61—74. Amsterdam-Oxford-Princeton: Excerpta Medica. 1982.
6. Tomita, M., Gotoh, F., Amano, T., Tanahashi, N., Kobari, M., Shinohara, T., Mihara, B.: Transfer function through regional cerebral cortex evaluated by a photoelectric method. Amer. J. Physiol. (in press).

Authors' address: M. Kobari, M.D., Department of Neurology, School of Medicine, Keio University, 35 Shinanomachi, Shinjuku-ku, Tokyo 160, Japan.

Centre of Neurosurgery, Justus-Liebig University of Giessen, Federal Republic of Germany

Morphological Analysis of Experimental Decerebration After Acute Epidural Compression

Z. M. RAP, G. CSÉCSEI, N. KLUG, J. ZIERSKI, and H. W. PIA

Summary

Alterations in the blood-brain barrier (BBB) and structural changes in the diencephalon and brain stem after decerebration produced by short-lasting cerebral compression were studied. Two linear primary ischaemic lesions with BBB damage localized parallel to the tentorium were found. The first zone was in the plana gyrus cinguli—massa intermedia of the thalamus and corpora mamillaria of the hypothalamus. The second zone extended from the superior and inferior colliculi of the lamina quadrigemina through the tegmentum to the border of pons. In electrophysiological tests, such as the blink reflex, R_2, disappeared earlier than R_1 and the auditory brain-stem evoked potentials (BAEP) suggesting that the first signs signs of decerebration could be caused by primary ischaemic lesions of the diencephalon whilst the mesencephalon and pons structures were damaged at a later stage.

Keywords: ICP; ischaemic lesions; decerebration.

Introduction

Most experimental models of decerebration have been produced by localized reproducible lesions, in order to obtain clinical or neurophysiological symptoms of brain stem dysfunction[4, 6–8].

The aim of the present study was the demonstration of structural changes connected with the occurrence of decerebration and the disturbance of the blink reflex (BR) and auditory brain stem-evoked potentials (BAEP) after repeated short-lasting cerebral compression[1, 5]. Early structural changes in the dien-mesencephalon, and pontine structures were revealed by means of alterations in the blood-brain barrier (BBB) and by fluoresescence-light microscopy and electron-microscopy studies.

Material and Methods

Fifteen adult cats were anaesthetized with Narcoren 2 ml/kg, tracheotomized and were breathing spontaneously during the experiments. Intracranial hypertension (IH) was produced by inflation of a balloon (0.1 ml/min) placed epidurally in the fronto-temporal region, repeated 2–3 times until the occurrence of a pupillary disturbance. The intracranial pressure (ICP), respiratory rate (RR), heart rate (HR), systemic blood pressure (SBP) and central venous pressure (CVP) were continuously monitored. GR and BAEP were tested repeatedly during the progressive increase in IVP[1, 5]. BBB was studied using 2 per cent Evans blue solution (2 ml/kg) injected i.v. prior to each experiment. Immediately after the last ICP-increase the heart was perfused with either a 10 per cent formalin solution or a 5 per cent glutaraldehyde solution. Specimens for fluorescence microscopy were frozen and cut using a microtome whilst those for light microscopy were embedded in parafin. Histological sections were stained using the haematoxylin-eosin, PAS and Heidenhein method.

The material for electron-microscopy was fixed in 5 per cent glutaraldehyde and post-fixed in 2 per cent OSO_4, then embedded in Durcupan-ACM.

Results

Pathophysiological Observations

Raising the ICP to 50–100 mm Hg was accompanied by an increase in HR, SBP, CVP and a drastic decrease in the RR, followed by respiratory arrest. At that time both pupils were dilated. 1–2 minutes after immediate decompression the spontaneous respiration returned and its rate as well as that of the heart (HR) the SBP and the CVP normalized. Neurophysiological studies showed that $BR-R_2$ disappeared earlier, after the first cerebral compression, whereas $BR-R_1$ and BAEP were affected after the second or third compression[1, 5].

Pathomorphological Studies

Sagittal sections through the cat's brain after two cerebral compressions (without Evans blue) showed no primary or secondary lesions in the brain stem (Fig. 1 a). The same procedure in cats with Evans blue injected i.v. revealed two primary ischaemic zones, with BBB damage localized in front of and behind the tentorium (Fig. 1 b). The first zone extended from the gyrus cinguli through the splenium of the corpus callosum, fornix, the massa intermedia of the thalamus to the corpora mamillaria of the hypothalamus. The second zone divided the mesencephalon, running from the superior and inferior colliculi of the lamina quadrigemina, through the tegmentum of the basal part of the pons. Sometimes after the ICP had been raised two or three times, secondary haemorrhages in the ischaemic areas were observed (Fig. 1 c). These morphological changes, associated with a supratentorial space-occupying process, were related mainly to the gyri lingualis and cinguli, the splenium of the corpus callosum, the thalamus and the lateral geniculate body (Figs. 2 a and b). Sometimes, herniation of the cerebellar tonsils was found (Fig. 2 c).

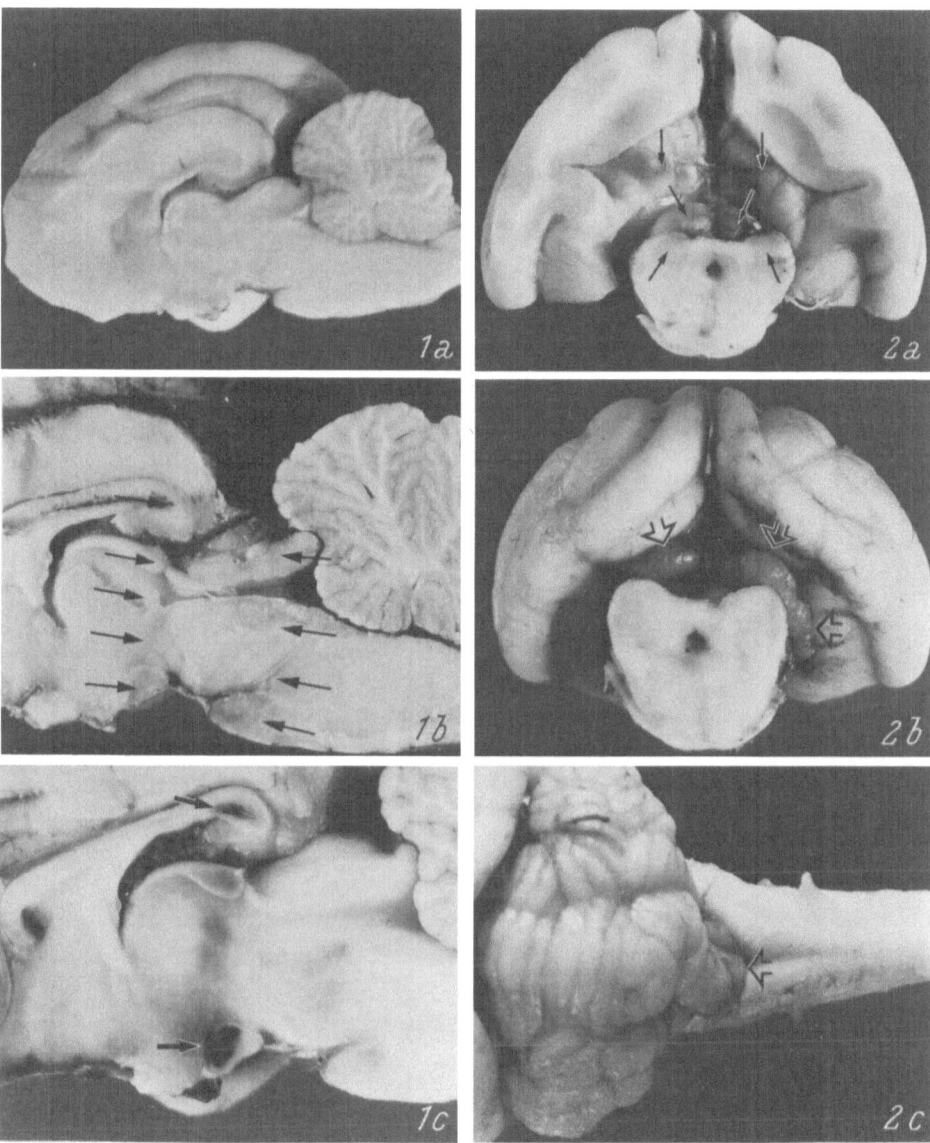

Fig. 1 a. Structural changes in the diencephalon and brain stem were not seen. Exp. without Evans blue. × 1.5. b. Primary ischaemic lesions with **BBB** damage localized in the dien- and mesencephalon (arrows). × 3. c. Primary ischaemic zones with secondary haemorrhage (arrows). × 6.

Fig. 2 a. **BBB** damage in the gyri cinguli and sup. and inf. colliculi (arrows). × 2. b. Herniation of the gyri lingualis, cinguli and corpus geniculatum lat. showing **BBB** damage (arrows). × 2. c. **BBB** damage in the herniated area of the cerebellar tonsil, (arrow). × 3

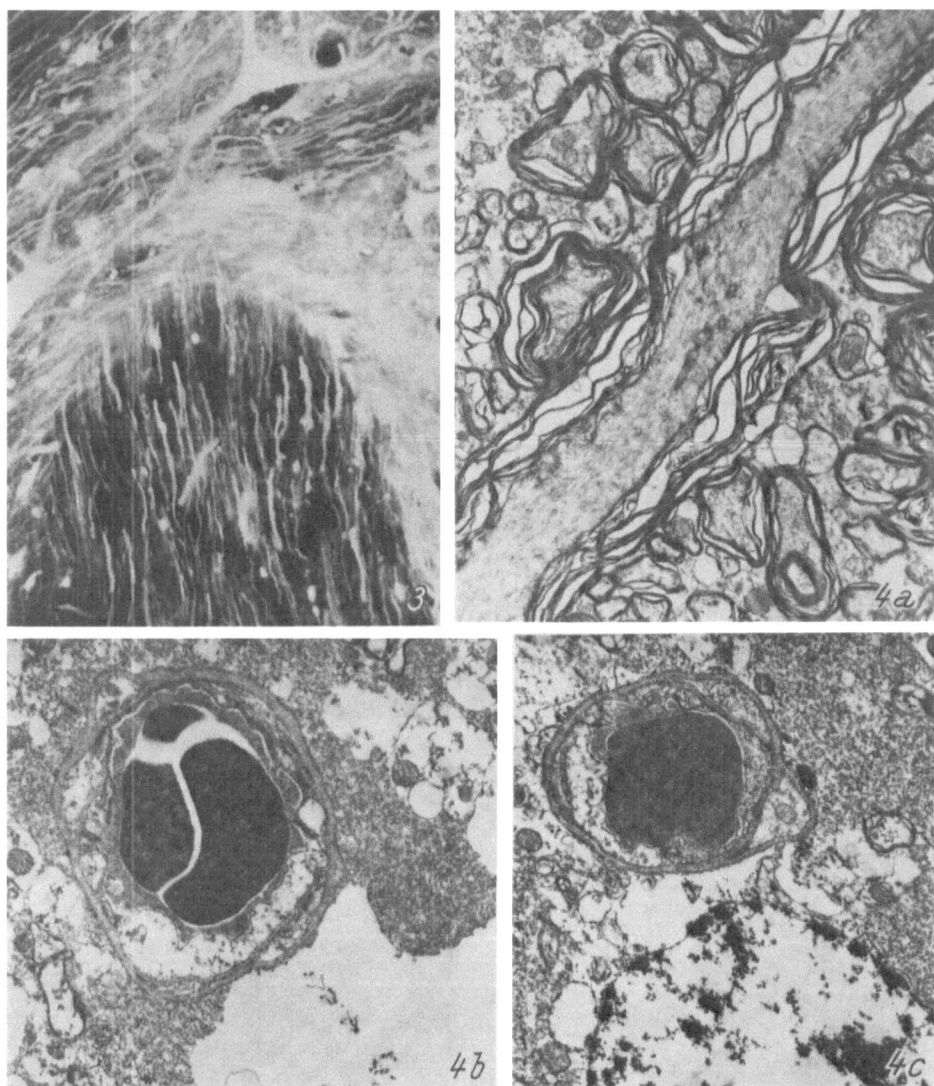

Fig. 3. Red fluorescence of Evans blue-albumin complex in the nervous fibres, nerves and glial cells at the mesencephalo-pontine border. × 250

Fig. 4 a. Delamination of the myelin sheaths at the border of the pons. × 9,000. b. Multi-layered capillary from ischaemic zone (thalamus), acute-swelling of the endothelial cell. Enlargement of the perivascular space. × 9,000. c. Small venule (lamina quadrigemina)—swelling of endothelial cell and perivascular glial cells. × 9,000

Fluorescence microscopy revealed the bright diffuse red fluorescence of the Evans blue—albumin complex in all the elements of the nervous tissue in ischaemic zones. The structures extending between these zones, as well as in the pons and the medulla oblongata, accumulated the tracer in the nervous fibres, including both the large and small neurons of the reticular formation and nuclei of the cranial nerves (Fig. 3). These changes were accompanied by PAS-positive substances accumulated mainly in the primary ischaemic zones around occluded vessels. Small haemorrhagic foci and splitting of myelin completed this picture.

The electron-microscopy studies showed swelling of the endothelial cytoplasm, with aggregates of red cells in collecting venules, capillaries and small veins, as well as enlargement of the perivascular space with swelling of the perivascular glial cells and their processes, shrinkage of nervous cells, and delamination of the myelin sheaths (Figs. 4 a–c).

Discussion

The short-lasting intermittent cerebral compression causing severe disturbance of cerebral blood circulation led to the ischaemic lesions of the nervous structures localized in front of and behind the tentorium. These limited ischaemic lesions resulted in a reduction in the cerebral perfusion pressure, mechanical compression of the cerebral vessels during transtentorial herniation, and with a hyperaemic phase after decompression.

Alternate compression and decompression of the brain caused slowing down of the blood flow and the formation of those blood cell aggregates which obstructed the small vessels, producing the so-called "no-reflow phenomena"[2,3]. The vascular obstruction, accompanying the changes in the endothelial cells were related to localized BBB damage and ischaemic brain oedema formation. The ischaemic lesions of the thalamo-hypothalamic areas appeared earlier than the mesencephalo-pontine damage. It was demonstrated mainly by neurophysiological studies; $BR-R_1$ disappeared earlier usually, after the first cerebral compression, whereas $BR-R_2$ and BAEP were affected after the second or third cerebral compression[1,5,6].

These morphological and neurophysiological observations were supported by CBF-studies following short-lived cerebral compression. The dynamics and the decrease in CBF correlated with the above-mentioned phenomena[9].

In conclusion, it could be said that decerebration symptoms after intermittent short-lasting cerebral compression were evoked by primary and secondary vasogenic lesions primarily localized in the diencephalon and then in the mesencephalo-pontine structures.

Acknowledgements

The skillful technical assistance of Mrs. C. Dambmann, Mrs. M. Knappe and Mrs. H. Rahje is gratefully acknowledged.

This work was supported by the Deutsche Forschungsgemeinschaft.

References

1. Csécsei, G., Klug, N., Rap, Z. M.: Effect of increased intracranial pressure on blink reflex in cats. Acta neurochir. (Wien) *68* (1983), 85—92.
2. Cuypers, J., Matakas, F.: The Effect of postischemic hyperemia on intracranial pressure and no reflow phenomenon. Acta Neuropath. (Berlin) *29* (1974), 73—84.
3. Hakmatpanah, J.: Irreversible damage and cerebral death in intracranial pressure. In: Advances in Neurosurgery (Schürmann, K., *et al.*, eds.), pp. 112—117. Berlin-Heidelberg-New York: Springer. 1973.
4. Hiraoka, M., Shimamura, M.: Neural mechanism of the corneal blinking reflex in cats. Brain Res. *125* (1977), 265—275.
5. Klug, N., Csécsei, G., Rap, Z. M.: Das Verhalten der frühen akustisch evozierten Potentiale in Abhängigkeit des intrakraniellen Druckes. Experimentelle Untersuchungen an der Katze. 27. Jahrestagung der Deutsch. EEG-Gesellschaft Freiburg vom 30. 9.—2. 10. 1982.
6. Nagao, S., Roccaforte, P., Moody, R.: Acute intracranial hypertension and auditory brain stem responses. Part 2: The effect of brain stem movement on the auditory brain stem responses due to transtentorial herniation. J. Neurosurg. *51* (1979), 846—851.
7. Tokunega, A., Oka, M., Murao, T., Sakai, H., Okumura, T., Hirate, T., Miyashita, Y., Yoshitatsu, S.: An experiment study on facial reflex by evoked electromyography. Med. J. Osaka Univ. *9* (1958), 397—411.
8. Tsementzis, S. A., Gillingham, F. J., Gordon, A., Hitchcock, E. R., Campbell, D., Medelow, A. D.: Decerebrate rigidity produced in cats by focal stereotactic radiofrequency lesions. Acta neurochir. (Wien) *59* (1981), 19—33.
9. Zierski, J., Kurzaj, E., Hoffmann, O., Winkler, B.: Cerebral blood flow in the brain stem during increased ICP. In: Intracranial Pressure V. Berlin-Heidelberg-New York-Tokyo: Springer (in press).

Author's address: Z. M. Rap, M.D., D.Sci., Neurosurgery Department of J. L. University Giessen, Klinikstrasse 29, D-6300 Giessen, Federal Republic of Germany.

Centre of Neurosurgery, Justus-Liebig University of Giessen, Federal Republic of Germany

The Role of Cerebral Veins in the Genesis of Secondary Lesions in the Occipital Lobe

W. E. Braunsdorf, P. Christophis, Z. M. Rap, and H. W. Pia

Summary

CT was undertaken in 17 patients with craniocerebral injury, in whom infarction of occipital lobe had occurred secondary to space-occupying, supratentorial lesions causing an increased intracranial pressure (ICP). The patients' findings were compared with the post-mortem findings from 5 non-surviving patients. Diffuse and locally blood-brain barrier (BBB) disturbances were demonstrated in the contrast enhanced CT pictures.

Congestion of vessels, stasis, local brain oedema and ischaemic-haemorrhagic lesions were all found in areas corresponding to hypodense zones.

Ischaemic and haemorrhagic lesions in the temporo-occipital areas were produced experimentally in 10 anaesthetized cats by repeated acute expansion of an extradural balloon. Light-, fluorescence- and electronmicroscopic studies showed congested vessels, local BBB damage, acute swelling of the endothelia of capillaries and venules and enlargement of the perivascular space which was characteristic of the early stage of brain oedema.

The pathogenesis of vasogenic lesions in the occipital lobe is discussed. Regional disturbances of cerebral venous outflow may be responsible for the occurrence of infarctions in the occipital lobe secondary to increased ICP.

Keywords: Occipital lobe infarction; increase in ICP.

Introduction

In the pathogenesis of occipital lobe infarction, two vasogenic factors have been discussed: the occlusion of the posterior cerebral artery or its branches[1,2,4,5,9,12], and the obstruction of venous blood outflow from occipital and basal veins[7,10,11]. We wanted to concentrate on some regional aspects of circulatory disturbances in the occipital lobe, secondary to the space occupying process, which are connected with BBB damage and with the early stage of brain oedema formation.

Material and Methods

1. In 17 patients with cranio-cerebral injury, occipital lobe infarctions were revealed in serial CT studies. 14 patients had extracerebral-extradural or subdural haematomas which were evacuated by surgery. In all of them a midline shift was present, which persisted after the operation. 3 patients had diffuse non-surgical brain injury with brainstem involvement. 12 patients survived (group 1) and 5 died (group 2). Brain sections were cut horizontally corresponding to the CT plane and stained with routine histological methods (Prof. Gerhard, Institute of Neuropathology, Essen).

2. Experimental studies were carried out on 10 anaesthetized (Narcoren 2 ml/kg) cats under spontaneous respiration. Intracranial hypertension was produced 2–3 times by short-lasting balloon inflation. In the *first group* of animals, (5) a balloon was placed in the fronto-temporal region of the brain, in the *second group, (5)* in the parieto-temporal region. Brain sections were studied using light-, fluorescence- and electronmicroscopy. Details of the methods have been described by Rap *et al.* 1982[8].

Results

1. CT Studies in Patients with Posttraumatic Occipital Lobe Infarction

In all the patients a diffuse zone of hypodensity localized in the medio-basal part of the occipital lobes. If not present in the CT scan on admission, it appeared in the CT soon after 12–24 hours (Figs. 1 a and b). In 5 cases hypodensity of the occipital lobe appeared on the side contralateral to the lesion. In 6 patients an enhancement in the hypodense area was present following contrast medium injection. In patients who survived, CT performed two weeks after the trauma showed only small defective areas in the previously diffuse hypodense occipital lobe.

From the 5 patients who died, 4 had unilateral and 1 bilateral lobe infarction as judged from the CT findings.

Pathomorphological studies revealed two ischaemic infarctions in the medio-basal part of the occipital lobe as a thin layer underlying the cortex and white matter subcortically. In one case a so-called "total infarction" and in another two cases ischaemic-haemorrhagic infarctions were observed (Figs. 2 a, b, e, and d). They were accompanied by secondary in 3 cases and primary (1 case) lesions in the midbrain and pons. In three cases primary and secondary haemorrhages were found in the cerebellum.

The light-microscopic studies showed morphological changes at different stages of tissue organization. Congestion of vessels, mainly veins in the pia mater and cortex, thrombosis and extravasation of blood into the perivascular space were noted. Disintegration of neurons and glial cells, and an increased number of polymorph leucocytes were observed during the first 24 hours after the lesion had occurred. This was accompanied by an increased number of macrophages, proliferation and hypertrophy of astrocytes, as well as proliferation of vessels at a later stage of tissue organization.

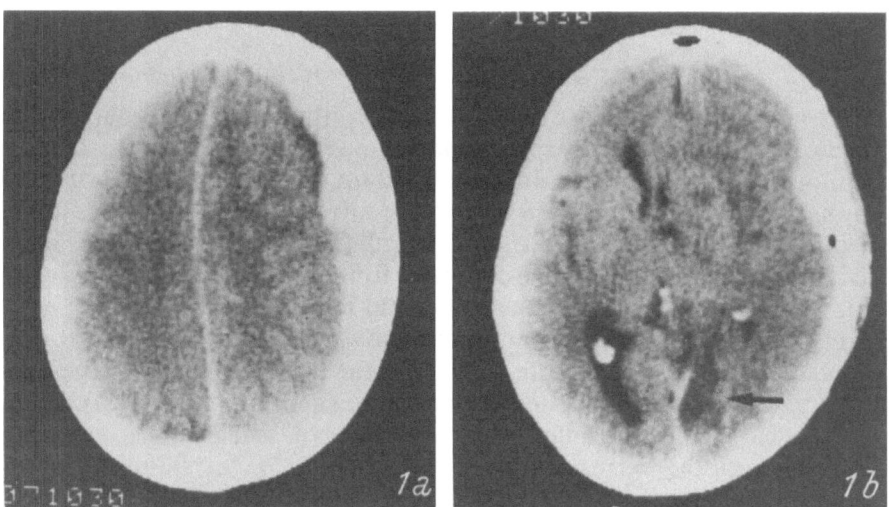

Fig. 1. CT scan in patient with acute subdural haematoma a) before, b) after operation occipital infarction (arrows)

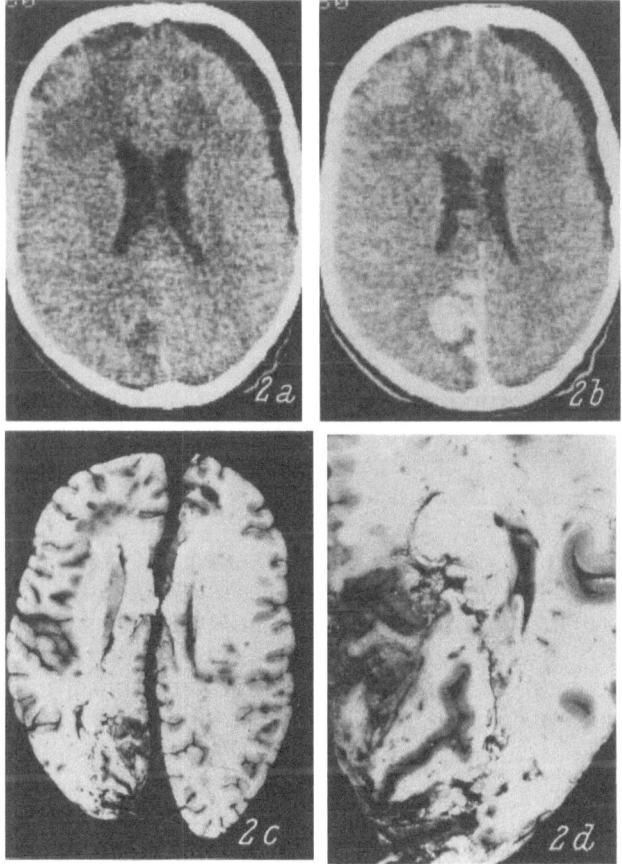

Fig. 2. CT scan without (a) and with contrast medium (b) in patient 12 days after severe head injury. Ipsilateral occipital infarction (c). The same at high magnification (d)

2. BBB and Structural Studies in Temporal-Occipital Lobes in Animals

Expansion of an epidural balloon in fronto-temporal (group 1) and parieto-temporal (group 2) regions determined the different localization of the morphological changes. In the first group the ischaemic-haemorrhagic lesions, with surrounding **BBB** damage, occurred mainly in the splenium of the corpus callosum and in the gyrus cinguli (Fig. 3 a). In the second group the ischaemic-haemorrhagic lesions were found in the grey and subcortical white matter of the temporal and occipital lobes (Figs. 3 b and c).

The fluorescent microscopic studies showed a diffuse spread in the red fluorescence of the Evans blue complex in the damaged area and the dye's accumulation mainly in the neurons and their axons (Fig. 4). Light microscopic studies revealed the congestion of small vessels and their occlusion with the extravasation of blood (Fig. 5), the shrinkage of the neurons, and a swelling of glial cells.

Swelling of the endothelial cells with aggregates of blood cells in vessels and swelling of the perivascular glial processes were typical of the early stage of brain oedema observed in the electron-microscopic study (Figs. 6 a and b).

Discussion

In all patients hypodense areas were seen in the occipital lobe 12–24 hours after the injury. Morphologically it corresponded with hyperaemia, brain oedema and ischaemic-haemorrhagic infarction[3, 6]. An increase in the density in this area after contrast medium injection, indicated severe alterations in the BBB accompanying secondary haemorrhage. Resolution of this local pathological process, primarily including brain oedema, after 2–3 weeks resulted in isodensity in the CT scan.

In 3 out of 5 cases examined at post-mortem showed that severe hyperaemia accompanied the occipital lobe infarction. In these cases the pole of the occipital lobe was compressed not only from above, but also from below through elevation of the tentorium caused by the ischaemic haemorrhagic infarction of the cerebellar hemisphere. The local obstruction of the venous drainage from the occipital lobe, both involving small veins as well as large veins (which are compressed between the tentorium and the splenium of corpus callosum[7, 10, 11]), causes hypoxia, an increase in vascular permeability. Experimental studies in cats showed that in the area of infarction in the temporo-occipital region there was congestion of the vessels, local stasis and aggregation of blood cells which were caused by the repeated extradural compression. The same mechanism lead to haemo-dynamic disturbances, which additionally complicate the acute swelling of endothelial cells of capillaries and venules. BBB damage and brain oedema formation were the consequences[8]. Probably secondary haemorrhagic foci were the results of intermittent intracranial hypertension. The same mechanism of secondary haemorrhage in the occipital lobe occurred in our group of operated patients, where evacuation of hematomas was made but not followed by a permanent reduction in the ICP. The assessment of the

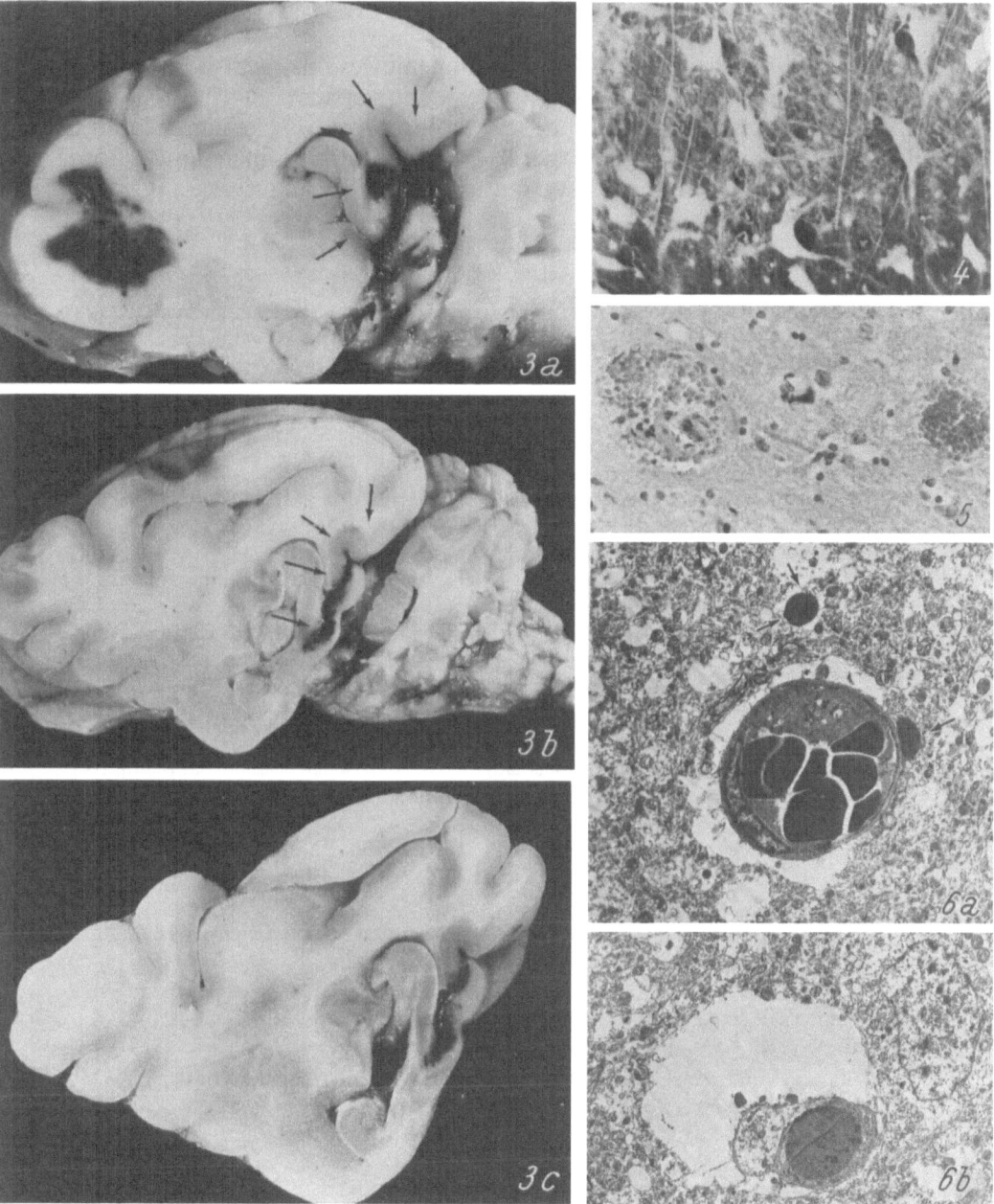

Fig. 3. a) Sagittal section through three cat brains (group 1): BBB damage in gyrus cinguli and in adjacent occipital lobe. Arrows: haemorrhagic foci. b) Cat brain from group 2, ischaemic haemorrhagic infarction in temporo-occipital lobes (arrows). c) The same cat, mirror section, high magnification, × 5

Fig. 4. Red fluorescence of Evans blue-albumin complex in neurons and their fibres. Data from occipital lobe, × 250

Fig. 5. Two small vessels in the occipital lobe. Thrombosis and extravasation of blood are seen. Haematoxylin-Eosin, × 150

Fig. 6. a) Aggregate of blood cells in the capillary in occipital lobe. Swelling of perivascular glial processes. Red cellular extravasation per diapedesis (arrows), × 9,000. b) Enlargement of perivascular space around the small multi-layered venules in occipital lobe. × 3,600

mechanism of the development of vasogenic lesions, with ischaemic or haemorrhagic infarction and brain oedema, by a space occupying process in the occipital lobe is still difficult. Morphological studies in humans and in experimental animals suggest that in the occipital lobe infarction by local disturbances in the venous blood outflow caused by an increased ICP may play an important role.

Acknowledgement

The authors would like to thank Prof. L. Gerhard for her co-operation, and also thank to Mrs. M. Knappe for the photography and the radiological assistants of the Department of Neuroradiology (Head: Priv.-Doz. Dr. Agnoli) for preparing the CT scans.

References

1. Adebahr, G., Schewe, G.: Sekundärschäden in der Area striata. Hefte zur Unfallheilkunde *94* (1968), 261—265.
2. Blackwood, W.: Vascular disease of the central nervous system. In: Neuropathology (Greenfield, J. G., *et al.*, eds.), pp. 96—103. London: E. Arnold (Publishers) Ltd. 1961.
3. Clasen, R. A., Huekman, M. S., van Roem, K. A., Pandof, S., Leinig, J., Clasen, J. R.: Time course of cerebral swelling in stroke, a correlative autopsy and CT study. Advances in Neurology, Vol. 28, Brain Edema, pp. 395—412. New York: Raven Press. 1980.
4. Kleihues, P. P., Hizawa, K.: Die Infarkte der A. cerebri posterior: Pathogenese und topographische Beziehungen zur Sehrinde. Arch. Psychiatrie Nervenhk. *208* (1966), 263—284.
5. Kleihues, P. P.: Über die doppelseitigen symmetrischen Occipitallappeninfarkte. Pathologie und klinisch-opthalmologische Befunde. Dtsch. Z. Nervenheilk. *188* (1966), 25—52.
6. Miller, J. D., Gudeom, S. K., Kishore, P. S., Becker, D. P.: Computed tomography in brain due to trauma. Advances in Neurology, Vol. 28, Brain Edema, pp. 414—421. New York: Raven Press. 1980.
7. Pia, H. W.: Die Schädigung des Hirnstamms bei den raumfordernden Prozessen fes Gehirns. Acta neurochir. (Wien) Suppl. IV (1957).
8. Rap, Z. M., Csécsei, G., Klug, N., Zierski, J., Pia, H. W.: Morphological analysis of experimental decerebration after acute epidural compression. In: The Cerebral Veins (Auer, L. M., Loew, F., eds.), pp. 293—298. Wien-New York: Springer. 1983.
9. Robinson, F., Porro, R. S., Scotliff, H. J.: Angiographic recognition of occipital lobe infarction. Neurology *16* (10) (1967), 1016—1021.
10. Spatz, H.: Pathologische Anatomie der Kreislaufstörungen des Gehirns. Z. ges. Neurol. Psychiat. *167* (1939), 301—357.
11. Stochdorph, O.: Über Verteilungsmuster von venösen Kreislaufstörungen des Gehirns. Arch. Psych. Nervenk. *208* (1966), 285—298.
12. Zülch, K. J.: Störungen des intrakraniellen Druckes. In: Handbuch der Neurochirurgie, Bd. 1/1 (Olivecrona, H., Tönnis, W., eds.), pp. 208—303. Berlin-Göttingen-Heidelberg: Springer. 1959.

Author's address: Z. M. Rap, M.D., D.Sci., Neurosurgery Department of J.L.University Giessen, Klinikstrasse 29, D-6300 Giessen, Federal Republic of Germany.

Department of Forensic Medicine, Lund University, Lund, Sweden

Superior Cerebral Vein Susceptibility to Injury in Head Trauma

C. G. P. Löwenhielm

Summary

Following disruption of the parasagittal bridging veins and venous bleedings near the superior margin of the cerebral hemispheres, situated in the cortex or in the subcortical white matter, these "gliding contusions" are common in trauma cases, when the head has suffered rotational acceleration. Such injuries are explained by a sudden head rotation which gives rise to slipping along the brain-skull interface and to deformation of the brain matter. Physical and mathematical models, cadaver tests and animal experiments have been utilized in order to clarify the genesis and to establish tolerance levels of these types of rotational injuries to the cerebral veins.

Keywords: Parasagittal veins; rotational acceleration; gliding contusion.

Introduction

Some of the most frequent traumatic injuries to the brain are disruption of the parasagittal bridging veins (superior cerebral veins), accompanied by a local subarachnoidal haemorrhage and/or cerebral lesions consisting of cortical and subcortical bleedings close to the superior margin of the brain, typically located in the posterior part of the superior frontal gyrus or in the central gyrus. Disruption of one or several bridging veins leads to a subdural haematoma. In case of a subdural haematoma it may be possible to save the patient's life by neurosurgical treatment, but in the other situation, when many or all of the bridging veins are disrupted, death occurs immediately, or after persistant unconsciousness[9]. Most striking in these cases is the absence of a noteworthy subdural haematoma, even when life has been maintained by respirator treatment. The brain shows acute swelling but without flattening of the gyri, and no midline dislocation. The brain weight is 200–300 g above normal[7,9].

In the fresh state the intracerebral lesions are characterized by subarachnoidal bleedings, cortical and subcortical perivascular rod-shaped bleedings. Lindenberg and Freytag[3] designated the subcortical lesions as "gliding contusions". The patho-anatomical outcome of gliding contusions has been described by Voigt et al.[7], who also discussed their differential diagnosis.

Angular acceleration of the head is considered to be the causative agent of these injuries. The original work in this area was presented by Holbourne[2], who argued that linear acceleration was an insignificant cause of brain damage. On the basis of the head angular acceleration concept, this paper comprises a summing up of different experimental studies on tolerance criteria for the frequently occurring bridging vein disruptions and gliding contusions, as well as the their genesis.

Materials and Methods

In 658 consecutive cases of blunt trauma fatalities, disruption of parasagittal bridging veins and gliding contusions were studied. At autopsy the brain was removed, ad modum Flechsig, i.e. the upper part of the cerebrum being cut off together with the skull and then removed in toto together with the dura, by using a spatula. This procedure allows examination of the parasagittal bridging veins by carefull folding back of dura. Autopsy of the brain was always performed after fixation in formalin.

The dynamic mechanical properties of human cadaver bridging veins were investigated by using special test equipment. Specimens of intact bridging veins were subjected to tensile tests in which the strain rate was varied and the ultimate elongation was recorded. Voigt and Lange[8] carried out acceleration sled experiments with cadavers to study the optimal design of vehicle instrument panels. In these experiments, bridging vein disruption was obtained in some cases. High-speed films from these tests were analysed regarding the angular displacement-time history of the head. The angular velocity change and the peak angular acceleration were then calculated by using smoothed cubic spline functions[1] in order to get high quality derivatives.

The genesis of the gliding contusions was similated by using a mathematical model given by Ljung[4]. The model, to which a sudden rotation could be applied, consists of a hemisphere of viscoelastic material which is enclosed in a rigid shell. The plane part of the shell corresponds to the falx cerebri and the spherical part of the shell to the skull. This model gives a good approximation of the motion of the brain close to its superior margin. For angular acceleration pulses of varying amplitude and duration, the deformation and the shear of the brain matter can be calculated. As a gliding contusion in the fresh state is characterized by haemorrhage, a simulation can be accomplished thus: assuming that the intracerebral branches of the superior cerebral veins passively follow the deformation of the brain, and taking the dynamic properties of the cerebral veins into consideration (then using the values obtained for bridging veins), the site of maximal stress can be obtained and the corresponding threshold levels for intra-cerebral venous disruption calculated.

Results and Discussion

Disruption of the parasagittal bridging veins was seen in 29% of the cases and gliding contusions in 11%. In 19% of the cases with bridging vein disruption there was no skull fracture. The corresponding figure for the gliding contusions was 25%. Slipping of the brain surface in relation to the inner surface of the skull was first shown by Pudenz and Shelden[5]. This slipping has been regarded as the causal mechanism of bridging vein disruption (cf.[6,9]).

The tensile tests on isolated cadaver bridging veins showed that ultimate strain is highly dependant on the strain rate. For low strain rates the ultimate strain is about 100%, whilst for very high strain rates the ultimate strain is reduced to about 15%. Strain rates valid for head impact in traffic accidents give a strain of about 40–50%.

Analysis of high speed films from the cadaver experiments yielded the tolerance criteria: both a critical angular acceleration ($\ddot{\Phi} = 4{,}500 \, \text{rad/sec}^2$) and a critical angular velocity change ($\Delta\dot{\Phi} = 30 \, \text{rad/sec}$) must be exceeded, for bridging vein disruption to occur.

Lindenberg and Freytag[3], who were the first to describe the gliding contusions, assumed that they resulted from stretching of the parasagittal bridging veins and the surrounding Pacchionian granulations. However, mathematical simulation of gliding contusions indicated that, when a sudden angular acceleration is applied to the head, the probable location for injury induced by the brain's own intrinsic deformation, is obtained some 8–10 millimetres below the surface of the brain, which corresponds to the site of the subcortical gliding contusions. The shear levels in the cortex were found to be fairly low compared with the levels obtained subcortically. This result is in good agreement with the fact that the cortex can be preserved uninjured despite the deeper lying brain tissue showing extensive lesions. The tolerance criteria for gliding contusions was found to be that both a critical angular acceleration ($\ddot{\Phi} = 4{,}500 \, \text{rad/sec}^2$) and a critical angular velocity change ($\Delta\dot{\Phi} = 70 \, \text{rad/sec}$) had to be exceeded if injury was to occur.

Sometimes gliding contusions occur without accompanying bridging vein disruption. This finding seems to be contradicted by the fact that the tolerance level for gliding contusions is higher than those for bridging vein disruption. However, if the bridging vein disruption/gliding contusion frequency quotient is studied as a function of age, it is found that the quotient is lower for the elderly and higher for the young. This may be explained by the proliferating connective tissue which, with time, will form a firmer attachment of the parasagittal surface of the brain to the dura. Thus, slipping may be said to be gradually prevented, and the bridging veins protected, whilst deformation of the brain is unaffected.

Disruption of parasagittal bridging veins and gliding contusions occur as a consequence of head rotation. Reconstruction of the course of events in motor vehicle accidents indicates that rotational acceleration occurs when the heads of car occupants hit the steering wheel, the dash-board or the wind shield. A driver runs a high risk of suffering from the injuries described since his head may often hit the steering wheel regardless of seat belt use or not.

An angular acceleration of the head may give rise to considerable strain in the connection between the cerebrum and the spinal cord, and hence to brain stem injuries. These lesions lead to immediate death or death following deep unconsciousness. The fact that these patients are already beyond medical help seems to be the reason why such cases have not attained clinical interest.

References

1. Berghaus, D. G., Cannon, J. P.: Obtaining derivative from experimental data using smoothed spline functions. Exp. Mechanics (1973), 38—42.
2. Holbourne, A. H. S.: Mechanics of brain injuries. Lancet *II* (1943), 438—441.
3. Lindenberg, R., Freytag, E.: The mechanism of cerebral contusions. Arch. Path. *69* (1960), 440—469.
4. Ljung, C. B. A.: A model for brain deformation due to rotation of the skull. J. Biomechanics *8* (1975), 263—274.
5. Pudenz, R. H., Shelden, C. H.: The lucite calvarium—A new method for direct observation of the brain. J. Neurosurg. *3* (1946), 487—505.
6. Unterharnscheidt, F., Sellier, K.: Vom Boxen, Mechanik, Pathomorphologie, und Klinik der traumatischen Schäden des ZNS bei Boxern. Fortschr. Neurol. Psychiat. *39* (1971), 109—151.
7. Voigt, G. E., Löwenhielm, C. G. P., Ljung, C. B. A.: Rotational cerebral injuries near the superior margin of the brain. Acta Neuropath. (1977), 201—209.
8. Voigt, G. E., Lange, W.: Simulation of head-on collision with unrestrained front seat passenger and different imstrument panels, Proc. 15th Stapp Car Crash Conference, 466—488, Society of Automotive Engineers, New York, 1971.
9. Voigt, G. E., Saldeen, T.: Über den Abriß zahlreicher oder sämtlicher Vv. cerebri sup. mit geringem Subduralhaematom und Hirnstammläsion. Dtsch. Z. ges. gerichtl. Med. *64* (1968), 9—20.

Author's address: C. G. P. Löwenhielm, M.D., Department of Forensic Medicine, Lund University, Sölvegatan 25, S-223 62 Lund, Sweden.

Diagnosis and Treatment of Cerebral Venous Diseases

Department of Neurosurgery, Hanwa Memorial Hospital, Osaka, Japan

The Significance of the Veins in the Determination of Cerebral Blood Volume in Clinical Cases

M. Taneda

Summary

The changes in the intracranial blood volume (ICBV) and the cerebral blood flow (CBF) caused by decompression were monitored in clinical cases. The methods for decompression were the intravenous injection of glycerol and the lowering of cerebrospinal fluid (CSF) pressure by ventricular drainage. Changes in the ICBV were monitored in 11 patients by continuous extracranial recording of gamma activity, while monitoring CSF pressure, after the intravenous administration of ^{99m}Tc-labelled red blood cells. The CBF was measured in 6 cases by the ^{133}Xe inhalation method. Using these methods the changes in the ICBV were not conclusive in 3 cases of CSF drainage and 6 cases with glycerol infusion. In the other cases, the ICBV increased as a result of CSF drainage and was decreased by glycerol infusion. This indicated that there was resistance to flow distal to the vascular bed which was diminished probably by the removal of extravascular fluid after glycerol infusion, and was not diminished by simple decompression following CSF drainage. The CBF generally increased following both methods of decompression. However, glycerol infusion increased the CBF much more than CSF drainage, despite the fact that the former decreased CSF pressure less than the latter. This greater increase in the CBF following glycerol infusion also suggested the occurrence of a decrease in resistance on the venous side. The presence of such a resistance indicates that veins may have an important role in the mechanism for the determination of the ICBV in clinical cases.

Keywords: Cerebral blood volume; cerebral blood flow; intracranial pressure; cerebral veins.

Introduction

The cerebrovascular bed is one of the compartments of which the intracranial cavity is composed. An increase in the intracranial blood volume (ICBV) may be clinically important, because it may raise the intracranial pressure (ICP) and lead to devastating results. However, the pathogenesis of the expansion of the vascular compartment has not been well investigated in clinical cases.

Theoretically, there are two mechanisms which would increase the cerebrovascular volume. One is vasoparalysis occurring on the arterial side. The other is stasis caused by obstruction or narrowing of the veins.

In this study decompression, by reducing the size of the non-vascular compartments, was performed to investigate the corresponding responses of the ICBV and the cerebral blood flow (CBF) to decompression, and to correlate these effects with the mechanisms which bring about an increase in the ICBV, especially in relation to the role of the veins in clinical cases.

Material and Methods

The patients included in this study were diagnosed as having either intracerebral haemorrhage or subarachnoid haemorrhage by computer tomography. They were all adult and clinically in a serious condition. Changes in the ICBV and the CBF caused by decompression were investigated in these cases.

The cerebrospinal fluid pressure (CSFP) was monitored by means of a catheter placed in the frontal horn of the lateral ventricle which was connected to the recording equipment.

Decompression was performed twice for every case as follows: Firstly, CSF was withdrawn to keep the CSFP at 0 mm Hg. Then, it was replaced into the ventricle to raise the CSFP again to the level before decompression. Secondary, after stabilization of the CSFP, 200 ml of glycerol 10% was infused intravenously over a period of between 5 and 10 minutes.

The changes in the ICBV were monitored in 11 cases using a modification of the method described by Risberg et al.[9]. The gamma activity was continuously recorded extracranially, while monitoring CSFP, following intravenous administration of 10 mCi of ^{99m}Tc-labelled red blood cells. Simultaneous changes in the ICBV accompanying the lowering of CSFP caused by decompression were reflected by changes in the gamma activity. A continuous measurement of the relative changes in gamma activity was obtained by external recording of the radiation from the chest, as well as from the head. The gamma activity from the chest was detected mainly from the heart and was presumed to reflect the gamma activity mainly in the blood. Therefore, its changes provided the basis on which the changes in the ICBV could be postulated. The data were plotted on the same chart by means of a X-Y recorder (Fig. 1).

The CBF was measured by the ^{133}Xe inhalation method[7] in six cases before decompression, several minutes after CSF drainage, and 15 minutes after the initiation of glycerol administration. The time interval between measurements was at least 30 minutes. The CBF was calculated using the initial-slope method. Immediately after each xenon inhalation, arterial blood was withdrawn and analyzed for CO_2 and O_2 tension as well as O_2 saturation.

Results

Infusion of glycerol was effective in reducing CSFP in all cases, although it did not lower CSFP to 0 mm Hg in the way that drainage of CSF did.

The ICBV increased as a result of drainage of CSF in eight out of 11 cases, whereas the changes in gamma activity were not clearly shown in the other three cases. The reinjection of the CSF returned the ICBV to the level before decompression in the cases with increased ICBV as a result of CSF drainage.

Glycerol infusion caused a decrease in the ICBV in five out of 11 cases, whereas changes in the ICBV were not clearly shown in the other six cases. In the former cases, the ICBV began to decrease before the intracranial pressure decreased. This was not caused by a decrease in radioactivity due to dilution of the blood by the rapid infusion of glycerol, because, as shown in Fig. 1, the reduction in the radioactivity from the chest (ChBV), (shown as a trace to which a thick line was added for the clear discrimination from the trace of the ICBV), was much smaller than the reduction in the ICBV.

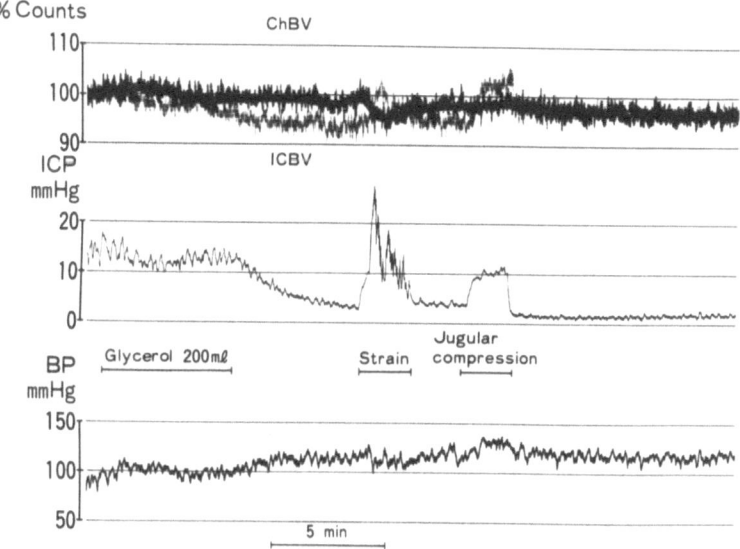

Fig. 1. Changes in intracranial blood volume (ICBV) caused by the intravenous administration of glycerol. Radioactivity from the chest (ChBV) (shown as a trace to which a thick line has been added for the clear discrimination from the trace of the ICBV), provided the basis on which the changes in ICBV could be postulated. Note the apparent decrease in ICBV as compared with the changes in ChBV before the decrease in the intracranial pressure (ICP)

Changes in the hemispheric CBF following decompression on the side of the lesion were expressed as percentage changes of the CBF. In five out of six cases, the CBF increased by 1.2, 4.1, 6.3, 11.2, and 15.9% respectively, as a result of CSF drainage, and by 22.2, 30.8, 37.8, 33.1, and 48.4% respectively, as a result of glycerol infusion. There were no marked differences in the arterial CO_2 tension during the measurements in every case. In one case, the CBF decreased by 15.2% following drainage of CSF, and increased by 11.1% after glycerol infusion. The drainage of CSF caused hyperventilation in this case, and this in its turn caused a decrease in the arterial CO_2 tension of 3 mm Hg.

Discussion

The role of the ICBV in the pathological brain is not absolutely clear. It was suggested by Langfitt et al.[3], that vascular engorgement by vasomotor paralysis rather than oedema was the primary cause of the brain swelling in the injured brain. In the experimental work of Lowell and Bloor[5], they found a somewhat diminished vascular compartment in monkeys with an increased ICP following trauma. There was no marked difference in the ICBV values determined by emission computed tomography between normal controls and patients with head injuries[2]. Unfortunately, difficulties in the measurement of the ICBV have not allowed extensive clinical research in this field. From the clinical point of view, changes in the ICBV are much more important than the net values of the ICBV, because, in a swollen brain, a small increase in the ICBV may cause a tremendous increase in the ICP.

Alteration in the ICBV may be caused both by physiological and pathological changes in the resistance to flow. In normal subjects, it is believed that there is a variable resistance on the arterial side of the vascular channel, to maintain constant blood flow (autoregulation). The ICBV and CBF may proportionally increase or decrease in response to the alterations of this resistance, caused by changes in arterial CO_2 tension[8] or in response to the alterations in cerebral perfusion pressure (CPP) in cases where there is a fall in the resistance because of vasoparalysis or vasodilatation. By means of an increase in the resistance on the venous side of the vascular channel, in response to an intracranial mass or brain swelling[4], stasis occurs. The ICBV may increase, whereas the CBF may decrease. If such venous obstruction diminishes, the ICBV may decrease and the CBF may increase. Thus, whether the volume-flow relationship is proportional or reciprocal seems to depend on which portion of the vessel is responsible for the changes in the circulation.

In this study, the CBF increased and the ICBV decreased by decompression, following the infusion of glycerol in five out of 11 cases. This indicated that there was a resistance to flow on the venous side, and that it was decreased after glycerol administration, which diminished the resistance to flow distal to the capillary bed by the removal of extravascular fluid.

The drainage of CSF increased the ICBV as well as the CBF in most the cases. In normal subjects with an intact autoregulation, decompression alone may not increase the CBF[1]. An increase in the CBF by increasing the CPP suggested the presence of a disturbed autoregulation in this series. The increase in the ICBV following drainage of CSF was probably caused not only by the increased intravascular pressure resulting from the increased CPP but also by a simple reciprocal increase in the volume of the vascular compartment to fill the space created by CSF drainage.

Intravenously administered glycerol was reported to be effective in increasing the CBF[6]. In this study, the infusion of glycerol increased the CBF much more than the drainage of CSF, although the glycerol infusion decreased the CSFP to a lesser extent than CSF drainage did. Similar observations were reported by Shenkin *et al*[10]. They found an increase in the CBF after the injection of hypertonic glucose, which is also a hyperosmolar solution, like glycerol or mannitol, but no change in the CBF after an acute reduction in the ICP following ventricular drainage. They suggested that the increase in CBF was caused by lowering the blood viscosity rather than by an increase in CPP. However, considering the fact that the ICBV decreased after glycerol infusion, it may be that a decrease in the resistance to flow might have a major role in the increase in CBF following the infusion of a hyperosmolar solution rather than decreased blood viscosity. Thus, the resistance to flow of blood on the venous side seems to exist in most of the clinically serious patients.

The absolute volume of the vascular compartment was not measured in this study, because the studies were confined to gross dynamic changes in the ICBV caused by decompression. The results obtained suggested the presence of a resistance on the venous side as well as a disturbance in the autoregulation, and their possible role in increasing the ICBV has been discussed. Particularly, the veins seemed to be significant with regard to their role in the mechanism which determines the ICBV. It is quite possible that the increase in the ICBV, caused by pathological conditions such as head injury, may lead to severe intracranial hypertension in some circumstances.

References

1. Häggendal, E., Löfgren, J., Nilsson, N. J., Zwetnow, N. N.: Effects of varied cerebrospinal fluid pressure on cerebral blood flow in dogs. Acta Physiol. Scand. *79* (1970), 262—271.
2. Kuhl, D. E., Alavi, A., Hoffman, E. J., Phelps, M. E., Zimmerman, R. A., Obrist, W. D., Bruce, D. A., Greenberg, J. H., Uzzell, B.: Local cerebral blood volume in head-injured patients. Determination by emission computed tomography of ^{99m}Tc-labelled red cells. J. Neurosurg. *52* (1980), 309—320.
3. Langfitt, T. W., Tannanbaum, H. M., Kassell, N. F.: The etiology of acute brain swelling in experimental head injury. J. Neurosurg. *24* (1966), 47—56.
4. Langfitt, T. W., Weinstein, J. D., Kassell, N. F., Gagliardi, L. J., Shapiro, H. M.: Compression of cerebral vessels by intracranial hypertension. I. Dural sinus pressure. Acta neurochir. (Wien) *15* (1966), 212—222.
5. Lowell, H. M., Bloor, B. M.: The effect of increased intracranial pressure on cerebrovascular hemodynamics. J. Neurosurg. *34* (1971), 760—769.
6. Mayer, J. S., Fukuuchi, Y., Shimazu, K., Ohuchi, T., Ericsson, A. D.: Effect of intravenous infusion of glycerol on hemispheric blood flow and metabolism in patients with acute cerebral infarction. Stroke *3* (1972), 168—180.
7. Obrist, W. D., Thompson, H. K., Wang, H. S., Wilkinson, W. E.: Regional cerebral blood flow estimated by Xenon^{133} inhalation. Stroke *6* (1975), 245—256.
8. Risberg, J., Ancri, D., Ingvar, D. H.: Correlation between cerebral blood volume and cerebral blood flow in the cat. Exp. Brain Res. *8* (1969), 321—326.
9. Risberg, J., Lundberg, N., Ingvar, D. H.: Regional cerebral blood volume during acute transient rises of the intracranial pressure (Plateau waves). J. Neurosurg. *31* (1969), 303—310.

10. Shenkin, H. A., Spitz, E. B., Grant, F. C., Kety, S. S.: The acute effects on the cerebral circulation of the reduction of increased intracranial pressure by means of intravenous glucose or ventricular drainage. J. Neurosurg. *5* (1948), 466—470.

Author's address: M. Taneda, M.D., Department of Neurosurgery, Hanwa Memorial Hospital, 11-11, Karita 7, Sumiyoshi-ku, Osaka 558, Japan.

Service de Neuroradiologie, C.H.U. Nancy, France

Spontaneous Dural Fistulas
Classification—Diagnosis—Endovascular Treatment

L. Picard, J. Roland, S. Bracard, J. Lepoire, and J. Montaut

Summary

The classification of the cranial dural fistulas depends from the topography of the pathological part of the sinus. The main and most frequent fistulas concern: cavernous and intercavernous sinus; superior sagittal sinus; lateral sinus. The fistulas of the cavernous sinus are always multipedicular. The embolization is the best treatment with very small pieces of dura, because Isobutyl is sometimes dangerous in such a localization. The most frequent fistulas are localized on the lateral sinus; the best treatment is embolization but the long-term results are irregular and a complete anatomical cure is rare. The fistulas of the superior sagittal sinus are less frequent and are often fistulas with cortical vein drainage. The best treatment seems embolization with IBC; when there is a cortical vein drainage, a neurosurgical intervention is better.

Keywords: Dural fistulas; embolization.

1. Definition

We call dural fistulas all the arterio-venous fistulas developed between the meningeal arterial branches providing the arterial supply to the wall of a sinus and the sinus itself. From a pathophysiological point of view, we think that the fistula is due to a lesion of the wall of the sinus and mainly of a portion of its wall (Figs. 1 and 2). On the angiographies, all the branches which supply a fistula are orientated towards the same pathological portion of the sinus.

2. Classification

The definition leads to a topographical classification, based on the pathological part of the sinus. This classification has not only a speculative interest but also a therapeutical one. The topographical classification is as follows:

1. Fistula of the cavernous sinus and inter-cavernous sinus.
2. Fistula of the superior sagittal sinus.
3. Fistula of the transverse sinus or of the sigmoid sinus.
4. Other fistulas which can be located on: inferior sagittal sinus; superior petrous sinus; inferior petrous sinus; rectus sinus; sinus of the tentorium; spheno-parietal sinus.

3. Dural Fistulas of the Cavernous and Inter-Cavernous Sinus

Among fifty-seven carotido-cavernous fistulas from our data, only nine have been selected because they were not post-traumatic. The lesion was located on the wall of the venous system, and lastly they were really or potentially supplied by all the regional meningeal branches.

The symptomatology of the 9 cases was:

Exophthalmos	8 cases.
Orbital pains	8 cases.
Diplopia	7 cases.
Murmurs	4 cases.

All these fistulas are multipedicular, mainly supplied by meningeal branches of the carotid siphon (posterior group—inferolateral trunk), internal maxillary artery (terminal branch—middle meningeal artery) and ascending pharyngeal artery (Fig. 1).

During the last few years, P. Lasjaunias and J. Moret have suggested a new classification:

Type 1 fistulas: including fistulas supplied by a complex nework of dural arteries without any systematization.

Type 2 fistulas: in the angiogram, it is possible to identify all the feeding arterial pedicles. These fistulas are perhaps due to a rupture of intra-cavernous collaterals of the carotid siphon.

The outcome is often favourable, mainly after embolization for the fistulas type 2, but the spontaneous or post-embolization outcome of the type 1 fistulas is less favourable.

Concerning our nine cases; all the patients were treated by embolization: seven with dura and two with isobutyl cyano-acrylate (IBC). One patient died because a small amount of isobutyl cyano-acrylate went through an anastomosis with the carotid siphon and invaded the middle cerebral artery. In seven cases the cure was complete and in one case, we obtained a clinical stabilization. In consequence, we think that embolization with dura is efficient and we have to avoid isobutyl cyano-acrylate in this territory because, due to the numerous anastomoses, it would seem to be too dangerous.

4. Dural Fistulas of the Transverse Sinus and/or Sigmoid Sinus

This is the most frequent location. We have embolized 21 cases. Twenty had a murmur, five a local pain, five dizziness and two had epileptic seizures.

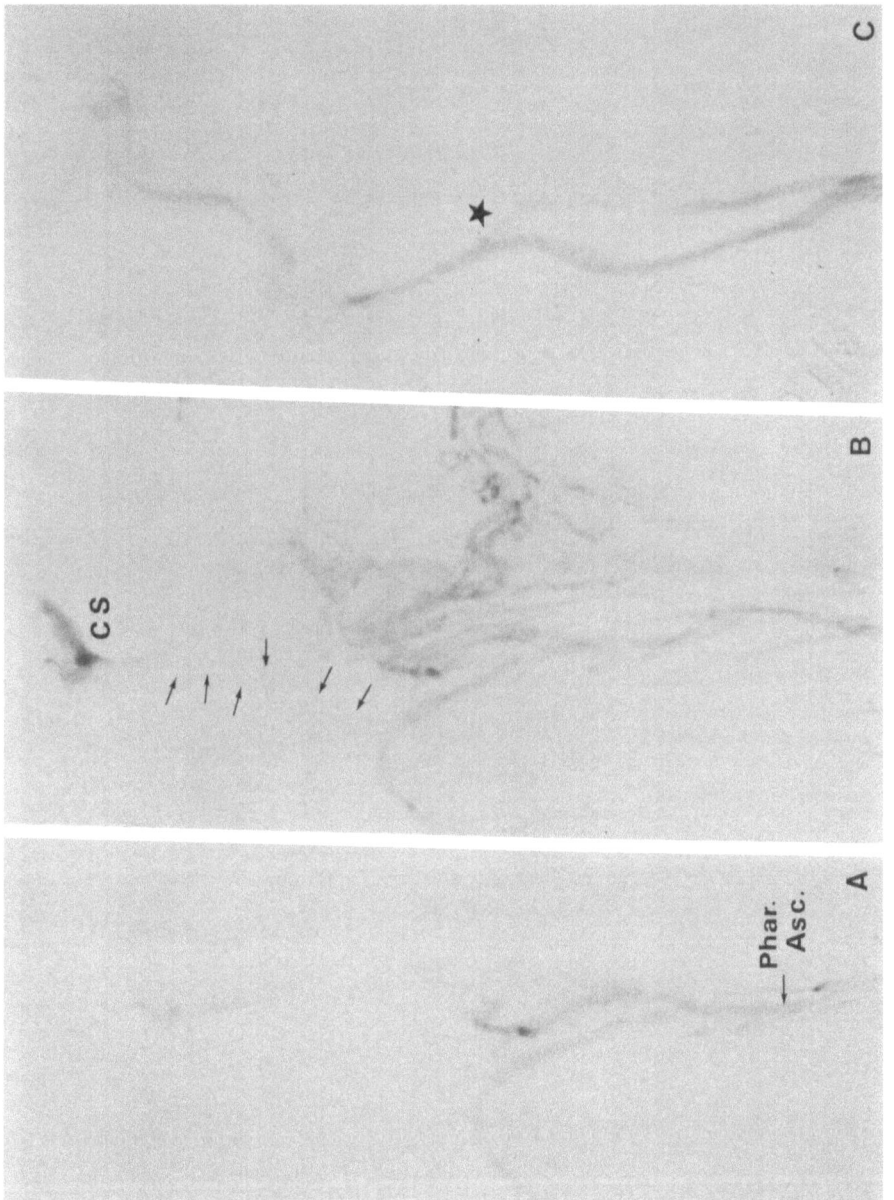

Fig. 1. Seventy-year-old man with spontaneous right exophthalmos. A) and B) hyperselective angiography of the right ascending pharyngeal artery (Phar. Asc.) of the right ascending pharyngeal artery. C) After hyperselective embolization with small pieces of dura, there is a complete obstruction (★) of the right ascending pharyngeal artery with reflux of contrast into the internal carotid artery leading to a complete anatomical and clinical cure

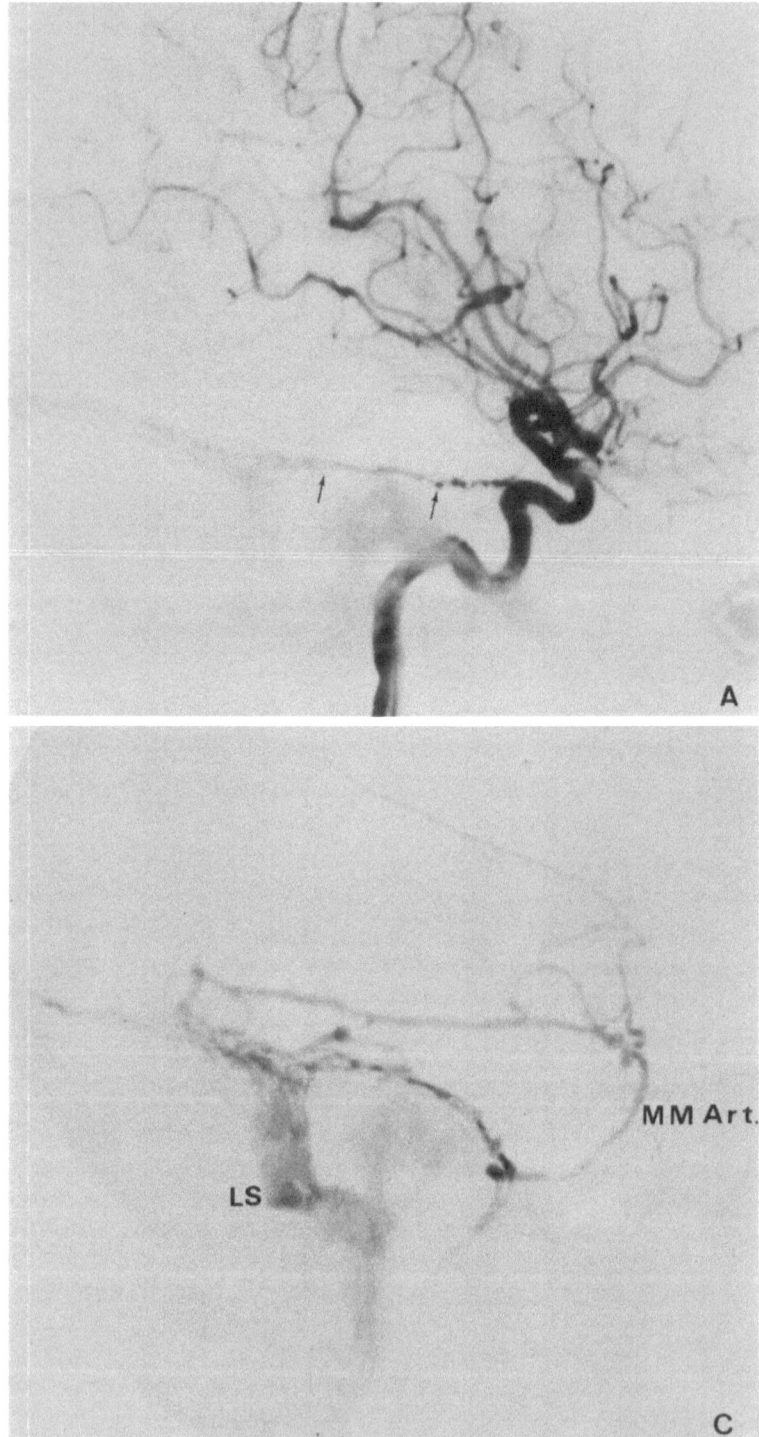

Fig. 2. Twenty-six-year-old woman with a spontaneous murmur in the right ear. A) Right internal carotid angiography (lateral view). The fistula is provided by an artery from the free margin of the tentorium (↗ ↗). B) Right hyperselective occipital angiography (lateral view). Many branches of occipital artery (*OCC*) supply the fistula in the lateral sinus (*LS*). C) Right

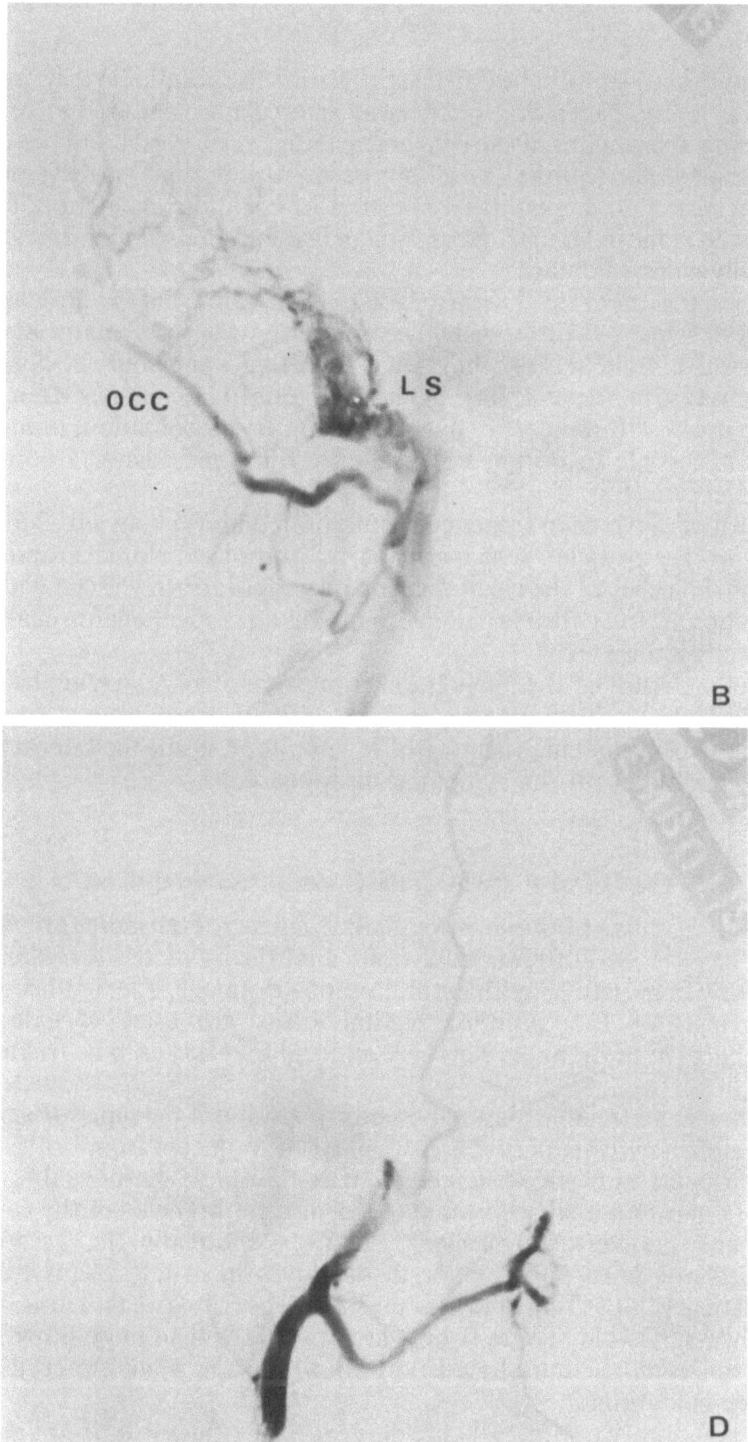

hyperselective middle meningeal angiography (lateral view). The fistula in the lateral sinus (*LS*) is also supplied by posterior branches of the middle meningeal artery (*M M Art.*). D) Right external carotid angiography after embolization of all the feeding pedicles of the fistula. Two years later, all the clinical symptoms had disappeared

Many arteries can supply these fistulas including the occipital artery, middle meningeal artery, ascending pharyngeal artery, auricular artery, cervical and meningeal branches of the vertebral arteries, meningeal branches of the carotid siphon and others. In all our cases, the occipital artery, middle meningeal artery and meningeal branches of carotid siphon supplied the fistula. The monopedicular dural fistulas are exceptional and are always potentially pluripedicular.

The best treatment for these fistulas is embolization, but the choice of the material has to be discussed. Theoretically, the best material is a polymerizable fluid (IBC) but IBC is often dangerous because of anastomoses with the vertebro-basilar or spinal circulation. Often, it is better to use very thin pieces of dura or Ivalon. By embolization, in most of cases, it is possible to obtain a clinical cure but almost never a complete anatomical cure (Fig. 2).

Some cases have been treated by combining embolization with surgery, by ligating the pedicles which cannot be embolized: for example, the meningeal branches of the carotid siphon or vertebral artery. Even with this combination, it is not always possible to obtain a complete anatomical cure, without any recurrence.

Direct occlusion of the sinus is sometimes possible. A sixty-eight-year-old man who had a dural fistula associated with an arterio-venous angioma involving the brain stem, was treated by a balloon inside the lateral sinus with the result that all the symptoms disappeared.

5. The Dural Fistulas of the Superior Sagittal Sinus

The dural fistulas of the superior sagittal sinus are even more rare. These fistulas have to be differentiated from aneurysms of the torcular and particularly from fistulas with cortical venous drainage. The fistulas of the anterior part of the superior sagittal sinus are often revealed by subarachnoid haemorrhage. This kind of dural fistulas has to be treated by surgery.

The fistulas of the middle and posterior portions of the superior sagittal sinus are often multipedicular and supplied by branches from both sides: particularly the superficial temporal artery, middle meningeal artery, occipital artery and also sometimes the meningeal branches of the carotid siphon and the vertebro-basilar system exceptionally the ascending pharyngeal artery. In our opinion, the best treatment for these fistulas is embolization with IBC injected near the site of the fistula on the sinus, but it is not always possible. These fistulas are more rare than those having the other localization. In our three cases, we obtained a complete cure after several embolizations.

6. It is necessary to individualize the group of fistulas with drainage into the cortical veins. These patients mainly had subarachnoidal or intra-cerebral haemorrhages due to the rupture of the drainage veins. The best treatment for them is surgical excision.

Conclusion

In conclusion, we can say that the short-term results after embolization are usually excellent. There is often a disappearance or a reduction in the murmur heard by the patient. It is possible that the patient does not hear a murmur anymore which is nevertheless always heard by auscultation. Immediately after embolization, the fistula often seems completely occluded in the angiogram.

The long term follow up shows modifications of the anatomical aspects of the fistula. Frequently, we have observed modification of the arterial feeding of the fistula, even if the subjective improvement of the clinical symptoms persists. The modifications of the arterial supply are the appearance of new arterial branches, particularly if the sinus remains permeable at the site of the initial fistula. From a perspective point of view, one might suggest that the best treatment for these fistulas is perhaps an occlusion of the sinus itself. We have carried out such an occlusion in a fistula of the lateral sinus in a sixty-eight-year-old man and the clinical result was excellent two years later.

We can select the best treatment according to the sinus concerned:

a) For the cavernous or paracavernous sinus, embolization with dura is the best, particularly for the type 2.

b) For the superior sagittal sinus, the anterior fistulas have to be treated by surgery and the middle and posterior fistulas have to be treated by embolization (if possible IBC).

c) For the transverse and sigmoid sinus, embolizations with dura and IBC have to be used.

The long-term clinical follow-up varies according to the location, the subjective complaint and the angiographic results: For the cavernous territory, the clinical results were good except for one dramatic complication which resulted in death. For the transverse or sigmoid sinus, the subjective results were often good but the anatomical results were usually imperfect.

The most important question is to define the limitations of the treatment. The best treatment has few risks and has to be able to cure the main clinical symptoms.

Authors' address: Prof. L. Picard, Service de Neuroradiologie, C.H.U. Nancy, France.

[1] Universitätsklinik für Radiologie, [2] Universitätskinderklinik, [3] Institut für Pathologische Anatomie, Graz, Austria

CT Visualization of Thrombosis of Dural Sinuses in Childhood

E. JUSTICH[1], G. SCHNEIDER[1], G. FRITSCH[2], W. MUNTEAN[2], and G. F. WALTER[3]

Summary

Based on 5 cases of cerebral sinovenous occlusion in childhood, the value of cranial computer tomography is discussed. Cranial CT, with both native and contrast enhanced scans, will provide the correct diagnosis in some cases. Characteristic CT findings are the visulization of thrombosed sinuses and veins, known as the "filled triangle" and the "cord sign" in the native scan and positive contrast enhancement signs, like the "empty triangle" with tentorial and gyral enhancement. If the CT scan is normal, or shows only noncharacteristic symptoms, cerebral angiography has to be performed to obtain the correct diagnosis as early as possible. Remarkable is the high incidence of thrombosed internal cerebral veins and superior cerebellar veins with sinus thrombosis in children. CT is the method of choice for planning of therapy and controlling studies.

Authors' address: Dr. E. Justich, Department of Radiology, Graz University, A-8036 Graz, Austria.

[1] Departments of Neurosurgery, School of Medicine, Keio University Tokyo, and [2] Institute of Brain and Blood Vessels, Mihara Memorial Hospital, Gunma, Japan

Cerebral Venous Thrombosis: Findings from Computer Tomography and Fluorescein Angiography

T. KAWASE[1], T. TAZAWA[2], and M. MIZUKAMI[2]

Summary

Four cases of cerebrovenous thrombosis showed a high density focus in simulating subcortical arterial haemorrhage in the computer-tomographic (CT) scan. The diagnosis was made by angiography and fluorescein microangiography (FCA) performed at operation. Frontal cortical veins, with or without the anterior part of the superior sagittal sinus, were involved in 3 cases and the vein of Labbé was occluded in one case. In the CT scan, marginal irregularity and cortical involvement of the haemorrhage with a wide perifocal low density area were the specific findings which suggested venous haemorrhage. Contrast medium enhancement was minimal in the acute stage. Marked subarachnoid haemorrhage was noted at operation. With FCA, the site of occluded veins, collateral flow and fluorescein extravasation were visualized. A pathological specimen taken from a high density focus demonstrated haemorrhagic infarction. All patients had a good recovery during in-patient treatment, but two had recurrent attacks after discharge.

A full radiological examination, including magnified angiography, and lumbal puncture contributes to the correct diagnosis of cerebral venous thrombosis, when CT scans show an irregulary shaped high density focus.

Keywords: Cerebral venous thrombosis; computerized tomography; fluorescein angiography.

Introduction

Cerebral sinovenous thrombosis has been considered an uncommon clinical occurrence[3] in that there are few reports in which a thrombus localized in a cortical vein have been described[4, 5]. However, some reports[18] suggested many cases might escape documentation clinically because they had been verified only by angiography or brain scanning[2, 8, 12].

The development of computer tomography (CT) offers new possibilities for the correct diagnosis of intracranial lesions, and CT findings in cases with sinovenous thrombosis have been described since the first case was reported in 1977[1]. However, the accurate diagnosis of sinovenous thrombosis is sometimes not easy even by CT, due to the variation in CT findings[4]. It is the purpose of this report to present the CT findings in 4 cases of cortical venous thrombosis simulating subcortical arterial haemorrhage. The accurate diagnosis was made by surgery and fluorescein microangiography (FCA) at operation.

Methods and Report of Cases

A Hitachi H-250, 256 × 256 matrices CT scanner was used. Contrast enhancement was performed by the intravenous infusion of 100 ml of 60% meglumine diatrizoate in all patients. Conventional and/or magnified angiography was done in each patient. All patients underwent surgery, and a pathological cerebral microcirculation was observed at the operation using the fluorescein cerebral microangiographic technique. For FCA, a Nikon motor driven camera equipped with a telephoto lens, stroboscopic light and two types of fluorescein filters was used. A bolus of 20 to 30 ml of 0.5% fluorescein dye was injected retrogradely via a catheter placed in the brachial or the superficial temporal artery. The detail of method of FCA has been reported elsewhere[10].

Case 1. A right-handed 41-year-old female had a sudden headache. The next morning she vomited and complained of photophobia, and was admitted into our hospital on the 5th day. Sensory aphasia and a stiff neck were found on neurological examination. Bloody cerebrospinal fluid, taken by lumbar puncture, supported the diagnosis of subarachnoid haemorrhage. The CT showed patchy haemorrhages located in the temporal lobe (Fig. 1), and no remarkable enhancement was demonstrated after contrast medium infusion. Angiography offered little information except for the probable obstruction of the vein of Labbé (Fig. 2). A diagnostic craniotomy, performed on 8th day, revealed thrombi in the cortical veins with recent subarachnoid haemorrhage. On FCA, the proximal portion of the vein of Labbé was found to be occluded, and tortuous pial veins in the territory formed collateral pathways (Fig. 3). The patients' clinical course was satisfactory, showing a full recovery from the neurological deficits at discharge.

Case 2. A right-handed 54-year-old male had a history of craniotomy 9 years previously due to right temporal subcortical haemorrhage. He had a generalized convulsive seizure following a continuous headache for several days. A deterioration of consciousness and progressive left hemiparesis were noted on his admission. A large solitary high density focus (HU 40–80) with a perifocal low density area was demonstrated in the right frontal lobe by CT (Fig. 4). Angiography showed the presence of a mass without an abnormal vascular network, but an oblique view was not taken pre-operatively. The follow up CT, performed one week later, showed a large low density area with a slight gyral enhancement. The "Empty delta sign"[4] was not seen. A surgical exploration, performed on 8th day, showed petechial haemorrhages in the cortex, marked subarachnoid haemorrhage and multiple thrombi, which were assumed to extend retrogradely from the

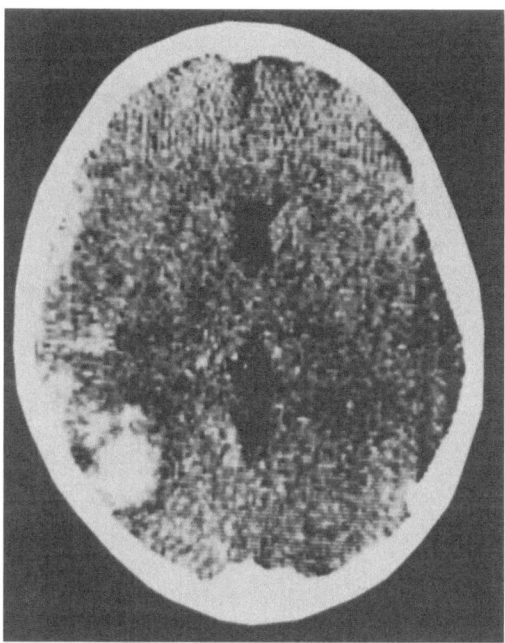

Fig. 1. Case 1. CT demonstrates patchy high density areas in the left temporal lobe

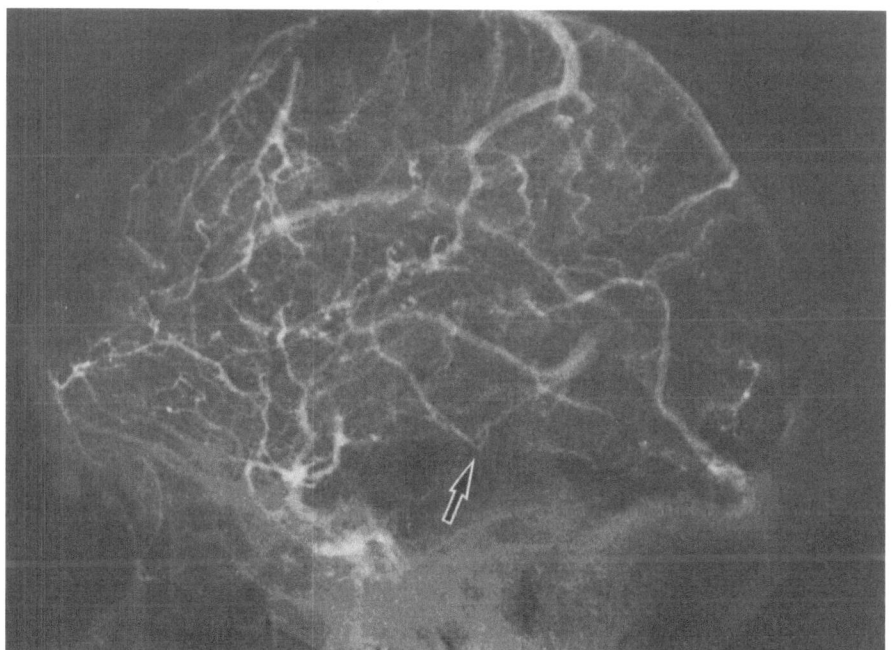

Fig. 2. Angiography in case 1. Occlusion of the vein of Labbé is suspected (arrow). The transverse sinus is opacified

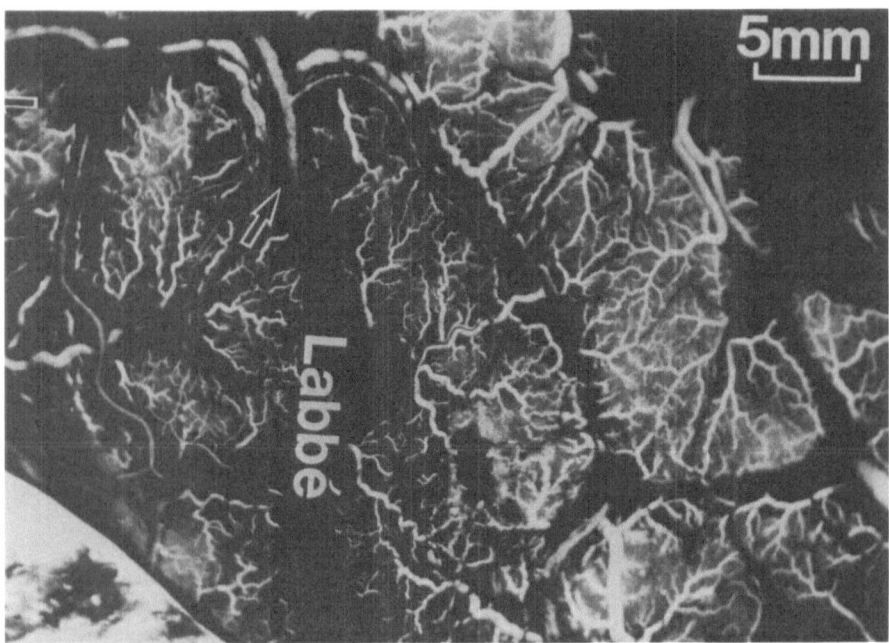

Fig. 3. Fluorescein cerebral angiography in case 1. Vein of Labbé is occluded (arrow). The pial veins in its territory are tortuous and congestive

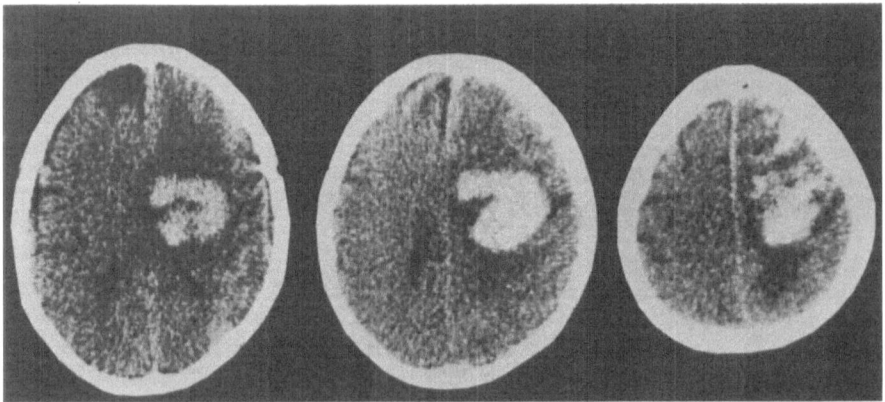

Fig. 4. Case 2. A solitary haemorrhage is demonstrated in the plain CT, with marginal petechial haemorrhage and a low density area. Petechial haemorrhages are observed in the cortex

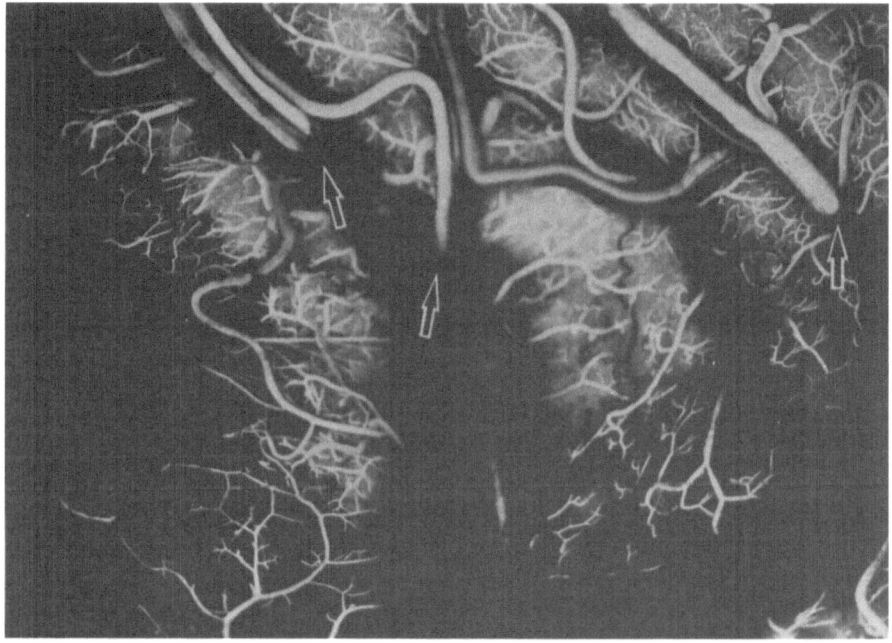

Fig. 5. Fluorescein cerebral angiography in case 2. Cortical veins are occluded (arrows)

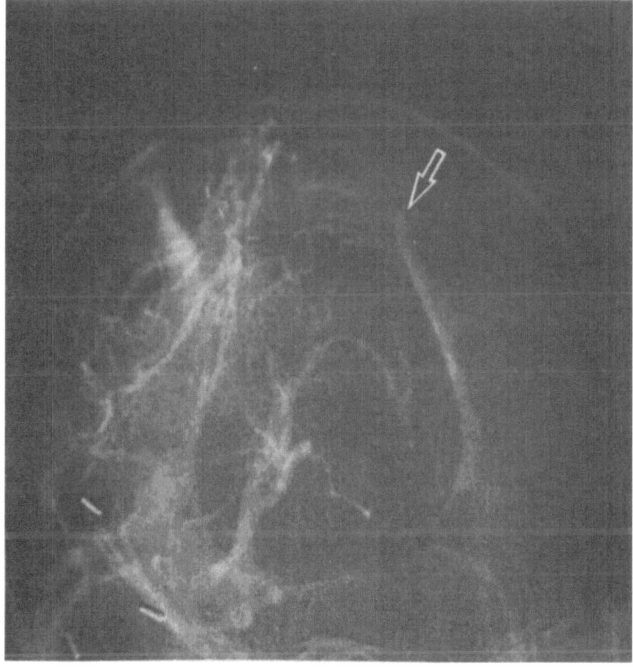

Fig. 6. Angiography in case 2. Anterior part of superior sagittal sinus is occluded (arrows)

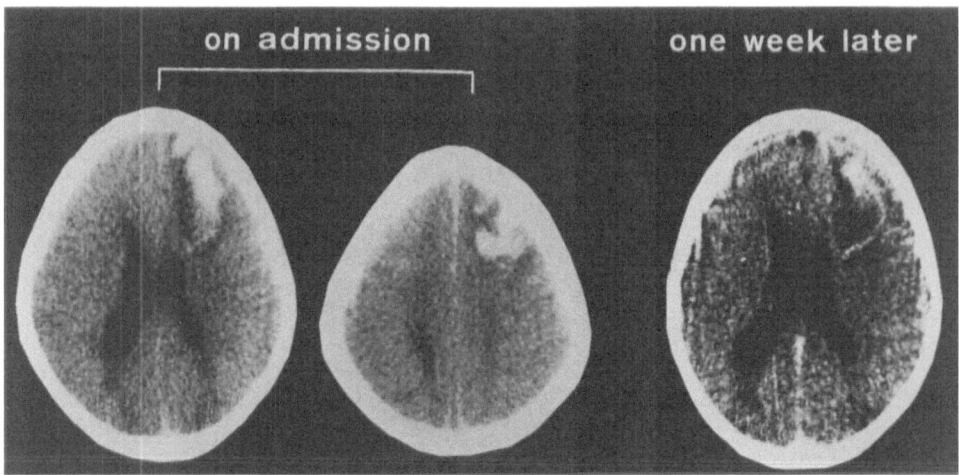

Fig. 7. Case 3. Irregularly shaped high density focus is demonstrated. A ring enhancement was visualized one week later

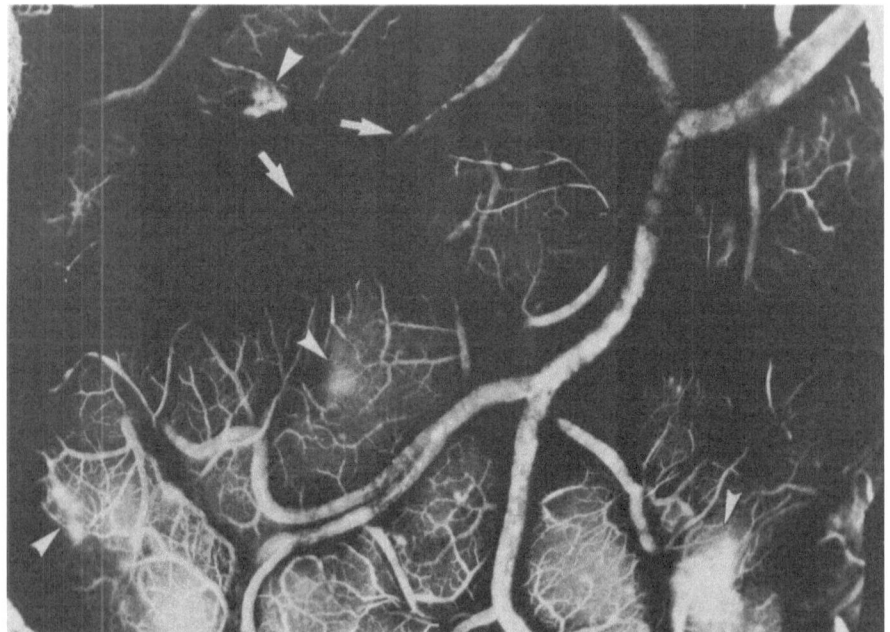

Fig. 8. Fluorescein cerebral angiography in case 3. Filling defect in the pial vein (arrowheads). Extravasation of fluorescein dye can be seen (arrows)

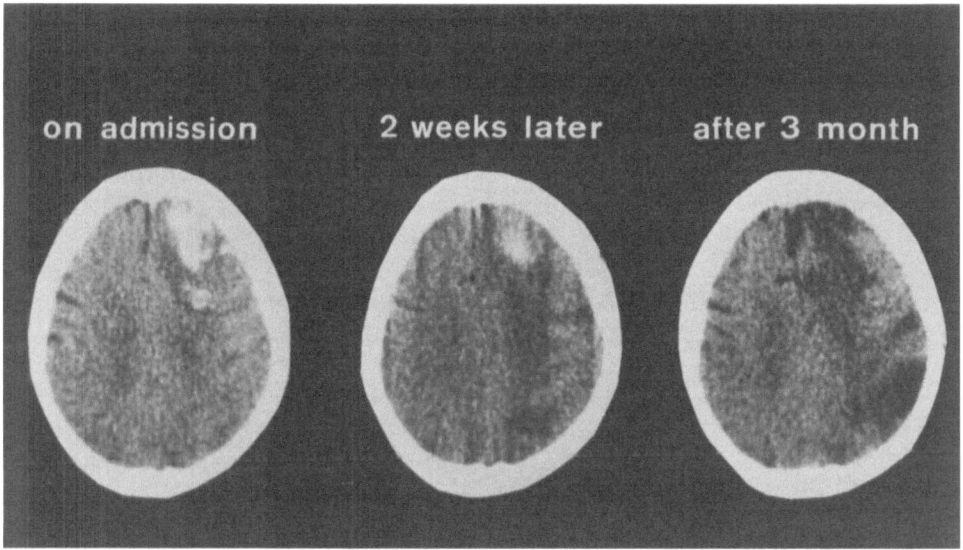

Fig. 9. Case 4. Irregularly shaped high density focus is visualized in the CT (left). A low density area spread 2 weeks later, and was marked after 3 months

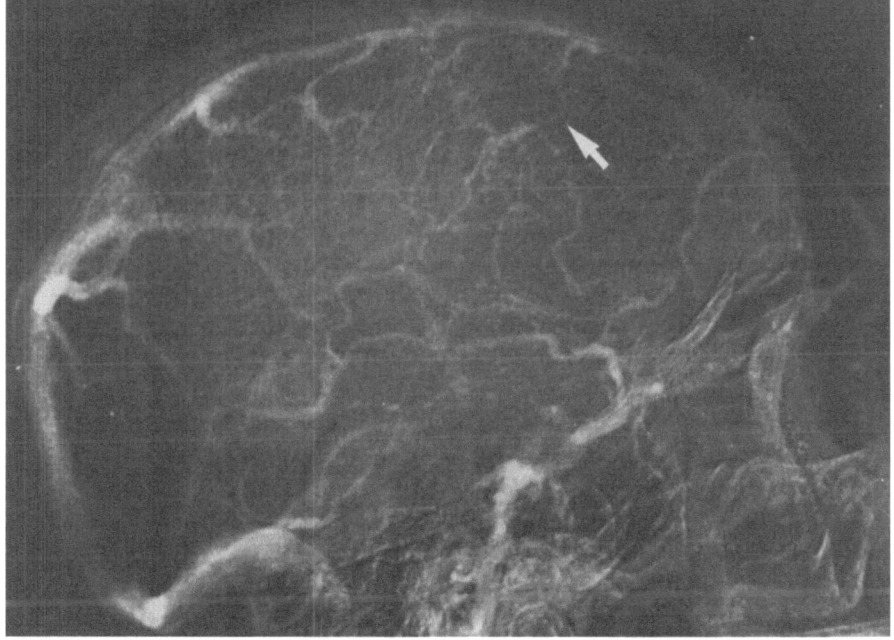

Fig. 10. Occlusion of frontal vein is suspected on angiography (arrow)

sagittal sinus to the cortical veins. The cortical veins were occluded at the portion 3–4 cm distal from the superior sagittal sinus on FCA (Fig. 5). A pathological specimen, taken from the area corresponding to the high density focus in the CT, showed haemorrhagic infarction. Postoperative angiography confirmed an occlusion of the anterior portion of the superior sagittal sinus (Fig. 6). The deterioration in the level of consciousness has been gradually improved after operation, and the patient's rehabilitation was successful.

Table 1. *Reported CT Findings in 30 Cases of "Primary" Cerebral Sinovenous Thrombosis*

	(acute stage)
Empty delta sign	6
Cord sign	1
Increased sinus enhancement	1
Small ventricules only	5
Focal haemorrhage	5
Multiple haemorrhages	2 (2)
Low density with petechial haemorrhages	6 (1)
Diffuse low density	2
Gyral enhancement	6
Others	1
Normal	2
Total	37

( ) Bilateral haemorrhages.

Case 3. A 67-year-old male complained of headache, nausea and mild fever, and was transferred to our hospital with a proven diagnosis of subarachnoid haemorrhage. Lumbar puncture showed bloody cerebrospinal fluid without a marked increase in white cells. An irregularly shaped high density focus was noted in the CT (Fig. 7). Angiography showed only an indication of a mass without an abnormal vascular lesion. Blood coagulability was normal. The operation revealed subrachnoid haemorrhage, and FCA at operation documented an occluded cortical vein with spotted extravasation of fluorescein dye in the corresponding territory (Fig. 8). The patient had a good clinical course during his hospital treatment. However, the same type of attack recurred after discharge. The CT scan made after the second admission showed a small high density focus in the opposite frontal lobe, whereas angiography could not give any definitive information of a venous thrombosis.

Case 4. A 56-year-old, right-handed male had a history of cerebral infarction from which he recovered after five years. He complained of progressive left hemiparesis and dysarthria, and mental retardation was noted when he was admitted. He had mild fever with a leucocytosis, but no focal infection was seen. The CT scan demonstrated an irregular high density focus in his right frontal lobe (Fig. 9). A low density area and slight gyral enhancement were shown in the subsequent contrast CT made two weeks later. Angiography showed a tortuous cortical vein with poor filling of the contrast medium in the proximal portion of the vein (Fig. 10). The patient had a good clinical course during his hospital treatment. However, hemiparesis progressed three months after his discharge. The development of cerebral infarction was noted in the CT. No marked change was noted in the second angiographic series. A surgical exploration was made for a definitive diagnosis, and a white thrombus was found in the cortical vein corresponding to the suspicious lesion shown in the angiograms.

Discussion

Positive computer-tomographic findings on sinovenous thrombosis has been described since 1977, and a total of 30 cases have been reported[1,2,4,6,9,11,13 − 17,19,20], including this report. The CT findings in the reported cases are classified into 11 types, as shown in Table 1. "Empty delta sign"[4], a low density area with petechial haemorrhages (haemorrhagic infarction) and "gyral enhancement"[4] are typical findings with a high incidence. Thirteen patients (43%) showed high density spots, and three of those had haemorrhages bilaterally. A focal haemorrhage was observed in 5 patients. Of five patients with a focal haemorrhage in the CT, four had a localized thrombus in the cortical veins. Three are described in this report, and the other case was reported by Buonanno *et al.*[4]. Other 24 patients had lesions mainly in the dural sinus, and demonstrated various CT findings. This implies that cortical venous thrombosis is closely related to focal haemorrhage.

The differentiation between arterial and venous haemorrhage by CT is not always easy, and the patient with a focal haemorrhage has a possibility of being diagnosed as a case of hypertensive or idiopathic intracerebral haemorrhage. Marginal irregularity with cortical haemorrhage and a relatively low X-ray attentuation of the high density focus, associated with a wide perifocal low density area, will support the diagnosis of venous haemorrhage. When bloody cerebrospinal fluid is obtained, this possibility is increased. Magnified angiography may give the most accurate information on a cortical venous thrombus. Obstruction of cortical veins with a slow circulation in the relevant territory and tortuosity of the affected veins can contribute to the diagnosis.

The clinical courses of patients with cortical venous thrombosis reported here were statisfactory over a certain period, however, recurrence of haemorrhage or progressing infarction occurred a few months later. Anticoagulant therapy, such as Warfarin or Heparin[7], should be given to prevent the spreading of the intravascular coagulatory process after the absorption of the focal haemorrhage.

References

1. Banna, M., Groves, J. T.: Deep vascular congestion in dural venous thrombosis on computed tomography. J. Comput. Assist. Tomogr. *3* (1979), 539—541.
2. Barnes, B. E., Winestock, D. P.: Dynamic radionuclide scanning in the diagnosis of thrombosis of the superior sagittal sinus. Neurology (Minneap.) *27* (1977), 656—661.
3. Barnett, H. J. M., Hyland, H. H.: Noninfective intracranial venous thrombosis. Brain *76* (1953), 36—49.
4. Buonanno, F. S., Moody, D. M., Baol, M. R., Laster, D. W.: Computed cranial tomographic findings in cerebral sino-venous occlusion. J. Comput. Assist. Tomogr. *2* (1978), 281—290.
5. Cambria, S.: Thrombosis of the vein of Labbé with haemorrhagic cerebral infarction. Rev. Neurol. (Paris) *136* (1980), 321—326.
6. Di Rocco, C., Lannelli, A., Leone, G., Moschini, M., Valori, V. M.: Heparin-urokinase treatment in aseptic dural sinus thrombosis. Arch. Neurol. *38* (1981), 431—435.
7. Gettelfinger, D. M., Kokmen, E.: Superior sagittal sinus thrombosis. Arch. Neurol. *34* (1977), 2—6.
8. Go, R. T., Chiu, C. L., Neumann, L. A.: Diagnosis of superior sagittal sinus thrombosis by dynamic and sequential brain scanning. Report of one case. Neurology (Minneap.) *23* (1973), 1199—1204.
9. Iijima, S., Nakamura, S., Shio, H., Kameyama, M., Takeuchi, J.: A case of superior sagittal sinus thrombosis with primary thrombocythemia. Clin. Neurol. *20* (1980), 333—338.
10. Kawase, T.: Study of cerebral microcirculation and haemodynamic with serial fluorescein cortical angiography. Part 1. Method for fluorescein angiography and measurements of local circulation time. Brain and Nerve *29* (1977), 181—189.
11. Kingsley, D. P. E., Kendall, B. E., Moseley, I. F.: Superior sagittal sinus thrombosis: an evaluation of the changes demonstrated on computed tomography. J. Neurol. Psychiat. *41* (1978), 1065—1068.
12. Krayenbühl, H.: Cerebral venous thrombosis: The diagnostic value of cerebral angiography. Schweiz. Arch. Neurol. Psychiat. *74* (1954), 261—287.
13. Lavin, P. J. M., Bone, I., Lamb, J. T., Swinburne, L. M.: Intracranial venous thrombosis in the first trimester of pregnancy. J. Neurol. Neurosurg. Psychiat. *41* (1978), 726—729.
14. Ohya, M., Sato, O.: A case of cerebral venous and superior sagittal sinus thrombosis, treated with decompressive craniectomy. Neurosurg. *8* (1980), 803—810.
15. Patronas, N. J., Duda, E. E., Mirfakhrall, M., Wollmann, R. L.: Superior sagittal sinus thrombosis diagnosed by computed tomography. Surg. Neurol. *15* (1981), 11—14.
16. Sayama, I., Kobayashi, T., Nakajima, K.: Cerebral sino-venous thrombosis, its clinical course and the study of repeated CT findings. Brain and Nerve *34* (1982), 547—554.
17. Seki, H., Sonobe, M., Higuchi, H.: A case of sinus thrombosis causing hemorrhagic infarction. Neurol. Med. Chir. (Tokyo) *19* (1979), 537—541.
18. Towbin, A.: The syndrome of latent cerebral venous thrombosis: Its frequency and relation to age and congestive heart failure. Stroke *4* (1970), 419—430.
19. Wendling, L. R.: Intracranial venous sinus thrombosis: Diagnosis suggested by computed tomography. Amer. J. Roentgenol. *130* (1978), 978—980.
20. Zilkha, A., Daiz, A. S.: Case report. Computed tomography in the diagnosis of superior sagittal sinus thrombosis. J. Comput. Assist. Tomogr. *4* (1980), 124—126.

Authors' addresses: T. Kawase, M.D., Department of Neurosurgery, School of Medicine, Keio University, 35 Shinanomachi, Shinjuku-ku, Tokyo, Japan 160, T. Tazawa, M.D., and M. Mizukami, M.D., Department of Neurosurgery, Institute of Brain and Blood Vessels, Mihara Memorial Hospital, 366 Ootamachi, Isesaki, Gunma 372, Japan.

Departments of Surgical Neurology and *Neurology, Research Institute for Brain and Blood Vessels, Akita, Japan

Cerebral Sino-Venous Thrombosis—A Clinical and CT Study for Its Treatment

I. Sayama, J. Kobayashi*, and K. Nakajima*

Summary

Three cases of cerebral sino-venous thrombosis (CSVT) are reported. Repeated CT findings were studied and discussed, with special reference to the sequential changes in the pathological condition.

The aetiology of these cases was considered to be non-infectious and was regarded to be of the primary type, with normal laboratory data including fibrinolytic activities.

A diagnosis was made with serial angiography and CT scans in every case, which revealed superior sagittal sinus thrombosis (SSST) in two, and cortical venous thrombosis (VT) in one, respectively.

Their CT findings were in agreement with those of other studies previously described but, in the case of SSST, an "Empty triangle sign", (term which was coined by Buonanno *et al.*), which was found in the post-contrast medium CT of this disease, could even be found in the native scan in our cases.

The CT findings should be classified as direct and indirect signs according to their pathological mechanisms, thus making it possible to select proper treatment.

Author's address: Dr. I. Sayama, Department of Surgical Neurology, Research Institute for Brain and Blood Vessels, 6-10 Senshu-kubotu-cho, Akita, Japan.

The University Clinic of Neurosurgery and the Neuroradiological Service of Marseille,
France

Microsurgical Treatment of Dural Arterio-Venous Fistula in Posterior Fossa

F. Grisoli, F. Vincentelli, and M. Baldini*

Summary

Four cases of dural fistula in the posterior fossa are described. The significance of the rare venous drainage with involvement of cortical veins is discussed and the value of the successful microsurgical treatment of this type of arteriovenous malformations is stressed.

Keywords: Arterio-venous fistula; posterior fossa; microsurgery

Introduction

Arteriovenous malformations of the dura mater involving the infratentorial compartment are uncommon [7].

Recent refinements in angiographical techniques have allowed the classification of such cases into pial, dural and mixed pial-dural types on the basis of their arterial supply. The dural types are rare [2, 5]. Such malformations may involve the tentorium and/or the dura covering the remainder of the posterior fossa. The arterial supply generally consists of branches from the middle meningeal artery, the temporal and superficial arteries, the meningohypophyseal trunk and the meningeal branches of the vertebral artery. The great majority of AV-malformations in the posterior fossa drain directly into the dural sinuses [5]. Castaigne *et al.* [1] report that amongst the pure dural malformations and the angiomas of the transverse sinus, the fistulas with cortical venous drainage represent a very special entity and their precise diagnosis assumes considerable importance since their surgical treatment has become possible.

* Present address: Dr. M. Baldini, Clinica Neurochirurgica dell'Università, Genova, Italy.

Clinical Material

Case 1. This 36-year-old man was admitted with a severe headache and signs of meningeal irritation following two episodes of subarachnoid haemorrhage in 12 days. The angiogram showed a dural fistulous communication in the middle part of the right superior petrosal sinus region, extending into Meckel's cavum. A branch of middle meningeal artery and the right Bernasconi-Cassinari artery represented the source of the arterial supply. The venous drainage entered the inferior vermian, lateral recessus and brain-stem veins.

Operation: After incision of the right free margin of the tentorium in the zone of Meckel's cavum, a large draining vein was observed and clipped at the base of the fistula. Bipolar coagulation of the tentorial feeding arteries completed the treatment. The control angiogram and the postoperative CT study confirmed the exclusion of an AV malformation. The neurological examination was normal.

Case 2. A 68-year-old woman suffered an ictus cerebri with diplopia and a right hemiparesis 9 years previously. At admission the patient had a cerebellar syndrome with nystagmus. The CT scan with contrast medium enhancement showed a hyperdense area in the right cerebello-pontine angle the *angiograms* showed that the meningeal and posterior auricular branches of the right external carotid artery were the blood supply. The right cerebello-pontine angle plexus drained directly into the mesencephalic and basal veins.

First operation: Ligature of the right external carotid artery.

A *second operation* was performed after one year because of the neurological symptoms and neuroradiological findings: The falcotentorial angle was exposed and the fistula was clipped at the base of the AV malformation. The cerebellopontine angle arterialized draining vein was clipped.

The control angiogram demonstrated the absence of a fistulous communication. The postoperative neurological examination was normal.

Case 3. This 45-year-old woman was admitted with a headache and meningeal signs following a subarachnoid haemorrhage.

CT scan: There was a spontaneous hyperdense area in the right and partially in the left tentorial regions.

The angiogram showed an infratentorial arteriovenous malformation. Feeding arteries were found to be the occipital, meningeal and vertebral arteries and both Bernasconi-Cassinari arteries in the tentorium. The venous drainage was represented by the posterior cerebellar vein without straight sinus opacification.

Operation: Exposure and bipolar coagulation of multiple feeding arteries in the tentorium was carried out. Simultaneously, clipping of the fistula and of the draining veins at the base of the fistula itself was performed. The draining arterialized vein was the cerebellar vein which was recognized at angiography.

The control angiogram showed the total exclusion of the dural AVM. The postoperative neurological control examination was negative.

Case 4. This 64-year-old man was admitted with headache, mental deterioration and a normal neurological examination.

CT scan showed the presence of a right haemorrhage at the level of the cistern of velum interpositum, within the choroid tissue.

The angiogram revealed a dural angioma. The arterial supply was from the right occipital artery, the tentorial artery of the free border and a meningeal branch from the ascending pharyngeal artery. Drainage was into the superior vermian vein and basal vein. Filling of the transverse sinus was delayed.

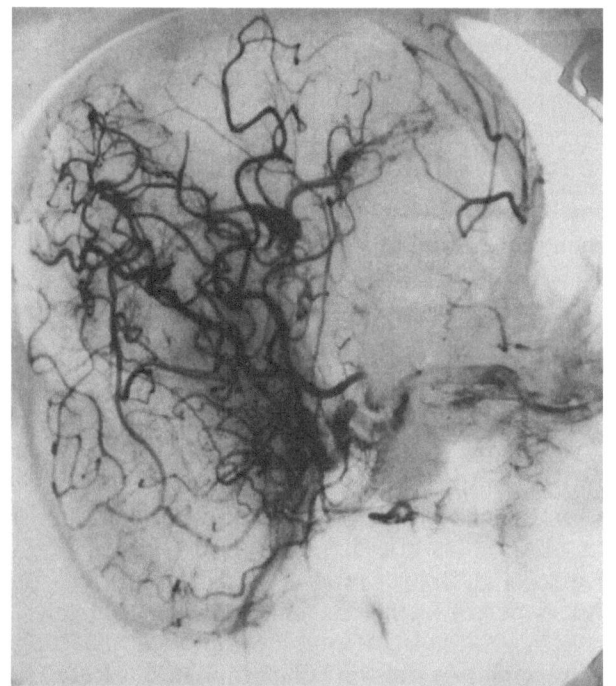

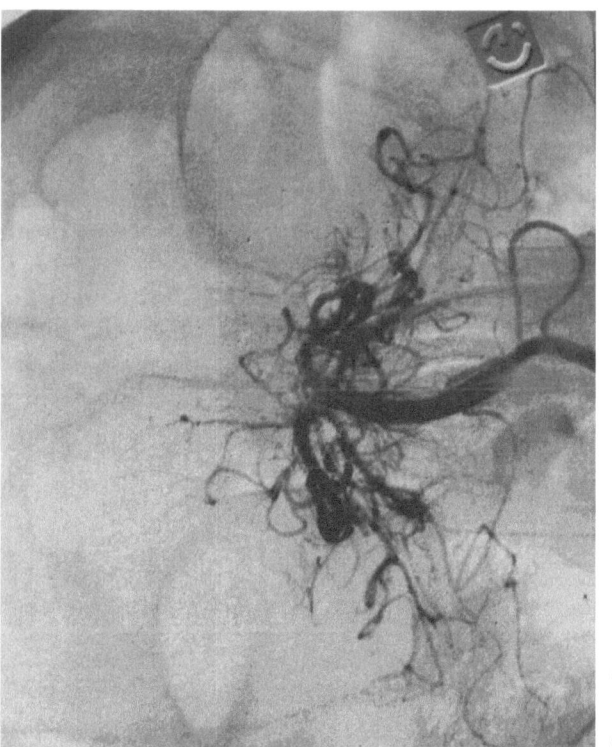

Fig. 1. Case 4. The fistula is at the level of the midportion of the straight sinus without opacification of the sinus itself, but filling of the superior vermian vein towards the internal cerebral vein, which probably caused the bleeding into the velum interpositum cistern (right). No arterial supply from the basilar system (left)

First operation. Dural arterial and venous vessels were preserved. Only bipolar accurate coagulation of the bridging tentorio-cerebellar veins was performed, and a large vein at the level of the tentorium was isolated and clipped.

The postoperative control angiogram confirmed the complete exclusion of the AV-communication.

Second operation: A ventriculo-peritoneal shunt was implanted due to the presence of a supratentorial hydrocephalus.

The CT control scan showed an evident regression of the ventricular dilatation. The neurological examination showed persistence of the mental deterioration.

Discussion

We have reported four cases of infratentorial dural AV malformations which showed venous drainage into the cortical veins. It is interesting that our cases were admitted to the department with meningeal signs and subarachnoid haemorrhage, which had occurred some time before the admission. Other most commonly occurring symptoms such as convulsive seizures, bruits and tinnitus, were not present, and traumatic aetiology was not reported in the clinical histories[8].

The venous drainage, moreover, was made up in all cases by cortical veins (inferior and superior veins of vermis, superficial and bridging tentorio-cerebellar veins) with only secondary and delayed participation of the dural sinuses. Amongst the dural fistulas in posterior fossa reported in literature, we have only encountered this special entity of fistolous drainage in the reports of Laine *et al.* (1963), of Pecker *et al.* (1964), of Houser *et al.* (1972), of Newton *et al.* (1968)[2, 4, 6].

This particular type of drainage is emphasized, because its obliteration allows the total exclusion of the dural fistula without the dissection and isolation of the sinus.

The operative exposure of the transverse and sigmoid sinuses and the dural excision of the fistula seemed not to be necessary in our cases, because the drainage into the sinus was only indirect[3].

The haemodynamic interruption of the dural fistula depends on the selective microsurgical interruption of the draining veins excluding the excision of the fistula.

References

1. Castaigne, P., Bories, J., Brunet, P., Merland, J. J., Meininger, V.: Les fistules artério-veineuses méningées pures à drainage veineux cortical. Rev. Neurol. (Paris) *132,* 3 (1976), 169—181.
2. Houser, O. H., Baker, H. L., Jr., Rhoton, A. L., Okazaki, H.: Intracranial dural malformations. Radiology *105* (1972), 55—64.
3. Kühner, A., Krastel, A., Stoll, W.: Arteriovenous malformations of the transverse sinus. J. Neurosurg. *45* (1974), 12—19.
4. Laine, E., Galibert, P., Lopez, C., Delahousse, J., Delantsheer, J. M., Christiaans, J. L.: Anévrysmes artérioveineux intraduraux de la fosse postérieure. Neuro-Chirurgie (Paris) *2* (1963), 147—158.
5. Newton, T. H., Weidner, W., Greitz, T.: Dural arteriovenous malformations in posterior fossa. Radiology *90* (1968), 27—35.

6. Pecker, J., Bonnal, J., Javalet, A.: Deux nouveaux cas d'anévrysmes artério-veineux intraduraux de la fosse postérieure alimentés par la carotide externe. Neuro-Chirurgie *5* (1964), 327—332.

7. Perret, G., Nishioka, H.: Arteriovenous malformations. J. Neurosurg. *25* (1966), 467—490.

8. Tönnis, W., Schiefer, W., Walter, W.: Signs and symptoms of supratentorial arteriovenous aneurysms. J. Neurosurg. *15* (1958), 471—480.

Author's address: Dr. M. Baldini, Clinica Neurochirurgia dell'Università, Genova, Italy.

University Department of Neurosurgery, University Medical Centre, Ljubljana, Yugoslavia

Disturbed Venous Flow Through the Cavernous and Transverse Sinuses

V. DOLENC

Summary

The report deals with the disturbed venous flow via the cavernous sinus (CS) and the transverse sinus (TS) observed in 15 and 6 patients, respectively. Surgical management of the intracavernous vascular lesions and tumours invading the CS by direct approach in the first group of patients required sacrifice of the superficial Sylvian veins and occlusion of at least the anterior third of the CS. Operative treatment for aneurysms situated on the intracavernous segment of the internal carotid artery (ICA) necessitated occlusion of barely a third of the CS. In carotid cavernous fistulas (CCF), however, quite a large portion of the CS had to be occluded. Tumours invading the CS accounted for a considerable preoperative, and even more extensive postoperative obliteration of the CS.

In the six patients with the tumour on the clivus extending around the tentorial edge to the middle and the posterior cranial fossa, the TS was surgically interrupted as laterally as possible to permit the best access to the lesion. Four patients out of 21 died: the first one operated for CCF died on the tenth postoperative day as a result of pulmonary embolism, the second patient, operated for a giant intracavernous aneurysm, died after three months from sepsis, which was the cause of death one month postoperatively also in the third case operated on for a large intracavernous tumour; the fourth patient operated on for a tumour of the CS, histologically diagnosed as a sarcoma, died six months after surgery due to multiple metastases.

Eleven out of the 21 operated patients are without neurological deficits at the time of this report. Five still show neurological signs but they are less pronounced than preoperatively. Two of the 6 patients, in whom the TS was interrupted, suffer from further neurological deficits in addition to the preoperative ones due to injury to the temporo-occipital veins. One patient operated for CCF developed transient hemiparesis and two epileptic fits due to the deranged venous flow resulting from the interruption of the superficial Sylvian veins.

Keywords: Cerebral veins; cavernous sinus; transverse sinus.

Introduction

Venous flow through the CS may be reduced or obstructed by thrombosis of the sinus or by the tumours, which may either compress it, invade it or originate from it, as well as by vascular lesions of the intracavernous part of the ICA: aneurysms and/or CCF[2]. The flow via the TS may be diminished or stopped by thrombosis of the sinus as well as by tumours in the sinus or around it.

Surgical procedures at the skull base usually necessitate sacrifice of the descending cerebral veins to the anterior part of CS and/or those draining into the TS. In the direct surgical approach to the CS lesions (aneurysms, CCFs, tumours) the CS itself has to be obliterated more or less extensively[2, 8], and the superficial Sylvian veins sacrificed. For surgical removal of the tumours sited on the clivus, extending from the posterior to the middle cranial fossa, a combined temporo-suboccipital approach has proved most suitable, though it involves discontinuance of the TS. The venous anastomosing network at the level of the insular pole, composed of the four large veins of Trolard, Rolando, Sylvius, and Labbé provides the rerouting of the venous flow, should the drainage into the CS or TS be blocked by a tumour or interrupted surgically. The blockage of drainage in one direction is usually satisfactorily compensated by drainage through the anastomotic veins. In very rare cases, though, an inadequate rerouting due to the insufficiently developed anastomoses may lead to, at least, temporary engorgement of the ligated veins and the resulting neurological deficits.

One of the 15 patients who had the superficial Sylvian veins sacrified and the CS partially or completely excluded, developed neurological deficits due to an impaired venous drainage. However, of the 6 patients in whom the TS was discontinued and the temporo-occipital veins injured, 2 developed the corresponding neurological signs due to insufficient venous drainage.

Patients, Surgery, Results

Our series of 21 patients was comprised of 7 females and 14 males, aged from 22 to 66 years.

Out of the 15 patients with intracavernous pathology, 4 had an aneurysm of the intracavernous part of the ICA, 4 unilateral CCF due to craniocerebral trauma and 7 had CS invading tumours. Six patients harboured large tumours located at the clivus and at the tentorial edge in the posterior and middle cranial fossa. In the first group of patients with intracavernous aneurysms of the ICA, CCFs and tumours situated both outside and inside, or only inside the CS, the pterional approach was used[11]. In cases requiring the exploration of the entire CS, the pterional approach was combined with the subtemporal approach[4], as well exposing the intrapetrous portion of the ICA[5]. In all cases of direct intracavernous surgery the superficial Sylvian veins were invariably sacrificed and some part of the CS was excluded due to package with surgicel[2].

The preoperative studies of the venous network for the existence of a sufficient number of anastomotic channels between the descending and the ascending superficial veins and between the superficial Sylvian veins and the insula and deep middle cerebral venous drainage systems did not lead to a prediction of what the ultimate consequences of the surgical interruption of the superficial Sylvian veins would be.

In four patients operated on for intracavernous aneurysms of the ICA, three aneurysms were clipped, whilst the fourth was resected and the ICA reconstructed by suturing. A control angiogram showed a satisfactory exclusion of the lesions and no evidence of oedema involving the hemisphere deprived of drainage via the superficial Sylvian veins into the

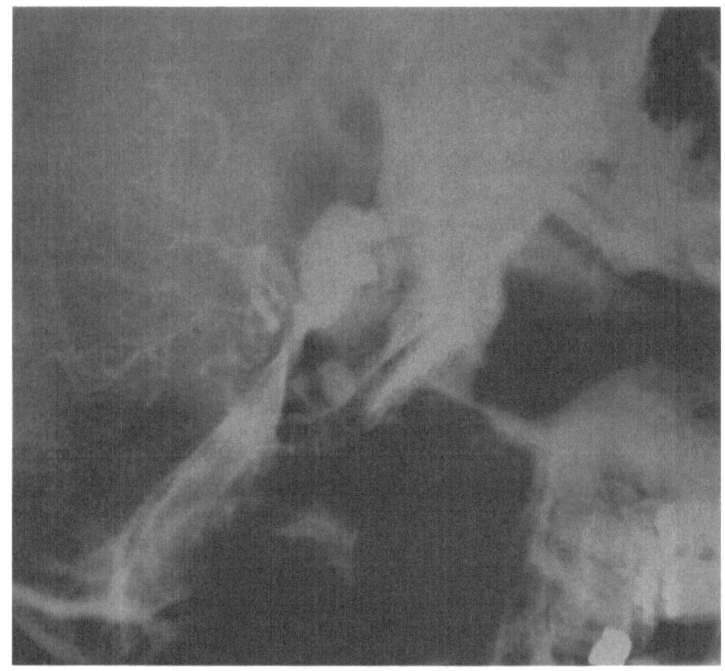

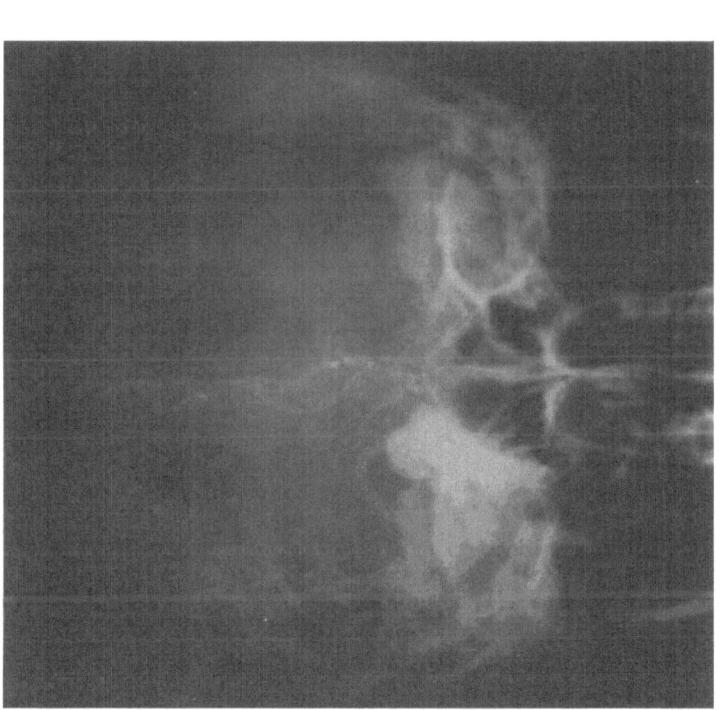

Fig. 1. Anteroposterior (left) and lateral (right) views of the right carotid angiogram showing a CCF

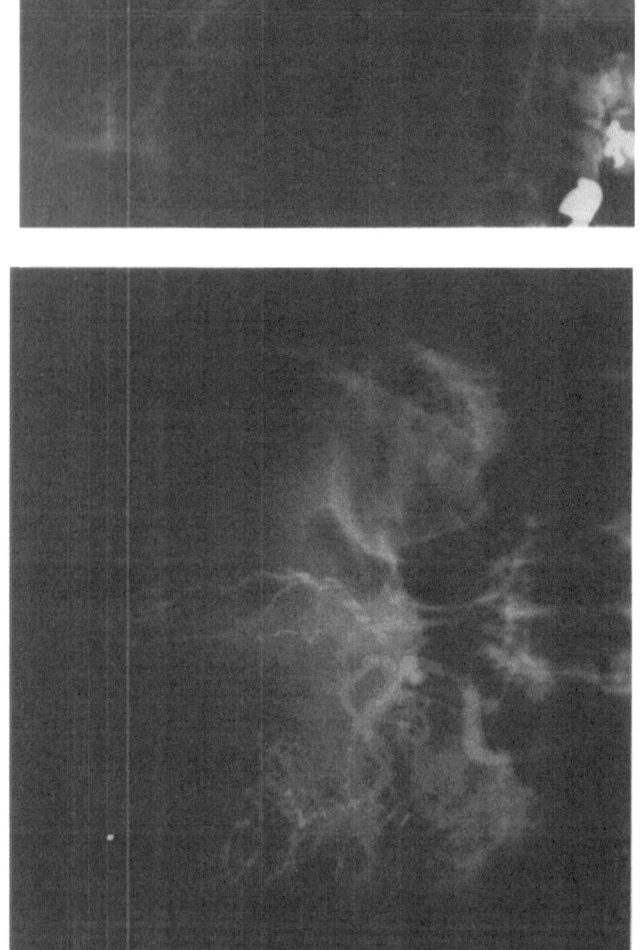

Fig. 2. Anteroposterior (left) and lateral (right) views of the right carotid angiogram following the operation. Note absence of CCF, normal ICA, but a marked shift of the ACA to the opposite side

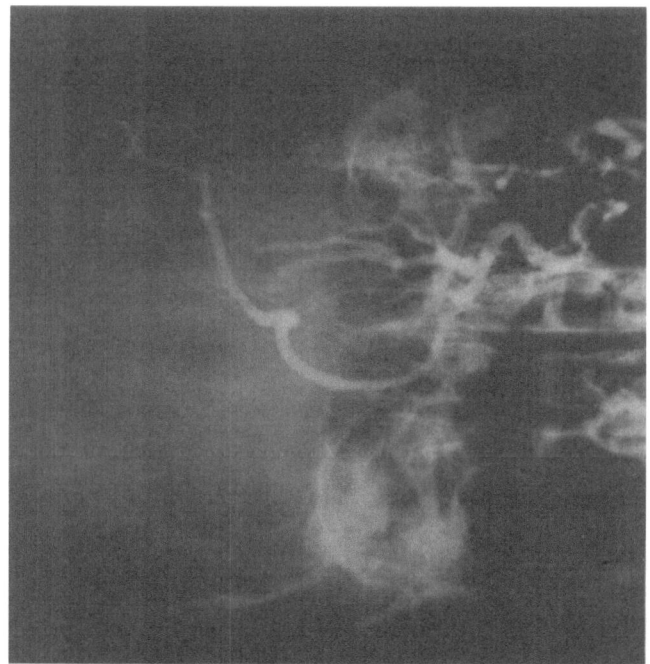

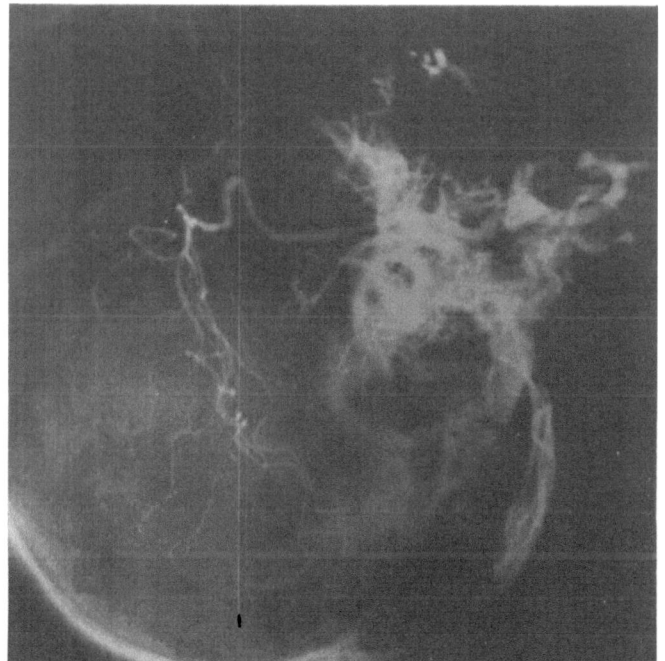

Fig. 3. Anteroposterior (left) and lateral (right) left carotid angiogram showing a marked displacement of ICA and the rich vascular supply of the tumour

V. Dolenc:

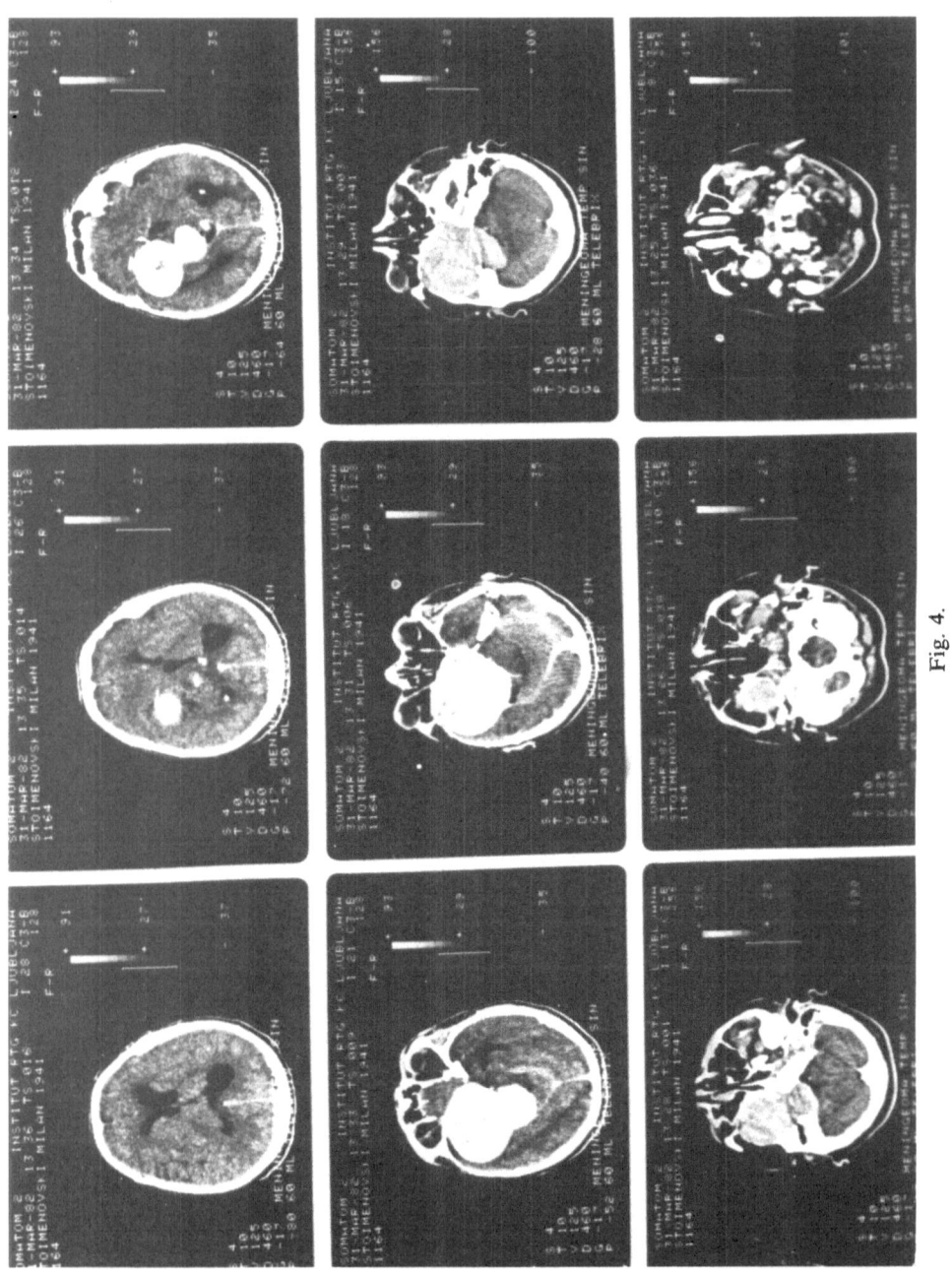

Fig. 4.

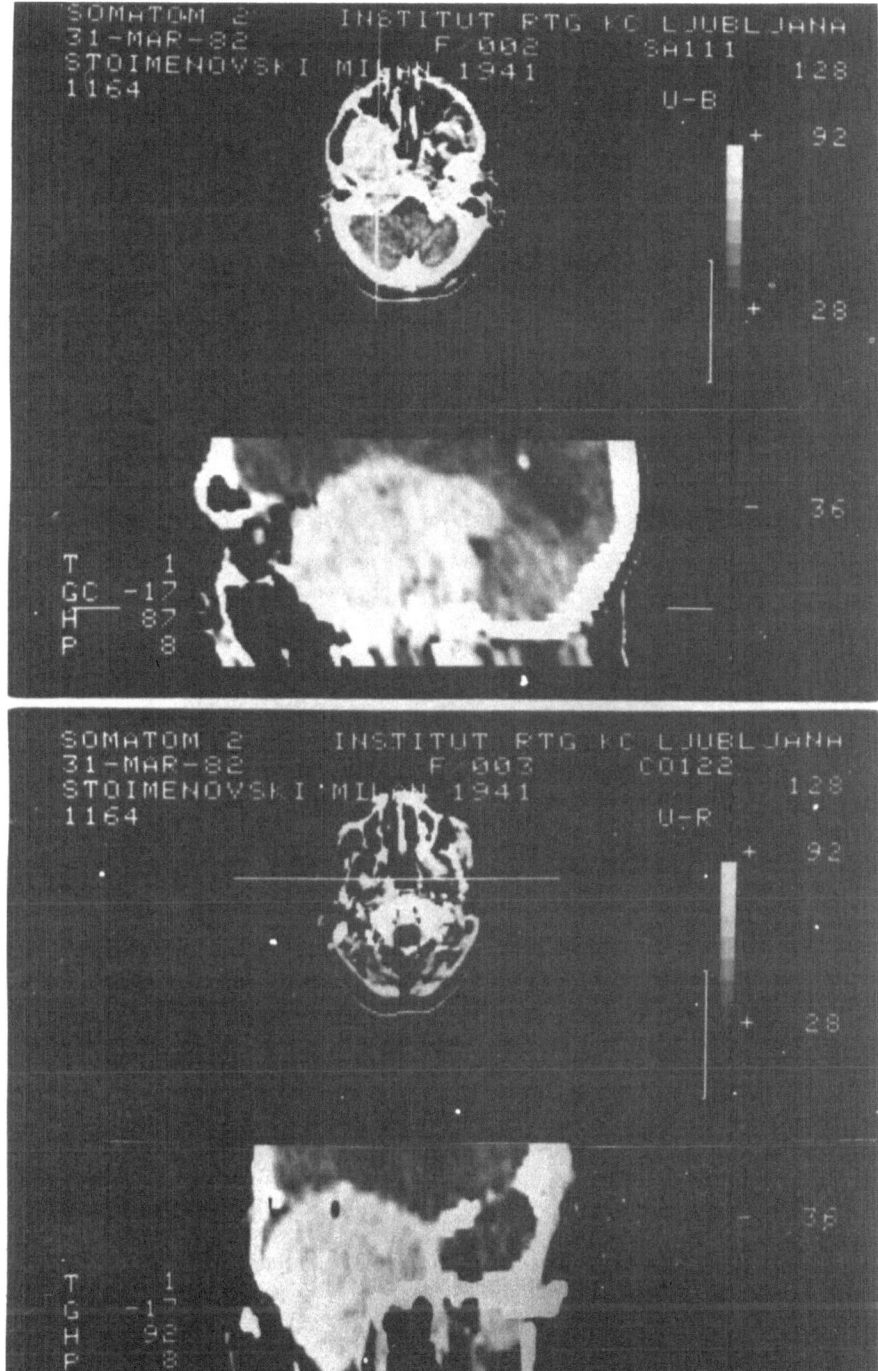

Fig. 4. CT scan showing an upward and downward expansion of the same huge tumour shown in Fig. 3, in horizontal levels and in parasagittal and coronal section

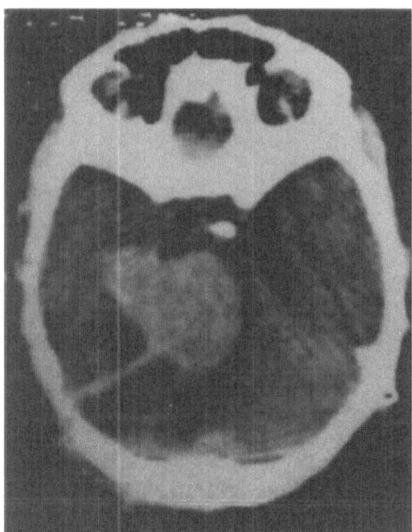 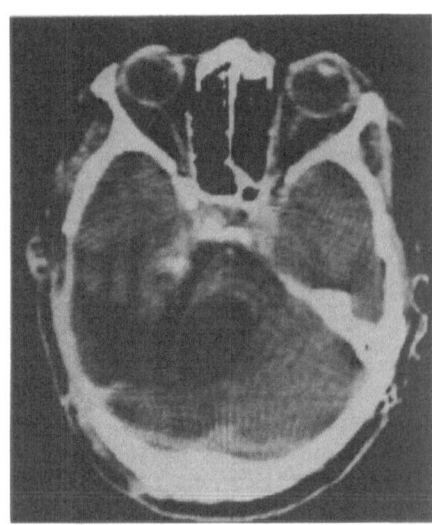

Fig. 5. CT scan showing a meningioma sited in the middle and posterior cranial fossa (left) and the condition two years after surgery (right)

anterior part of the CS, which had been interrupted at operation. Surgery produced no additional neurological deficits.

In two of the four patients operated for an unilateral CCF the lesion was excluded by clipping the fistula and in the other two by suturing the ICA wall. Postoperative carotid angiograms confirmed the complete exclusion of the CCF in all four cases. One week post-operatively, the ICA was not patent in one case and patent in the other three patients. The patient with CCF, in whom the ICA wall was sutured, developed a slight hemiparesis and had two Jacksonian epileptic fits involving his face and upper limb, contralaterally. A carotid angiogram revealed oedema of the brain on the operated side with a marked shift of the ACA to the opposite side (Figs. 1 and 2). Except for the above-mentioned complication, in all the patients operated surgery induced no additional neurological deficits in.

The extrasinusal portion of the lesion was removed in all seven of the cases operated for tumours invading the CS. Four cases had a complete and three an incomplete removal of the intrasinusal portion. The former was feasible in the patients with a neuroma of the Vth nerve, in a pituitary tumour extending into the CS and in two cases with an intracavernous meningioma. Incomplete tumour removal from the CS was done in one case of sarcoma, in one case with a huge neuroma of the Vth nerve (Figs. 3 and 4) and in one case with a very large meningioma. In the last three cases the tumours already extended into the bone compartments of the skull base below the CS. Surgery produced no further neurological deficits as compared to the preoperative signs due to the tumour, nor did it create any problems related to the venous drainage via the superficial Sylvian veins, which had been invariably divided at operation.

In the second group of patients with tumours sited at the clivus and around the tentorial edge in the middle and the posterior cranial fossa (Fig. 5), a combined subtemporal and suboccipital approach was used in all cases. For this operative procedure the patients were placed in a sitting position with necessary precautions for possible air embolism: A Doppler was placed precordially and an intravenous catheter was inserted through the subclavian vein into the right atrium. Preoperative angiographic and phlebographic studies of the affected side were done and the functional existence of the contralateral TS was proved in all cases. The TS was invariably ligated and discontinued at the most lateral corner, *i.e.*, at the point of its turning downwards. This lateral ligation of the TS was necessary to preserve the temporo-occipital veins. Six tumours, histologically diagnosed as benign meningiomas, were completely extirpated by a one-step procedure. An attempt was made in all cases to preserve the temporo-occipital veins, which were invariably well developed. In the first two cases in this series we failed to preserve the veins. In the remaining four cases we managed to preserve the temporo-occipital veins by dissecting at least a part of the sigmoid sinus (SS) and removing the bone encroaching on it. The SS was ligated twice and the ligatures placed 5–6 millimetres apart, with the proximal one being applied at least 1 cm laterally from the junction of the temporo-occipital veins and the TS. Ligating and dividing the superior petrous sinus permitted a safe interruption of the TS. The incision into the dura at the base of the temporal lobe, horizontally, and by the cerebellum, vertically, along the dorsal aspect to the SS, and horizontally, below and along the TS to the vicinity of the torcular Herophili, permitted the inspection of the attachment of the tentorium to the petrous bone from the middle and posterior cranial fossa, thus enabling a risk-free incision of the tentorium. Retraction of the temporal and occipital lobes with the tentorium and the TS upwards and the cerebellum backwards, makes a perfect visualization of the lesion possible. A lot of patience had been required to preserve the petrous vein. We succeeded only in 3 cases out of 6. Two patients with injury to the left temporo-occipital veins developed a postoperative right-sided paresis and dysphasia. The patience whose petrous vein was sacrificed showed no neurological deficits in the respective drainage area.

Discussion

The anatomy of the venous system of the brain has been studied by many authors [3, 6, 7, 10]. Many variations in the normal venous pattern were observed: *e.g.*, at the level of the insular pole [7]. From this particular region the venous drainage may run in two main directions: descending and ascending. The descending venous drainage is divided again into the anterior, posterior and inferior systems. The anastomoses between the superficial and deep middle cerebral veins [10] and between the middle cerebral veins and other superficial veins (*e.g.*, the vein of Trolard) may create an efficient rerouting of the venous drainage from the area of the non-functioning superficial Sylvian veins.

In our series of 21 patients with lesions at the skull base, the anterior and/or posterior descending venous drainage was damaged both by tumours and surgery. Of the 15 patients with interrupted superficial Sylvian veins, only one developed transient neurological deficits due to the insufficient venous drainage producing oedema (Fig. 2). According to the above observation, adequate venous drainage may be provided by the existing anastomoses in the great majority of cases.

In all 15 patients in the group with the intracavernous lesion, the sinus was opened and at least a third of it occluded by surgicel package. In no case was the drainage via the superior and/or inferior ophthalmic vein hindered. The only explanation would be that at operation the flow through the CS was never completely blocked and that the CS has many possibilites of rerouting the venous flow through the intercavernous sinuses[1,9] or the pterigoid plexus[3]. In four out of the six patients in whom the superior petrosal sinus and the TS were discontinued and the temporo-occipital veins preserved, the rerouting of the venous flow from the temporo-occipital vein via the TS through the torcular Herophili to the other TS seemed most probable. Two of the six patients with injury to the temporo-occipital veins showed insufficient venous drainage through other superficial veins.

Conclusions

According to our limited experience in the field the temporo-occipital veins running into the TS could not be interrupted by any means without neurological deficits resulting. Surgical ligation and interruption of the TS has to be preceded by preoperative studies to confirm the function of the contralateral TS and the patency of the torcular Herophili for flow in the transverse direction. Due to a large number of anastomoses between the superficial Sylvian veins and other venous systems, the ligation of these veins entails a lesser risk of engorgement as compared with the ligation of the temporo-occipital veins. The CS can tolerate exclusion of a large part of its volume unless the venous flow through its most minute portion is completely obstructed.

References

1. De Divitiis, E., Spaziante, R., Iaccarino, V., et al.: Phlebography of the cavernous and intercavernous sinuses. Surg. Neurol. 15 (1980), 306—312.
2. Dolenc, V.: Direct microsurgical repair of intracavernous vascular lesions. J. Neurosurg. 58 (1983), 824—831.
3. Doyon, D. L., Aron-Rosa, D. S., Ramée, A.: Orbital veins and cavernous sinus. In: Radiology of the Skull and Brain. Angiography (Newton, T. H., Potts, D. G., eds.), Vol. 2, Book 3, Veins, pp. 2220—2254. Saint Louis: The C. V. Mosby Company. 1974.
4. Drake, C. G.: Treatment of aneurysms of the posterior cranial fossa. In: Prog. neurol. Surg. 9, 122—194. Basel: Karger. 1978.
5. Glasscock, M. E.: Exposure of the intra-petrous portion of the carotid artery. In: Disorders of the skull base region. Proceedings of the tenth Nobel symposium (Hamburger, C. A., Wesäll, J., eds.), pp. 135—143. Stockholm: AWE Almquist-Wiksell, 1968.
6. Hacker, H.: Normal supratentorial veins and dural sinuses. In: Radiology of the Skull and Brain. Angiography (Newton, T. H., Potts, D. G., eds.), Vol. 2, Book 3, Veins, pp. 1851—1877. Saint Louis: The C. V. Mosby Company. 1974.
7. Krayenbühl, H., Yaşargil, M. G.: Radiological anatomy and topography of the cerebral veins. In: Handbook of Clinical Neurology. Vascular Diseases of the Nervous System, Part I (Vinken, P. J., Bruyn, G. W., eds.), Vol. 11, pp. 102—117. Amsterdam: North-Holland Publishing Company. New York: American Elsevier Publishing Co., Inc. 1975.
8. Parkinson, D.: Carotid cavernous fistula: direct repair with preservation of the carotid artery. Technical note. J. Neurosurg. 38 (1973), 99—106.
9. Sasaki, T., Ueda, Y., Saito, I., et al.: Cavernous sinus venography by transbasilic catheter technique. Neurol. Med. Chir. (Tokyo) 14, pt 1 (1974), 11—17.

10. Wolf, B. S., Huang, Y. P.: The insula and deep middle cerebral venous drainage system: normal anatomy and angiography. Am. J. Roentgenol., Rad. Therapy and Nuclear Med. *90*, N. 3 (1963), 472—489.
11. Yaşargil, M. G., Fox, J. L., Ray, M. W.: The operative approach to aneurysms of the anterior communicating artery. In: Advances and Technical Standards in Neurosurgery (Krayenbühl, H., *et al.,* eds.), Vol. 2, pp. 113—170. Wien-New York: Springer. 1975.

Author's address: V. Dolenc, M.D., University Department of Neurosurgery, University Medical Centre, Ljubljana, Yugoslavia.

National Institute of Neurosurgery, Budapest, Hungary

Analysis of 177 Operated Cases of Parasagittal Meningioma from the Point of View of Venous Disturbances

G. Benoist and E. Pásztor

Summary

A total of 177 cases was analysed, which were operated on in the years 1972–1981 with parasagittal meningioma. Since damage of arterial supply did not occur during the operation, some postoperative complications could well be attributed to disturbances of the venous circulation. Surgical mortality, caused either by cerebral oedema or emollition, arose in 2 cases (6%) with visible disturbances of venous circulation that were also noticeable in 6 cases (4%) after operations free of complications.

The occurrence and severity of postoperative oedema were not affected by the age of patients or by the size of tumour. Clinical signs due to postoperative oedema could be observed in 53% of those cases which had either total occlusion of the superior sagittal sinus by a tumour, or resection of the partially permeating sinus, or damage of the entering superficial veins running by or over the tumour.

The occurrence of postoperative oedema was 40% in cases without venous circulatory disturbance observed at surgery. The duration and severity of the oedematous period, and the degree and frequency of permanent neurological sequelae did not differ between the two groups of patients. Postoperative oedema occurred most frequently and in the severest form in cases of centrally localized tumours. Tumours adhering to the wall of the sinus or the lower part of the falx did not show any difference on this account, but oedema occurred less frequently and had a milder course in cases with parasagittal tumours of the convexity.

Out of 26 recurrences observed, 16 had histological signs of malignancy. Seven tumours out of the remaining 10 were localized in the central region, and 5 had had an attachment to the wall of the superior sagittal sinus.

Total removal of the tumour-matrix or only its coagulation did not have any noticeable effect with respect to recurrences.

Keywords: Parasagittal meningiomas; postoperative oedema; recurrences.

Introduction

The clinical significance of ensuing cerebral oedema following the damage or occlusion either of the superior sagittal sinus itself or of major veins entering the sinus is well-known. However, there is a decisive difference in the consequences, depending on whether an abrupt venous occlusion or a slowly developing obstruction occurred, since the latter condition allows the development of compensating routes for venous blood flow.

A significant number of parasagittal meningiomas infiltrate the wall of the superior sagittal sinus or even invade its lumen. A 40% rate of sinus infiltration was observed by Simpson[8]. Recently, mainly Bonnal et al.[1-4] suggested reconstruction of the superior sagittal sinus in order to extend the surgical radicality.

In this retrospective study, the surgical strategy and the postoperative course were both analyzed in cases presenting with parasagittal meningiomas, in order to establish the relationship between the severity and the duration of postoperative oedema, or permanent neurological sequelae and the condition of the superior sagittal sinus and entering veins as found at exploration and how possible intraoperative damage may have affected the oedema. We also investigated, how recurrences were influenced by surgical radicality.

Methods and Results

A total of 177 cases presenting with parasagittal meningiomas and operated on in our Institute during the last ten years were analyzed. Tumours, showing only slight histological signs of malignancy, such as frequent mitosis or the presence of irregular nuclei, were included, whilst meningosarcomas and definitely malignant meningiomas were excluded from this study (Table 1.).

Table 1. *Parasagittal Meningiomas*

	Number	Histologically malignant signs	Recurrence
Convexity	56	5	5 (4 mal.)
S S Sinus	69	7	9 (4 mal.)
Falx	52	10	12 (8 mal.)
Frontal	49	10	10 (9 mal.)
Central	89	7	10 (3 mal.)
Parietooccipital	39	5	6 (4 mal.)
	(177)	(22)	(26) (16 mal.)

As one part of our study, the examination of the postoperative course was aimed at by our analysis. Since in our cases, any damage of cortical tissue, including brain arteries, was meticulously avoided, postoperative circulatory calamities could well be attributed to the disturbance of the venous circulation. As the second part of our study, the influence of the total excision of the tumour matrix, or only its coagulation, on the frequency of recurrences was taken into account.

From the point of view of localization, tumours were classified according to two criteria. First: the attachment to the superior sagittal sinus. A.: Meningioma of the convexity (though the tumour reached and probably compressed the sinus, infiltration of the sinus wall was not present), *i.e.,* group 1 "adhesion only" by Logue[6]. B.: Tumour of the sinus, where the tumour-matrix involved and infiltrated the wall to a varying degree, the tumour entering the cavity of sinus and occluding it totally or partially, *i.e.,* groups 2–4 of Logue[6]. C.: Tumours attached to the falx without infiltrating the sinus wall. Second: according to their location along the superior sagittal sinus: frontal, central and parieto-occipital tumours were distinguished.

Slight histological signs of malignancy were detected in 22 of the 177 cases. There were 26 recurrences; 16 of them indicated a tendency toward malignant proliferation.

Total or partial occlusion of the superior sagittal sinus and the functional state of the cortical and bridge-veins were evaluated either by angiography or by means of the surgical report. Sinography was only carried out in a single case.

Postoperative disturbance of the venous circulation was assumed when the sinus had been resected during surgery or preservation of the adjacent cortical veins failed.

The onset of postoperative oedema was diagnosed by the examination of consciousness and neurological signs such as hemiparesis, and in several cases by postoperative angiography or CT, in addition (Table 2.). Oedema usually caused clinical signs on the first or second postoperative day and became resolved within a week.

Table 2. *Operative Findings and Postoperative Course*

Location of tumours	Number	Cases Occlusion of Veins	Sinus	Postoperative oedema	
Convexity	56	7	—	17 (3 v)	(30%)
S S Sinus	69	3	19	31 (2 v, 8 s)	(45%)
Falx	52	3	2	29 (3 v, 2 s)	(56%)
Frontal	49	2	6	17 (2 v, 3 s)	(35%)
Central	89	7	10	44 (4 v, 5 s)	(49%)
Parieto-occipital	39	4	5	16 (2 v, 2 s)	(41%)
	177	13	21	77 (8 v, 10 s)	(44%)

v = occlusion of veins, s = occlusion of sinus.

Surgical mortality accounted for eleven patients (6.2%) in our hands. The cause of death was postoperative haemorrhage in two cases, pulmonary embolism in another case, and brain swelling and emollition in the remaining ones, respectively. Of the patients, who died as a likely consequence of cerebral swelling, two also presented with occlusion of superior sagittal sinus due to the tumour whilst, in the six remaining cases, noticeable alteration of the venous bed could not be observed during surgery. Thus the mortality rate due to oedema was 6% with visible disturbances of the venous circulation, whereas it was 4% amongst patients without those complications mentioned above.

Postoperative oedema occurred most frequently in the central region, after the operation of tumours attached to the falx, and was most rarely present in cases of tumours of the frontal convexity.

Neurological deterioration of various degrees observed during the oedematous period was followed by permanent sequelae in eleven patients, whereas a rapid complete improvement could be seen in the others. Later, all of the patients could walk, then became independent of medical care and were eventually able to work. A permanent neurological deficit was only found in cases with tumours of the central region.

The age of the patients or the size of the tumours did not have any influence on the development of postoperative oedema, nor on its severity, nor on the duration of the damage (Tables 3 and 4).

Table 3. *Age of Patients and Postoperative Course*

Age in years	Convexity No.	oedema (%)	S S Sinus No.	oedema (%)	Falx No.	oedema (%)	Total	oedema (%)
—40	6	1 (17%)	10	5 (50%)	7	4 (57%)	23	10 (43%)
41—60	39	12 (31%)	47	22 (47%)	36	19 (53%)	122	53 (43%)
61—	11	4 (36%)	12	4 (33%)	9	6 (67%)	32	14 (44%)

Table 4. *Size of Tumours and Postoperative Course*

Size in cm³	Convexity No.	oedema (%)	S S sinus No.	oedema (%)	Falx No.	oedema (%)	Total	oedema (%)
—250	38	10 (26%)	51	24 (48%)	40	25 (63%)	129	59 (46%)
251—	18	7 (39%)	18	7 (39%)	12	4 (33%)	48	18 (38%)

The occurrence of postoperative brain swelling was 43–44% in each age-group. In the cases of tumours of smaller size, postoperative swelling occurred somewhat more frequently, but this correlation was not significant.

Transient postoperative oedema developed in 53% of the patients where either total occlusion of the superior sagittal sinus by the tumour or resection of the partially permeating sinus were also present, or there was damage of the cortical veins running by or over the tumour (Table 5). The occurrence of oedema was 41% in cases without venous circulatory disturbances observed at surgery. The duration and severity of the oedematous period and the quality and frequency of the permanent neurological sequelae did not differ between the two groups.

Altogether 26 recurrences were observed during a ten-year period, 16 cases presented with slight histological signs of malignancy, namely, a tendency towards quick tumour-growth could be traced. On the other hand, in six cases, although presenting with the same histological picture, recurrence has not been observed so far.

Table 5. *Operative Findings and Postoperative Course*

No.	Postoperative oedema	Permanent neurological deficit
Operative finding: venous disturbances	18 n = 34 (53%)	2 n = 34 (6%)
Without venous disturbances	59 n = 143 (41%)	9 n = 143 (6%)
In all cases	77 n = 177 (44%)	11 n = 177 (6%)

Tumour recurrences were operated again in 1–14 years after the initial surgery. Within the first 6-year period 20 patients underwent surgery, and during the following 8 years there were 6 patients who underwent surgery. The time elapsed until the first recurrence did not differ between the groups presenting with or without histological signs of malignancy. The over-all average was 5.5 years (range: 1–14 years).

It was also investigated whether total excision of the dura infiltrated by the tumour gave more success in avoiding recurrences than just coagulation of the tumour matrix (Tables 6 and 7).

Table 6. *Operative Technique and Recurrence*
(Except tumours with histologically malignant signs)

	No.	Coagulation of matrix	Recurrence	Excision of matrix	Recurrence
Convexity	47	5	0	42	1
S S sinus	57	47	3	10	2
Falx	40	29	3	11	1
	144	81	6 (7,4%)	63	4 (6,4%)
Malignant:	22	18	12	4	4
Postop. exit.:	11				

Table 7. *Operative Technique and Recurrence*
Recurrence: 26/177 (histologically malignant signs: 16/26)

	Coagulation of matrix	Excision of matrix
Time of recurrence in years (average)	6 years (1–14) (n = 18) (M = 12)	4.5 years (2–7) (n = 8) (M = 4)

M = histologically malignant signs.

In the majority of cases with meningiomas of the convexity, the excised dura was replaced by preserved fascia. The majority of tumours originating from the wall of the sinus, were treated by coagulation of the matrix, the sagittal sinus being excised in only ten cases, where a total or almost total occlusion, mostly by bilateral tumour growth, was already present at surgery. The tumour-matrix with adherence to the falx, but no attachment to the sinus, was dissected and removed only in a third of all the cases.

Excision of tumour matrix, or coagulation of its supply only, showed no detectable influence either on the frequency or the time-course of recurrences.

In the cases of tumours infiltrating the sinus wall, though several of them had invaded the lumen of the sinus, only coagulation was carried out. Amongst tumours without any histological malignancy, 3 recurrences occurred out of 47 cases following coagulation of the matrix, in contrast to the 2 recurrences which developed after the 10 cases of sinus resection. The majority of tumours presented with only slight histological signs of malignancy recurred, namely 16 out of 22. Amongst these cases, 12 out of 18 tumours after the matrix coagulation and all of the 4 tumours following the total matrix excision had recurrences. Of course, the solid intradural part of the tumours was totally removed in each case.

Timing of the recurrences was independent on the location of the tumour and occurred on an average after 4–7 years, respectively. Tumour recurrences could occur in every location after 1–2 years or also even after 10 years.

Discussion

According to our observations, it seems that the development and severity of cerebral oedema following the operation of parasagittal meningiomas is not correlated either with the age of patients or with the size of the tumours, but it might be affected first of all, by the location of the tumour and by noticeable damage of the venous circulation. The role of these latter factors, though, seems to have no particular significance, but the uncommon frequency of oedema occurring in connection with the operation of the central region might indicate the decisive importance of the vena Rolandica.

In the study of Hoessly and Olivecrona[5] mortality was 10.4% after excision of the occluded sinus, 10.0% after marginal resection of the partially occluded sinus and 18.7% following block-resection, respectively. In 30% of the lethal cases, oedema was the cause of death. In accordance with these findings, the authors adopted a conservative standpoint regarding sinus resection and they suggested it should be performed in the first third of the sinus.

Logue[6] stipulated, when an important great vein entered the partially occluded part of the sinus, it was most practical to limit the operation and be satisfied with the removal of the tumour and, only in case of recurrence, might the totally occluded sinus be resectioned. Simpson has the same opinion[8].

More recently, Bonnal et al.[3] tend to be more radical in each case. Sinus excision of a still patent sinus might cause oedema, after the infiltrated sinus as well as the intrasinusal part of the tumour and if necessary, after whole segments of the sinus had been removed, a subsequent sinus-reconstruction or substitution was also carried out. As the region of vena Rolandica bears paramount importance in the genesis of postoperative oedema, these authors' technique for sinus reconstruction involves preservation and replacement of the major veins, as well.

Recurrences of parasagittal meningiomas are determined by two factors: the histological character of tumour and the extent of its removal. A definite prognosis cannot be established by microscopic characterization of the tumour. Simpson[8] observed, that tumours with a highly differentiated tissue architecture might be extremely infiltrative, and very anaplastic ones might not recur.

In accordance with the opinion of Bonnal and Brotchi[3], Simpson[8] also regards recurrences as a consequence of insufficient surgery. Bonnal and Brotchi maintain that the chief causes of recurrence appear to the unnoticed invasion of dural venous sinuses and unsuspected spread across free dural septa.

Recurrences arose in 5–9% of cases after most radical surgery, in 16–17% following coagulation of the dura and in 29% after just removal of the solid tumour without coagulation of the matrix, as observed by Simpson[8]. Schafer[7] found an 8% recurrence-rate after macroscopically total extirpation of the tumour. Hoessly and Olivecrona[5] having a conservative standpoint, gave a 7% incidence of recurrences.

Analysis of our own material indicates that the recurrence of parasagittal meningiomas is barely influenced either by excision or coagulation of the matrix. A careful coagulation of the matrix and severing of the blood supply of the tumour might give the same results as the macroscopically total excision of the tumour-attachment, since the microscopically infiltrated edge of a virtually intact dura might not be taken into consideration.

The high incidence of recurrences in tumours presenting histological signs of rapid growth proves that in such cases, invasion of the tumour into more remote areas or into deeper layers of the dura might have wider implications than expected on the basis of the surgical situation.

Simpson[8] suggested that the routine block resection of the great venous sinuses would greatly decrease the incidence of recurrences, but at a prohibitive price in immediate disability or even in death. On the basis of our experience, it is questionable whether resection and reconstruction of the sinus with possible complications of the surgical solution would pay the expected dividends in every case.

References

1. Bonnal, J., Brotchi, J., Stevenaert, A., Petrov, V. T., Mouchette, R.: L'ablation de la portion intrasinusale des méningiomes parasagittaux rolandiques, suivie de plastique du sinus longitudinal supérieur. Neurochirurgie (Paris) *17* (1971), 341—354.
2. Bonnal, J., Buduba, C.: Surgery of the central third of the superior sagittal sinus. Acta neurochir. (Wien) *30* (1974), 207—215.
3. Bonnal, J., Brotchi, J.: Surgery of the superior sagittal sinus in parasagittal meningiomas. J. Neurosurg. *48* (1978), 935—945.

4. Bonnal, J.: La chirurgia conservatrice et réparatrice du sinus longitudinal supérieur. Neurochirurgie (Paris) *28* (1982), 147—172.
5. Hoessly, G. F., Olivecrona, H.: Report on 280 cases of verified parasagittal meningioma. J. Neurosurg. *12* (1955), 614—626.
6. Logue, V.: Parasagittal meningiomas. In: Advances and Technical Standards in Neurosurgery, Vol. 2 (Krayenbühl, H., *et al.,* eds.), pp. 171—198. Wien-New York: Springer. 1975.
7. Schafer, E. R.: Recidivhäufigkeit bei Meningeomen. Acta neurochir. (Wien) *13* (1965), 186—195.
8. Simpson, D.: The recurrence of intracranial meningiomas after surgical treatment. J. Neurol. Neurosurg. Psychiatr. *20* (1957), 22—40.

Authors' address: G. Benoist, M.D., and Prof. E. Pásztor, Director, National Institute of Neurosurgery, Amerikai ut 57, H-1145 Budapest, Hungary.

Subject Index

Aneurysm Surgery in the Acute Stage

Edited by
L. M. Auer, F. Heppner, and L. Symon

Symposium Graz, July 19—21, 1981

1982. 76 figures. VIII, 306 pages.
(Acta Neurochirurgica, Vol. 63/1—4, 1982)

This volume gives an excellent survey not only of the possibilities and risks of aneurysm operations during the first few days after subarachnoid haemorrhage but also of new knowledge and treatment directed to possible prevention of cerebral vasospasm.

Noted specialists from the United Kingdom, U.S.A., Canada, Federal Republic of Germany, Sweden, Italy, The Netherlands, Austria, Yugoslavia and especially Japan were attending this symposium to present their most recent findings and to exchange their experiences on the following main topics: Diagnosis, Spontaneous Course, Aspects of Surgical Techniques, Timing of Surgery, Surgical Results, Conservative Treatment, Experimental Data on Cerebral Vasospasm, Treatment of Vasospasm — Clinical Data.

The volume also includes a Honorary Lecture by Lindsay Symon on Perspectives in Aneurysm Surgery.

Springer-Verlag Wien New York